WASTE PROBLEMS AND MANAGEMENT IN DEVELOPING COUNTRIES

Edited by

Umair Riaz, PhD

Shazia Iqbal, PhD

Moazzam Jamil, PhD

APPLE ACADEMIC PRESS

Contributors

Ali Ahmad
Department of Agronomy, University of Agriculture, Agriculture University Road, Faisalabad, Pakistan

Muhammad Irfan Ahamad
Shaanxi Key Laboratory of Earth Surface System and Environmental Carrying Capacity, College of Urban and Environmental Sciences, Northwest University, Xi'an 710127, China

Fatima Akmal
Institute of Soil and Environmental Sciences, University of Agriculture, Faisalabad 38040, Pakistan

Syed Mustafa Ali
Health Informatics, University of Manchester, UK

Muhammad Ameen
Department of Soil Science, Faculty of Agriculture and Environment, The Islamia University of Bahawalpur, 63100, Pakistan; E-mail: soilchemist.ameen033@gmail.com

Aqsa
Institute of Soil and Environmental Sciences, University of Agriculture, Faisalabad 38000, Pakistan

Anam Ashraf
School of Environment, Tsinghua University, Beijing 100084, China

Muhammad Azhar
Institute of Soil and Environmental Sciences, University of Agriculture, Faisalabad 38040, Pakistan

Muhammad Mohsin Azim
Department of Chemistry, Norwegian University of Science and Technology, Trondheim, Norway

Zubair Aslam
Department of Agronomy, University of Agriculture, Agriculture University Road, Faisalabad, Pakistan; E-mail: zauaf@hotmail.com

Tabinda Athar
Institute of Soil and Environmental Sciences, University of Agriculture, Faisalabad 38000, Pakistan; E-mail: athartabinda@gmail.com

Safdar Bashir
Institute of Soil and Environmental Sciences, University of Agriculture, Agriculture University Road, Faisalabad, Pakistan

Korkmaz Belliturk
Faculty of Agriculture, Department of Soil Science and Plant Nutrition, Namık Kemal Üniversitesi, Kampüs Cad No:1, 59030 Süleymanpaşa/Tekirdağ, Turkey

Saira Bibi
Pak-Austria Fachhochschule, Institute of Applied Science and Technology, Mang, Haripur, Khyber Pakhtunkhwa 24421, Pakistan

Swarnavo Chakraborty
Department of Biotechnology, St. Xavier's College (Autonomous), 30, Mother Teresa Sarani, Kolkata 700016, West Bengal, India

Naseer Ahmad Dar
Department of Environmental Sciences, Govt. Degree College Shopian 192303, Jammu and Kashmir, India

Afzal Ahmed Dar
School of Environment Science and Engineering, Shaanxi University of Science and Technology, Xian 710000, China

Mukkaram Ejaz
School of Environmental and Municipal Engineering, Lanzhou Jiaotong University, Lanzhou 730070, Gansu, People's Republic of China

Muhammad Farhan
Department of Geodesy and Survey Engineering, College of Earth Science and Engineering, Jianging Campus Hohai University, Nanjing, People's Republic of China

Umer Farooq
Institute of Soil and Environmental Sciences, University of Agriculture, Faisalabad 38000, Pakistan

Zia Ur Rahman Farooqi
Institute of Soil and Environmental Sciences, University of Agriculture, Faisalabad 38040, Pakistan

Sait Geizgin
Department of Soil Science and Plant Nutrition, Selcuk University, Konya 42130, Turkey

Muhammad Ashir Hameed
Institute of Soil and Environmental Sciences, University of Agriculture, Faisalabad 38040, Pakistan

Yasir Hameed
Ministry of Education, Key Laboratory of Environmental Remediation and Ecosystem Health, College of Environmental and Resources Science, Zhejiang University, Hangzhou 310058, People's Republic of China

Mehmet Hamurcu
Department of Soil Science and Plant Nutrition, Selcuk University, Konya 42130, Turkey

Muhammad Anwar-Ul-Haq
Institute of Soil and Environmental Sciences, University of Agriculture, Faisalabad 38040, Pakistan

Muhammad Umer Hayyat
Sustainable Development Study Center, GC University Katchery Rd, Lahore 54000, Pakistan

Muhammad Mahroz Hussain
Institute of Soil and Environmental Sciences, University of Agriculture, Faisalabad 38040, Pakistan

Hafiz Abdul Kareem
College of Grassland Agricultural Sciences, Oregon State University, Corvallis, OR, United States; abdulkareem@nwafu.edu.cn

Samina Khalid
Department of Environmental Sciences, COMSATS University Islamabad, Vehari Campus, Punjab, Pakistan; E-mail: saminakhalid@cuivehari.edu.pk

Mohd. Kamran Khan
Department of Soil Science and Plant Nutrition, Selcuk University, Konya 42130, Turkey

Aman Ullah Malik
Postharvest Research and Training center, Institute of Horticultural Sciences, University of Agriculture Faisalabad, Punjab, Pakistan

Asma Mansoor
Visiting Lecturer Environmental Science NUML, Lahore, Pakistan

Rizwan Mehmood
Stored Grain Management Cell, Department of Entomology, University of Agriculture Faisalabad 38040, Pakistan

Tariq Mehmood
College of Environment, Hohai University Nanjing 210098, China; E-mail: tariqslamat@hhu.edu.cn

Anamika Pandey
Department of Soil Science and Plant Nutrition, Selcuk University, Konya 42130, Turkey

Abdul Qadir
Institute of Soil and Environmental Sciences, University of Agriculture, Faisalabad 38040, Pakistan

Sami Ullah Qadir
Department of Environmental Sciences, Govt. Degree College, Kokernag 192221, Jammu and Kashmir, India; E-mail: samievs84@gmail.com

Vaseem Raja
Department of Botany, Govt. Degree College for Women's Pulwama 192301, Jammu and Kashmir, India

Ghulam Rasool
College of Hydrology and Water Resources, Hohai University, Nanjing 210098, China

Adnan ul Rehman
Shaanxi Key Laboratory of Earth Surface System and Environmental Carrying Capacity, College of Urban and Environmental Sciences, Northwest University, Xi'an 710127, China

Muhammad Saif ur Rehman
Department of Chemical Engineering, Khwaja Fareed University of Engineering and Information Technology, Abu Dhabi Road, Rahim Yar Khan, Pakistan

Sami-Ur-Rehman
College of Life Science, Northwest A&F University, 3 Taicheng Rd, Yangling District, Xianyang, Shaanxi, China

Sobia Riaz
Institute of Soil and Environmental Sciences, University of Agriculture Faisalabad 38040, Pakistan; E-mail: sobiariazuaf@outlook.com

Umair Riaz
MNS-University of Agriculture, Multan, Pakistan

Aryadeep Roychoudhury
Department of Biotechnology, St. Xavier's College (Autonomous), 30, Mother Teresa Sarani, Kolkata 700016, West Bengal, India; E-mail:aryadeep.rc@gmail.com

Haleema Sadia
Department of Botany, University of Agriculture, Faisalabad 38040, Pakistan

Muhammad Sajid
Department of Materials and Chemical Engineering, Yibin University, Yibin 644000, Sichuan China; E-mail: engr.sajid80@gmail.com; drsajid@yibinu.edu.cn

Zulfiqar Ahmad Saqib
Institute of Soil and Environmental Sciences, University of Agriculture, Faisalabad 38000, Pakistan

Muhammad Salman Sarwar
School of Chemical and Materials Engineering, National University of Science and Technology, Islamabad 44000, Pakistan

Laila Shahzad
Sustainable Development Study Center, GC University Katchery Rd, Lahore 54000, Pakistan;
E-mail: lailashahzad@gcu.edu.pk

Awais Shakoor
Department of Environment and Soil Science, University of Lleida, Avinguda Alcalde Rovira Roure 191, 25198, Lleida, Spain

Faiza Sharif
Sustainable Development Study Center, GC University Katchery Rd, Lahore 54000, Pakistan

Muhammad Sohaib
College of Food and Agricultural Sciences, King Saud University, Riyadh 11451, KSA

Muhammad Irfan Sohail
Institute of Soil and Environmental Sciences, University of Agriculture, Faisalabad 38040, Pakistan;
E-mail: irfansohail9@gmail.com

Ayesha Siddiqui
Department of Botany, University of Agriculture, Faisalabad 38040, Pakistan

Muhammad Aammar Tufail
Department of Civil, Environmental and Mechanical Engineering, University of Trento, Trento 38123, Italy

Muhammad Irfan Ullah
Department of Entomology, College of Agriculture, University of Sargodha, Punjab, Pakistan

Jibbing Xiong
Jiangsu key Laboratory of Resources and Environment information Engineering,
China University of Mining and Technology, Xuzhou 221116, China

Asma Yasin
Sustainable Development Study Center, GC University Katchery Rd, Lahore 54000, Pakistan

Fazila Younas
Institute of Soil and Environmental Sciences, University of Agriculture, Faisalabad 38040, Pakistan;
E-mail: fazila.younas@gmail.com

Sadia Younas
Department of Chemistry, The Government Sadiq College Women University, Bahawalpur 63100, Pakistan

Qamar Uz Zaman
Department of Environmental Sciences, University of Lahore, 1-Km Raiwind Rd, Sultan Town, Lahore, Punjab, Pakistan

Lixin Zhang
College of Life Science, Northwest A&F University, 3 Taicheng Rd, Yangling District, Xianyang, Shaanxi, China

Abbreviations

AD	anaerobic digestion
AEC	American Earthworm Company
AIDS	acquired immunodeficiency syndrome
ANP	Ayubia National Park
API	American Petroleum Institute
BEP	best environmental practices
BOD	biological oxygen demand
Ca	calcium
CBMWTFs	common BMW treatment facilities
CIEL	Center for International Environmental Law
C:N ratio	carbon: nitrogen ratio
CO	carbon monoxide
CO_2	carbon di oxide
COD	chemical oxygen demand
CPCB	Central Pollution Control Boards
CWs	constructed wetlands
DAFs	dissolved air flotations
DCs	developing countries
DEFRA	Department for Environment Food and Rural Affairs
EC	electrical conductivity
EDCs	endocrine disrupting compounds
EEA	European environment agency
EFW	energy-from-wastes
EML	executive management level
EPA	Environmental Protection Agency
EU	European Union
FAO	Food and Agricultural Organization
FHL	formate hydrogen lyase
FLW	Food loss and waste
FSC	food supply chain
FUSIONS	Food Use for Social Innovation by Optimizing Waste Prevention Strategies
FYM	farmyard manure
g	gram

GDP	gross domestic product
GHGs	greenhouse gases
GISs	Geographic Information Systems
GIZ	Gesellschaft für Internationale Zusammenarbeit
GNI	Gross National Income
GPRS	General Packet Radio Service
GSM	Global System for Mobile Communications
GWI	Global Waste Index
GWMO	Global Waste Management Outlook
H2	hydrogen gas
HCl	hydrochloric acid
HCUs	health care units
HCW	health care waste
HDI	Human Development Index
HDPE	high-density polyethene
HEC	Higher Education Commission of Pakistan
HLPE	High-Level Panel of Experts on Food Security and Nutrition
HLW	high level wastes
HRT	hydraulic retention time
HTG	high-temperature gasification
INWAMI	United Nations Integrated Waste Management for Improved Livelihoods
IPCC	Intergovernmental Panel on Climate Change
ISWA	International Solid Waste Association
IUCN	International Union for Conservation of Nature
IWW	industrial waste water
K	potassium
KCCA	Kampala Capital City Authority
km^{-2}	kilo per square meter
L	liter
LCA	life cycle analysis
LDPE	low-density polyethylene
LFGR	landfill gas recovery
LLW	low level wastes
LPG	liquefied petroleum gas
LTG	low-temperature gasification
m	meter
MBT	mechanical biological treatment
Mg	magnesium

mm	millimeter
MoEF	Ministry of Environment and Forests
MRCO	Mandatory Recycling and Composting Ordinance
MRSA	methicillin resistant *staphylococcus aureus*
MSW	municipal solid waste
MSWM	municipal solid waste management
MT	million metric ton
MW	municipal waste
N	nitrogen
NBP	National Biosolids Partnership
NEP	National Environmental Policy
NORM	naturally occurring radioactive materials
NOx	nitrogen oxides
NSSWM	National Strategy for Solid Waste Management
ODA	ocean dumping act
ODS	ozone depleting substances
OECD	Organization for Economic Co-operation and Development
OSPAR	Oslo/Paris convention (for the Protection of the Marine Environment of the North-East Atlantic)
P	phosphorus
PAH	polycyclic aromatic hydrocarbons
PBDE	polybrominated diphenyl ethers
PCBs	polychlorinated biphenyl
PCDFs	polychlorinated dibenzofurans
PCDDs	polychlorinated dibenzo-p-dioxins
PET	polyethylene terephthalate
PF	phytoremediation factor
POS	point of sale
POPs	persistent organic pollutants
PP	polypropylene
PPE	personal protection equipment
PS	polystyrene or styrofoam
PVC	polyvinyl chloride
RBC	rotating biological contactor
RCRA	Resources Conservation Recovery Act
RDF	refused-derived fuel
REFRESH	Resource Efficient Food and Drink for the Entire Supply Chain
RFID	radio-frequency identification
S	sulfur

SAGD	steam-assisted gravity drainage
SARS	severe acute respiratory syndrome
SBRs	sequencing batch reactors
SMEs	small to medium-sized enterprises
SNF	spent nuclear fuel
SOPs	standard operating procedure
SOx	sulfur oxides
SS	suspended solids
SW	solid waste
SWG	solid waste generation
SWM	solid waste management
TDS	total dissolved solids
TF	transfer factor
TRUW	transuranic wastes
TWM	thermal waste management
UBCs	used beverage cartons
UNDP	United Nations Development Programme
UNEP	United Nation Environment Programme
US	United States
USD	United States Dollar
USDA-ERS	United States Department of Agriculture Economic Research Service
USEPA	United States Environmental Protection Agency
VOCs	volatile organic compounds
WAS	waste activated sludge
WCED	World Commission on Environment and Development
WFD	Waste Food Directive
WM	waste management
WRAP	Waste and Resources Action Programme
WRI	World Resource Institute
WtE	waste-to-energy
WWF	World Wilde Fund for Nature
ZWS	zero waste strategy
μm	micrometer
°C	degree centigrade
m^{-3}	per meter cube
m^{-2}	per square meter

Preface

The world population is increasing day by day and reportedly will rise to nine billion by 2050. More population means a greater need for more daily life necessities that will lead to more waste production. Waste is a serious problem of today's world as it is affecting all the biological life on planet Earth. Environmental issues triggered by wastes are water, air, and soil contamination, which pose threats to human health. The contemporary research works have elaborated the hazardous physiochemical effects of wastes on biota especially in fresh water and terrestrial living organism.

With the latest advancement in science and technology, huge amount of biosolid waste is produced throughout the globe and is posing serious threats to human beings, the environment, and agricultural lands. Biosolids are rich sources of energy and nutrients, and their proper management can help to reduce the burden on landfills, along with various other advantages. The livestock industry provides food and livelihood and contributes to the economy. Meanwhile it is a main cause of persistent organic compounds, odors, and harmful gases.

Industries like tannery, pulp and paper, sugar, fertilizer, and textiles produce heavy metal pollution that adversely affects the environment, especially the water streams and cause serious diseases. Wastes released from healthcare departments from patients suffering from contagious diseases spread this infection to humans directly or indirectly through different environmental segments if not properly handled. These infections become widespread due to the haphazard disposal.

Radioactive wastes are produced by exploitation of radioactive materials for the production of nuclear power in nuclear reactors, generation of nuclear weapons. Utilities of nuclear fission reactions and other forms of nuclear applications in the medical and research fields pose serious threats to the environment and different forms of life existing on this planet. The use of radioactive materials has increased every year since the last few decades, leading to a massive accumulation of radioactive wastes in the environment. Due to improper disposal of these wastes, exposure to humans and other living beings to harmful irradiation from these radioactive wastes has led to the progressive rise in health issues and other potential dangers in society. It has become necessary to ponder upon the management and proper handling of these wastes to control the rise in harmful effects on biological life due to exposure to these wastes.

In many parts of the world, the problem of large amounts of waste is solved through proper waste management. But waste management has always been a serious problem for the last few decades. Unlike developed countries, waste management is a serious issue in developing countries, including Pakistan. This is because of a lack in waste collection strategies and processes as well as awareness of its environmental impacts. One can think of the need of sustainability in waste management; however, it provides an opportunity of handling the waste before its production. Composting of waste and then using this compost as a nutrient source for agricultural land leads to dual benefits but it must be established at a large commercial scale. Minimizing or changing the consumption patterns of waste can lead to sustainable long-term benefits of resource management of developing countries. Solutions of waste management lay in proper collection, segregation, recycling and reusing, and creating secondary use of collected waste. In addition to this, handling waste properly has an opportunity to provide employment on a large scale and will definitely generate revenue. Waste has to be seen as an opportunity rather as a burden.

This book covers the characterization and problems, issues and regulatory development of waste management, and the management of municipal solid wastes, focusing on integrating the technical and regulatory complexities of waste management, particularly in developing countries. It also addresses hazardous wastes and their management from the perspectives of identification, transportation, and requirements for generators as well as the treatment, storage, and disposal facilities. This book describes all the main categories of wastes under regulation in developing countries as compared with the developed world. It also incorporates an extensive set of problems presented and includes numerous datasets obtained from different surveys. Special categories of waste that may not fit precisely into either RCRA Subtitle D (solid wastes) or Subtitle C (hazardous wastes) are highlighted in some chapters.

Academicians, researchers, and students will find this a comprehensible volume about waste management and its diversity, exploration, exploitation, and management strategies, and thus they will find this book to meet the requirements of training, teaching, and research.

We are extremely grateful to the authors who have contributed chapters in this book. We express our thanks to Apple Academic Press for their cooperation and publication of this book.

—Umair Riaz
Shazia Iqbal
Moazzam Jamil

PART I

Waste Generation in Developing Countries

CHAPTER 1

Solid Waste Generation and Its Characteristics

LAILA SHAHZAD[1*], ASMA MANSOOR[2], and SYED MUSTAFA ALI[3]

[1]*Sustainable Development Study Center, GC University Katchery Rd, Lahore 54000, Pakistan*

[2]*Environmental Science NUML, Lahore, Pakistan*

[3]*Informatics, University of Manchester, UK*

[*]*Corresponding author. E-mail: lailashahzad@gcu.edu.pk*

ABSTRACT

This chapter provides an overview of waste generation and its characteristics in different developing countries. An increase in waste generation and variety of characteristics is one of the consequences of global urbanization. Over 90% of the generated waste is often disposed of in open dump spaces or mostly burned in low-income countries. An uncontrolled waste generation without proper management has led to diverse environmental and human health problems. Managing solid waste is a serious challenge to the administration of small towns and metropolitan areas. Waste characteristics are explained which is essential to the long-term efficient and successful waste management policy. It is the initial step to estimate potential recovery of materials, identify sources of waste generation, facilitate the design of processing machineries, analyze physiochemical properties of the wastes, and sustain the compliance with guidelines. Both qualitative and quantitative characteristics of generated waste depend on seasonal changes, people lifestyle, population dynamics, geographic, and local laws. Therefore, this

Waste Problems and Management in Developing Countries. Umair Riaz, Shazia Iqbal, & Moazzam Jamil (Eds.)

chapter covers the depth of characteristics of solid waste generated in different developing countries and their possible health impacts and solutions.

1.1 INTRODUCTION

The definition of solid waste undergoes many changes over the time. The framework was approved for hazardous and nonhazardous waste management programs in 1976. According to this, "solid waste" means any garbage that includes solids, nonsoluble materials, semisolids, comprising gases and liquids in containers resulting from domestic, mining, agriculture, industrial, commercial, and operations activities. It is imperative to note that the solid waste is not only limited to physical solid. Some solid wastes exist in liquid form, others are found in semisolid, or gaseous form. In the 21st century, some discarded things have been eliminated from the class of solid waste like radioactive waste, domestic sewage, point source discharge, in-situ mining, hazardous secondary waste from the petroleum industry, and coke products (World Bank, 2020).

The waste problem is called the reflection of a society. The status of a society is related to its economic condition, historical background, cultural values, and environment. A complete knowledge regarding the status of society provides a direction to resolve the waste issues. For example, seashell mounds or kitchen waste by shellfish eaters in past give evidence to how people in particular society lived. With the passage of time, society becomes relatively wealthy and modern, it starts to rely on the extraction and metabolism of large quantities of resources including energy, in order to support the continuing population. An inevitable consequence of more consumption of resources led to the high quantity of solid waste production. All processes include extraction of input material, manufacture of products, consumption of material to generate the solid wastes. As we know the laws of thermodynamics, it stated that materials and energy can be changed from one form to another but never demolished. In the ancient world, there is basically zero waste produced, the reason is that the wastes of one organism become the food for another. This natural process of recycling of materials follows the principle of sustainability (Miller and Spoolman, 2011).

The global total magnitude of solid waste is 1.7–1.9 billion metric tonnes (Gichamo and Gökçekuş, 2019). The generation of solid waste is anticipated to increase to approximately 3.40 billion tonnes by the half of the 21st century (World Bank, 2020). This quantity is high than the doubling of population growth rate over the same time-lapse. Explicitly, waste generation and its characteristics like physical and chemical may fluctuate at the country level, province, city, and even within the various regions of the same city. It also changes from high-income countries to low-income countries as the increasing amount and complexity of solid waste linked with economy, industries growth, and urban population has drastic problems, particularly for developing nations (Table 1.1).

TABLE 1.1 Developing Economies[a]: Rates of Growth of Real GDP.

Country	2016	2017	2018	2019[b]	2020[c]	2021[c]
Bangladesh[d]	7.1	7.3	7.9	8.1	7.8	7.1
Pakistan[d]	5.6	5.8	3.3	3.3	2.1	3.3
India[d]	8.2	7.2	6.8	5.7	6.6	6.3
Philippines[d]	6.9	6.7	6.2	5.9	6.2	6.3
Kenya	5.9	4.9	6.3	5.6	5.5	5.7
Afghanistan	2.2	2.7	2.7	3	2.7	4.3
Nepal[d]	0.6	8.2	6.7	7.1	6.3	5.3
Ethiopia	8.5	8.1	6.8	7.3	7.5	7.4
Yemen	−14.8	−5.9	2.7	1.2	3.6	4.3
Indonesia	5	5.1	5.2	5	5.1	5.2

[a]*World Bank data (World Bank, 2020).*

[b]*Partly estimated.*

[c]*Measured by UN world economic forecasting model.*

[d]*On the basis of fiscal year.*

For example, collected works have reported that the waste generation has a positive correlation with the income level of a country. According to a statistic, waste generation in developed countries (daily per capita) is likely predicted to grow by 19% in 2050 than developing countries where it is expected to grow between 40 and 50%. Studies show that the generation of waste initially decreases as the income levels drop and then starts rising at a faster rate with high income. The fastest-growing continents are Africa and South Asia where total waste generation will be increasing to more than triple and double, respectively, in 2050. In these regions, most of the solid waste is openly discarded, and the pattern of its generation has wide consequences for the society health, and economy, thus needing serious actions (World Bank, 2020). The country of Nigeria, Africa's most populous nation, is facing challenges related to the management of piles of waste generation. Its rapidly emerging capital, Lagos, a city of 20 million population, is known as the "garbage capital of the world." The citizens of this city throw away 11,000 tonnes of solid waste regularly. Meanwhile, the city struggles to produce electrical power, sometimes they survive with only 3–4 h/day of electricity (Ike et al., 2018).

In the modern world, in developed countries, a waste generation might be limited because of the research in material sciences, adoption of waste reduction approaches, and advancement in technology. On the other hand, as we know, developing countries are lacking in financial resources, good governance, research, and development. This may also be the reason for difference in magnitude and characteristics of waste generation in the developed and developing countries. We are currently experiencing it, need to change since the solid waste is growing world problem with dramatic social and environmental impacts (Gichamo and Gökçekuş, 2019).

This chapter explores the generation of solid waste, its characteristics, and composition in developing countries, and also highlights the factors determining the rate of waste generation and ways to reduce its harmful effects.

1.2 WASTE GENERATION: A BIG PROBLEM

Solid waste generation (SWG) is emerging in every corner of the world, particularly in all urbanized areas. SWG is known as the most challenging problem encountered by developing nations that are suffering from drastic environmental hazards due to high SWG. In urban cities, high generation of solid waste stimulated the sanitary issues like water facilities, waste management, and infrastructure (Liyala, 2011)

In developing countries, the waste generation and its management by burgeoning metropolitan cities are exploiting functions of the municipalities and national governments. Developing countries do not have adequate waste management regulations such as trash collection services, local, and national organizations to handle the solid wastes. There are three reasons that explain the failure; firstly, lack of real stakeholder participation in taking efficient decisions and planning processes. Because their involvement is crucial for understanding the continuously varied relationships among government authorities, decision-makers, and social dynamics (Gichamo and Gökçekuş, 2019).

As it is a difficult process that demands cooperation between a wide range of stakeholders. Like, government bodies are failed to implement the waste minimization practices and collaborate with municipal bodies in determining the sites for waste management. It is the dilemma that municipalities are not having a proper organizational system with well-trained staff. Politicians are not leading their role in promoting waste management campaigns. Improper role of academia influence the culture of solid waste management. Public participation is hardly seen because of the lack of awareness and their unwillingness. NGOs and private institutions in developing countries have sufficient funds to resolve solid waste issues (Guan, 2011).

Secondly, the lack of knowledge that the system is composed of different level hierarchy like generation points, trash collection, transport, transfer, treatment, and final disposal destination. Thirdly, identifying that there are external factors influencing the system such as financial, institutional, environmental, technical, sociocultural, and legal (Dhokhikah and Trihadiningrum, 2012). Developing countries are characterized by poor information, inadequate data, and difficulties obtaining real figures on their waste quantitative analysis (Friedland et al., 2011). There are multiple reasons such as reduced funding, shortage of management skills, priorities to be solved, and inefficient local authorities (Khair et al., 2018). Data reliability and gathering from developing countries are generally difficult to attain due to limited open data sources, mismanagement of waste collection, and rural–urban migration inflow at the national level (Kawai and Tasaki, 2016).

1.2.1 PHILIPPINES

Philippines, a developing country, is ranked as the third-largest producer of solid waste among the countries of Southeast Asian. Approximately, every individual will be making five extra kilograms of waste by 2030 (Romero, 2020). Manila is the capital of the Philippines and is setting the worst example among developing countries. Its citizens generate an estimated 8000 tonnes of garbage on daily basis. Unfortunately, the government being a crucial stakeholder was unable to collect the waste, aware the people about waste reduction techniques. This resulted the piles of waste at many dumping sites, which were home for insects (flies, etc.) and animals (rats, other vermin, etc.). The dumping sites also encouraged needy people to collect the items from trash to earn the money for their survival. As poorest people even lived on the dumping site in shanties that emitted methane fumes and various toxins. Two decades ago, a bulk of trash called Payatas hit by typhoon disaster and 219 people were killed. Right after the incident, the government cleared the site, unfortunately, a new dumping site was operated that continued to provide a source of earning money for many scavengers (Duru et al., 2019).

1.2.2 SWEDEN

Among developed countries, Sweden set a successful example in solid waste management as they are using garbage as a useful resource for community development. Sweden recycles or reuses about 99% of all household waste and converts it into energy and various products. For example, heating the homes in winter and biogas is used in transportation and electricity generation. The reason behind their success is the public participation that is very appreciable. Further, they are educating children about the adoption of recycling from the childhood. The manufacturing companies in many areas take the responsibility for the proper handling of waste produced from their products. The Swedish government encourages the producers to generate more efficient goods that can be easily recycled. Being an EU member, Sweden considers the environment as serious concern and also has effective legislation for efficient resources management in the food market. At present, food chain industry stakeholders signed an agreement as an initiative to reduce 50% food garbage by 2030, according to the sustainable development agenda (Swedish Cleantech, 2020). That is why Sweden has been declared as a global leader in waste management.

1.3 CHARACTERISTICS AND COMPOSITION OF SOLID WASTE

Solid waste is categorized on the basis of its composition and sources. The composition of solid waste includes a plastic material, paper, rubber, leather, kitchen waste, glass, metal, garden waste, etc. The sources of the waste encompass markets, municipal, factories, agriculture, and demolition sites (Zhou et al., 2014). Often industrial waste may be governed by environmental concerns depending on its hazardous nature like ignitable, corrosivity, and toxicity.

1.3.1 COMPOSITION OF SOLID WASTE

The composition of waste is the physical existence of different types of materials in solid waste. It is usually identified by conducting a standard field survey of target areas. And samples of waste are extracted from generators or final disposal locations and separated into categories of composition. The physical composition varies across economy scenarios demonstrating various trends of composition. High-income or developed nations produce relatively low organic waste like 32% of total waste and also produce bulk of dry waste that is easily recyclable such as E-waste, plastic, metal, tires, and textiles, cover almost 51% of total waste (Sharma and Jain, 2020).

Developing countries produce 53% food waste and 56% green waste as the GDP level decreases (Figure 1.1) (Baawain et al., 2017). Developing countries are unable to manage rapidly changing waste composition without adequate systems. These variations in the composition bring light on consumption patterns, living standard, and financial status of people living in developing countries. For example, downward shift of organic waste from 64 to 56% can be seen in Figure 1.1 (Kumar and Samadder, 2017).

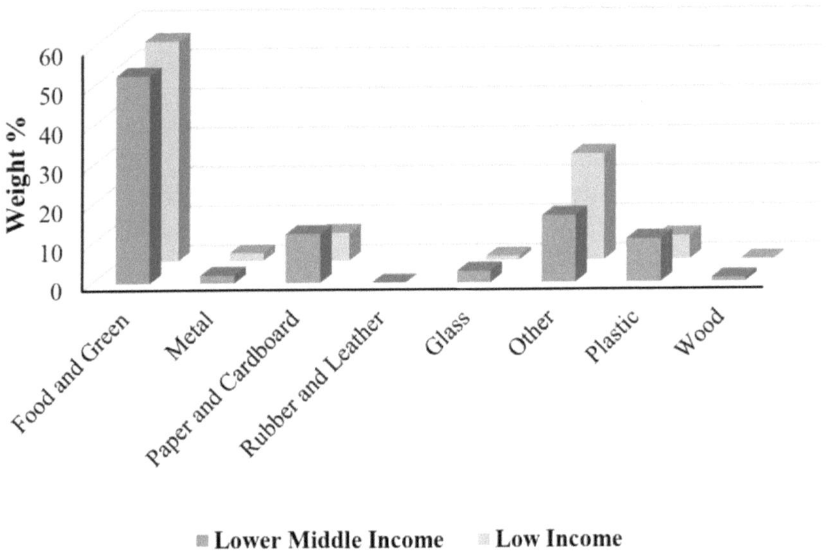

FIGURE 1.1 Waste composition by income level (%).

In India, daily total waste generation is about 91.01 g per capita and only organic waste is 74 g per capita. According to a recent study, average household waste generation had a positive correlation with household income status and education level, moreover, depicted a negative association with the number of family members. Organic waste material comprises almost 82%, which has a high recovery rate. The volume of total organic waste is about 232 Gg per year (Ramachandra et al., 2018).

1.3.2 SOURCES OF SOLID WASTE

Most of the solid waste arises from major anthropogenic activities and animal actions that are thrown away as unwanted material. These can be both organic and inorganic waste given by a society, which do not pose any benefits to the first hand (Olukoju, 2018).

1.3.2.1 MUNICIPAL SOLID WASTE

Municipal solid waste (MSW) entails the combined discarded material originated by residential areas and workplaces excluding factories, examples are cardboard, food wastes, steel, iron, pet bottles, yard wastes, kitchen unnecessary residue, plastics, metals, glass, and E-waste. Solid waste in the municipality, comprises diverse and uniform wastes, determined by their original sources such as urban, periurban regions. The generation rate of MSW is quicker than that of urban population growth due to the increasing demand for goods and services. In developed countries, most of MSW is dumped in landfill sites and some hazardous waste is burned in incinerators to kill pathogens. Many studies stated that developing countries are mainly generated households MSW (55–80%), followed by workplaces or marketable areas (10–30%) (Nabegu, 2017). Much of it ends up in open dumps where scavenger finds the items which they can sell for recycling and earn money. Mostly biodegradable waste is a major fraction having economic benefit and accounts for 54% with a moisture content of 60%. This proportion of waste is similar to the fraction noted in many countries such as India 40–60% (Lahiry, 2019), Nigeria 60–80% (Sridhar and Hammed, 2014), and Philippines 61–70% (World Bank, 2020).

1.3.2.1.1 *Case Study 1: Pakistan, Bangladesh, Yemen*

The population of the developing country, Pakistan, is growing annually by 2.4%. Currently, Pakistan is the home for 212.2 million people and is among the 10th most populous countries of the world. Lahore is the capital of the province Punjab and is considered as a second metropolitan city in Pakistan. This city has been modernized by means of an urbanization shift for standard lifestyle and economy upgradation. The daily SWG would reach 7150 tonnes daily with 0.65 kg per capita (Figure 1.2). The physical components of MSW are biodegradable, plastic, textile, diaper, and paper material (Azam et al., 2020). The majority of big cities of Pakistan like Karachi, Lahore, and Islamabad bear the same seasons, and some factors like geography, industry sector, infrastructure, and living culture influence the production of solid waste (Korai et al., 2017).

FIGURE 1.2 Solid waste generation in Lakhodair landfill, Lahore Pakistan.

Bangladesh is the country representing 160 million populations, and 29.4% out of it lives in the urban regions. They generate nearly 23,688 tonnes of MSW on daily basis and almost 70% covers organic waste (Alam and Qiao, 2020). According to World Bank, food waste is highest in developing countries like Pakistan, Bangladesh, and Yemen (Figure 1.3).

1.3.2.2 INDUSTRIAL WASTE

The industrial waste is comprised of mines material, factories, poultry farms, construction material, manufacturing processes, and supply of goods and services. This type of waste may be solid, liquid, or gaseous including dirt, masonry, oil, chemicals, concrete, and gravel (Azam et al., 2020).

Industrial waste is divided into two categories: nonhazardous and hazardous waste. It is stated that nonhazardous industrial wastes have characteristics between municipal and hazardous waste, which does not cause a threat to public health or environment, such as cartons, plastic, metals, glass, rock, and organic waste. As hazardous industrial waste possesses the hazard potential, If it is managed in improper manner, depends on its concentration, chemical, or infectious features that may contribute to or cause death or an increase the incapacitating reversible, illness, or ecological damage. As it is poisonous, chemically reactive, flammable, and toxic, such as medical discarded material, dried batteries, pesticide spray, dry-cell batteries, and ash. There are two largest groups of hazardous wastes, such as organic compounds (pesticides, Polychlorinated biphenyls, and dioxins) and nondegradable toxic metals (lead, mercury, and arsenic). The four chemical characteristics are used as means for detecting hazardous potential and include ignitability, reactivity, and toxicity (Miller and Spoolman, 2011).

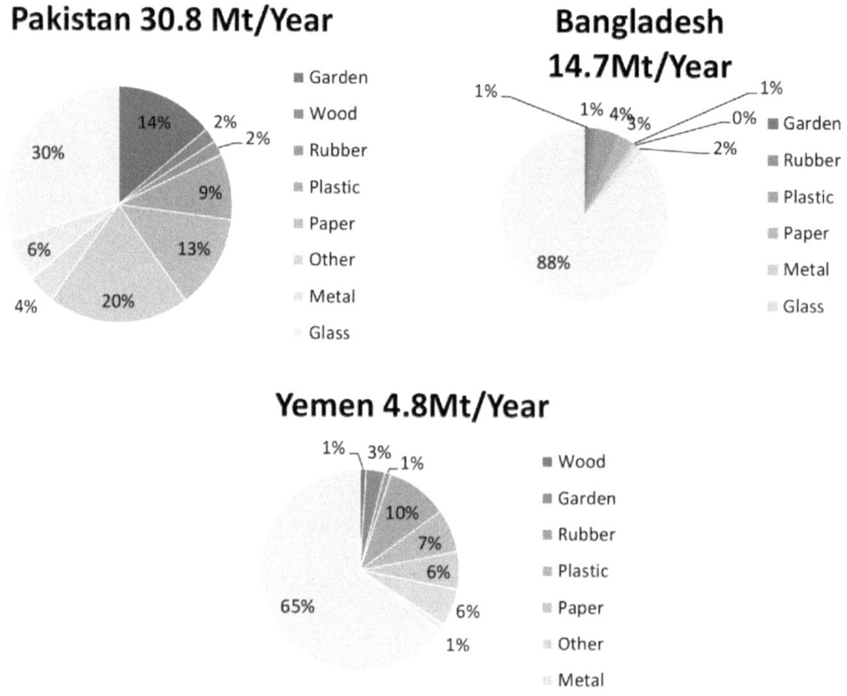

FIGURE 1.3 Comparison of municipal solid waste composition.

Source: Data based on 2016: World Bank.

1.3.2.2.1 Ignitable

This kind of waste is combustible and leads to ignition. For example, solvent-based paints, petroleum products like gasoline, detergents, and other wastes are flammable. Mostly these are available in both liquid and solid forms. Wood and paper can easily catch flame and erupt the combustion (World Bank, 2020). Solid waste can be hazardous due to its ignitability potential (Swedish Cleantech, 2020).

1.3.2.2.2 Corrosivity

These wastes are acidic or basic in nature which have pH < 2.0 or >12 and are present in aqueous that can easily soluble flesh and metals, for example, hydrocarbons, electroplating, sulfuric acid, oil, pesticides (Supplit et al., 2007).

1.3.2.2.3 Reactivity

A reactive waste undergoes violent chemical reactions under normal conditions like lithium, sulfur batteries, and explosives. This reactive waste is dangerous and can be explosive in reaction with water at high temperatures or normal optimum conditions (Weltens et al., 2012).

1.3.2.2.4 Toxicity

It is the ability of a chemical to cause a living organism to undergo adverse effects upon exposure. Toxic compounds or chemicals are found in industrial wastes that are leached into water tables and deteriorate the drinking water quality. These chemicals are arsenic, chromium, cadmium, and mercury (Guan, 2011).

1.3.2.3 AGRICULTURAL WASTE

Agricultural waste products such as bagasse, corn, cotton, rice, and wheat straw are mainly caused by the usage of agricultural practices. These by-products have a high calorific value. Moreover, these waste products also consist of chemical components such as ash, carbon, oxygen, and nitrogen. However, the waste material that is primarily produced by farm shops or vegetable packing plants is not a part of an agricultural waste.

1.3.2.4 MEDICAL WASTE

The diagnostic testing, treatment, and biological products used for humans and animals are considered to be the leading causes for generating medical waste. These products mainly include laboratory samples, media cultures, lancets, syringes, body parts, and fluids. Medical wastes can have detrimental effects on human health. However, these are not the only waste products, microbiological and biotechnology waste, human anatomical waste, and animal waste are also fall in the category of medical waste (Miller and Spoolman, 2011).

1.3.2.5 E-WASTE

In developing countries, disposal of E-waste is informal that causes harmful public health and environmental pollution problems. On the other hand, recycling and processing of electronic waste has a great risk to employees in developed nations. Proper precautions should be adopted to handle risks during recycling procedures. E-wastes include tetrabromo-bisphenol A (TBBA), chromium VI, polyvinyl chloride (PVC), lead lithium, mercury, toner dust, and other radioactive materials. Before disposal, the above-mentioned chemical compounds need to be treated. The diseases such as cancer, bronchitis, liver, and kidney damage are caused by these E-waste chemicals. Humans could be more vulnerable to environmental hazards like soil degradation and water contamination caused by these compounds (GIZ, 2019).

1.4 FACTOR AFFECTING RATE OF WASTE GENERATION AND CHARACTERISTICS

The quality and quantity of solid waste can be defined by various determinants such as income level, education status, population density, and human development (Figure 1.4). Moreover, the ever-rising population is causing immense pressure on demand for shelter, food, and other natural resources. The rise in community lifestyle standards, increasing population, and drastic have greatly accelerated SWG.

1.4.1 POPULATION DENSITY

The swift rise of population in most of the developing countries has posed challenges to respective governments to provide decent and quality of life to its citizens as stated in their constitutions. According to recent data, 83% of the population on planet earth is habituated in low-income countries. African and Asian regions experience dramatic increases in urban population. This dramatic increase in urban population eventually causes a radical increase in SWG.

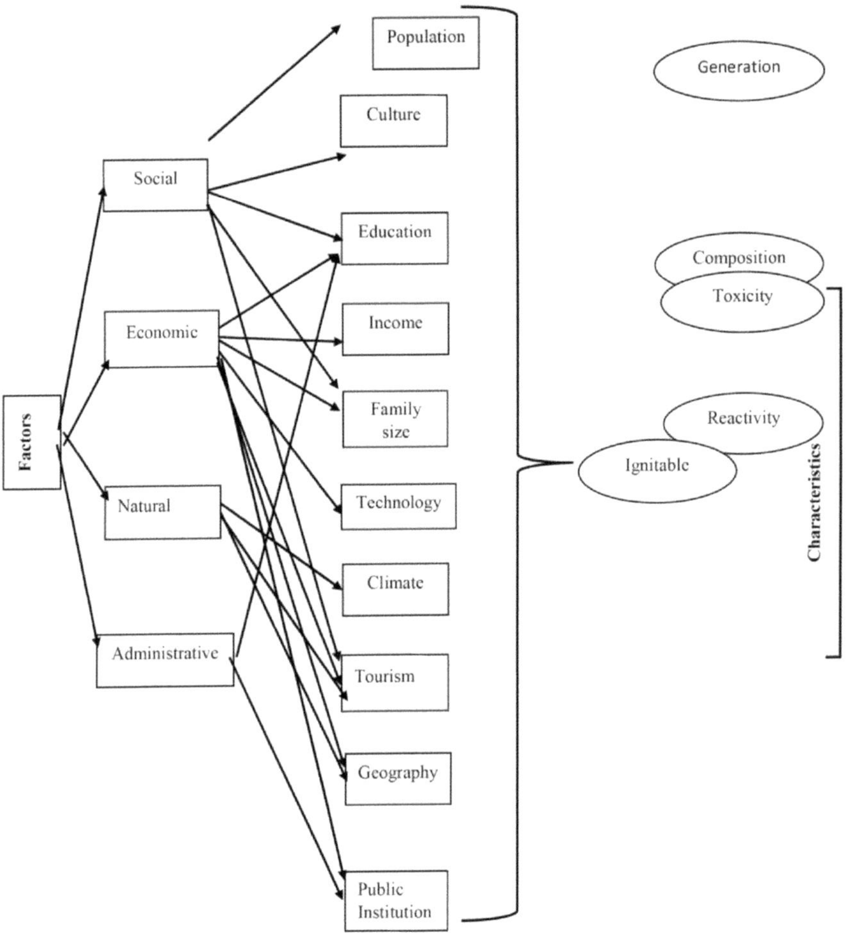

FIGURE 1.4 Factors influencing solid waste generation, composition, and characteristics.

1.4.2 CULTURE

The increasing SWG is mainly caused due to human behavior and its solution that can only be achieved by changing that behavior. Moreover, public awareness and attitudes have about waste, which can affect the whole process of SWM. The study revealed that national traditions and living style of people have also caused the variations in the composition of solid waste (Olukoju, 2018).

1.4.3 CONSUMPTION AND LIVING HABITS

Researchers have additionally located a hike in SWG because of the boom withinside the growing call for meals worldwide and different necessities, there was an escalation in quantity of waste being generated on daily basis through each family member. Consumption and dwelling behavior in exclusive areas and seasons are together a number of the maximum essential elements influencing the traits of home waste. The quantity of generated stable waste commonly accelerated in summertime season and has caused apparent variations in waste characteristics (Guan, 2011).

1.4.4 EDUCATION

Citizens having low education levels are supposed to throw their waste on streets and other undesignated locations set by the municipalities. Uneducated people discourage the waste sorting and waste collection services. Public awareness and participation are the main factors in reducing waste generation. Public involvement is a necessary tool to make the society cleaner and healthy. Without public participation, the utilization of resources will become less efficient and will lead to poor sustainability. A study investigated that public education on waste separation can develop the proper waste management system and decrease the cost of its disposal (Iraia et al., 2015).

1.4.5 MICROECONOMICS AND FAMILY SIZE

SWG is influenced by sociodemographic and economic aspects; average family size and structure, employment status, and monthly income. The study reported that the composition of waste and the social activities have direct relation and influence each other in a particular society. Research was conducted in cities of Indonesia. It disclosed that the city contains high waste generation has high population density and high economic growth characteristics. Many environmental problems are essential parts of society where households play an important role. For example, a family that has babies generates more waste comprised of diapers. Also, households with only older people show the low quantity of domestic waste as compared to mixed-age households (Abdel-Shafy and Mansour, 2018).

1.4.5.1 CASE STUDY 2: ETHIOPIA

Ethiopia is one of the developing regions of Africa poses many environmental challenges due to industrial activities and rapid population growth. A study reported that family size was positively associated with waste generated by each household. On the other hand, educational status had a negative correlation. Other studies reported the same results in Bangladesh and Nigeria. A possibility for this could be the effect of education on the attitude of individuals toward waste generation.

Scientists also found an encouraging correlation between SWG and people characteristics of income, education, and cultural patterns. Each family member's income and waste generation rate has no significant association due to cultural background. In Ethiopia, citizens regardless of their socioeconomic status, spend their income on Injera (local bread).

On the daily basis, total waste generated was 88,000 kg in Jimma city of Ethiopia and the solid waste production was 0.55 kg (average per capita), similar to Malaysia waste generation which is 0.5 kg (average per day). Almost domestic and institution waste were 87% and 13% produced, respectively. Approximately, 70% of population is engaged in agriculture occupation (Duru et al., 2019).

1.4.6 INFRASTRUCTURE AND TECHNOLOGY

The rapid growth of population in developing countries has created number of opportunities of infrastructural challenges and land use planning that collapse the capability of national and municipal governments. Unfortunately, lacking in infrastructure and technology development increases the high dependence on raw material and limits the utilization of waste reduction activities such as recycling. Moreover, inadequate landfill infrastructure amplifies the mismanagement of SWG (Al-Khatib et al., 2010).

1.4.7 PUBLIC INSTITUTION

It is dilemma that both within nations and between nations, waste is often transferred from higher-income to lower-income regions where there are fewer environmental regulations and less protection for those who work with waste (Duru et al., 2019).

1.4.8 CLIMATIC FACTORS

Climatic factors such as rainfall, humidity, and temperature in different seasons influence the both characteristics and quantity of solid waste. Humidity and rainfall fluctuate the water concentration of solid waste directly. Flooding produces piles of household waste which intensifies the load on nearby waste dumping sites. Heatwaves of high energy in the hot summer season, increase odor and dust from arable sites. Also, increased frequency in rainfall during summer causes the high risk of flooding affecting access to health services, facilities, water quality, and use of mobile waste management plants. After glacier melting, rising sea levels will lead to increase the erosion of coastal dumping sites causing pollution of coastal waters (Gichamo and Gökçekuş, 2019). Usually, coal and wood are used for the heating purposes of homes in winter in the most rural area of developing countries, both of these materials yield ash that is thrown away with domestic waste and disturb the composition of ambient atmospheric gases (Aldrin, 2017).

1.4.9 GEOGRAPHY

Area of living has effects on waste composition, solid waste, and extent of awareness. A study was conducted in Indonesia that shows residents of urban area have marginally higher awareness rate about solid waste management than the people in the suburb. The city center is responsible for less waste; 0.18 kg per capita on a daily basis and suburban area generates high waste 0.295 kg per capita (Khair et al., 2018).

1.4.10 TOURISM

Tourism is fast growing industry and a driving force for an increase in MSW generation in developing countries like Kenya, India, Pakistan, and

Philippines (Murava and Korobeinykova, 2016). Yet, systematical waste generated from the tourism sector remains uncounted and hidden behind residential waste flows. Study shows that an average tourist may generate more MSW than a local resident. Unsurprisingly, mostly waste is generated by the hospitality industry. Usually, the tourist spots are located in remote areas deprived of proper infrastructure and services for the management of MSW. This amalgamation can lead to a vicious cycle of tourism that extinguishes natural environment and host communities (Diaz-Farina et al., 2020). For example, Ayubia National Park is one of the highly fascinating and beautiful destinations in Pakistan and it maintains three types of forest ecozones; subalpine meadows, moist temperate coniferous, and subtropical chir pine forest, forest. Almost every year in the summer season, more than 120k tourists visit the park. Since the last few years, hotels and restaurants have been generating 2940–3225 kg/day waste due to tourists' pressure. This is an alarming situation as approximately, tourists generate 3.38–3.84 kg/capita/day solid wastes. Growing solid waste is constantly disturbing the biodiversity and also badly distressing aquatic resources.

1.5 HARMFUL EFFECTS ASSOCIATED WITH SOLID WASTE GENERATION

As we discussed earlier in this chapter, the rate of hazardous material is rapidly increasing in terms of total quantity and per capita average. In developing countries, rapid increase in urbanization, overproduction, irregular waste collection, improper sorting systems, open dumping, burning, unpleasant odors, and inappropriate disposal of MSW are serious threats to public health and environmental safety (Azam et al., 2020).

1.5.1 LAND USE

A large area is allocated for waste dumping site which is a tremendous opportunity, especially in a country that has limited land space. Those land areas could also be utilized for sustainable development that would generate higher GDP for the country. While the governments of developing countries are bound to spend substantial budget on recycling and expansion of landfill sites. There is an alarming situation to investigate the root causes behind the piles of solid waste and to adopt measures to mitigate the generation rate. Moreover, the increasing rate of malnutrition, the extreme humanitarian crises, and death from starvation in many countries of the developing world also require the awareness and consciousness campaigns in food wastage (Diaz-Farina et al., 2020).

1.5.2 IMPACTS ON ENVIRONMENT

Solid waste releases greenhouse gases (GHGs) that contribute in climate change. Approximately 5% of global GHG in the form of CO_2, CH_4, and N_2O are caused by emissions from dumping sites (Gichamo and Gökçekuş, 2019). As we mentioned earlier, more than 70% of solid waste is degradable and is found in low-income economies which majorly account for GHG emissions. The landfill is the third biggest emission source of methane gas around the globe. A study was conducted which stated that Pakistan stands as 135th country for global methane emissions, adding almost 0.8% global GHG (Zuberi and Ali, 2015).

1.5.3 IMPACTS ON HUMAN

Solid waste pollution has drastic impacts on human health. Uncontrolled SW increases the spreading and breeding of dengue mosquitoes (Khalid and Ghaffar, 2015). For example, a decade ago, 40,000 people were affected by dengue fever in Pakistan and almost 17,256 cases were reported in city Lahore (Khan and Abbas, 2014). The spread of dengue fever has suffered 10,000 people in Rawalpindi, another city of Pakistan (Farmer, 2019).

1.5.3.1 CASE STUDY 3: KIBERA SLUM, KENYA

Increased population has supplanted the existing social amenities such as houses, schools, and hospitals as strategies of urban planners. This has resulted in the explosion of informal settlements, whose geographical locations are surrounded by a poor environment that has unhealthy living conditions to mankind. The study elucidates that limited knowledge, pessimistic approach from the locals, and bad governance are the factors for the continued hostile status of African' biggest slum "Kibera" territory in Nairobi, Kenya. Kibera forms the part of Nairobi 6 million population (Figure 1.5). An estimate of the total population in the 225-hectare of area settlement between 500,000 and 700,000 inhabitants. According to government data, 82% of Kenyan people live in an informal and poor community and use charcoal material for energy. Their livelihood dependence on fuelwood rapidly decline the Kenya's forests and cause the adverse effects on the local climatic condition, wild animals and plants, water resources, and forest inhabitants. Most of Kibera's residents subsist on less than a dollar a day. Half of them are unemployed. Kibera generates approximately an average of 205 metric tonnes (226 tonnes) of waste per day and 75,000 tonnes per year. There was no proper waste collection system, half of the waste ends up in public spaces which was an eyesore and a menace to public health (Ouma, 2020). In Kibera, the prevalence of diarrhea mostly among children under age three is 40%, more than three times higher than rest of the city (The New Humanitarian, 2012).

FIGURE 1.5 Burning and human contact with solid waste in Kibera, Nairobi Kenya.

1.6 WAYS TO REDUCE WASTE GENERATION

Since the 1990s, people in the developed regions of the world commenced to encourage the concept of diverting waste materials with a famous slogan "Reduce, Reuse, Recycle," also known as the 3Rs. The adoption of this phrase incorporates a practical approach to the waste management because each technique gives benefits to the environment from the high priority to the low. From an environmental viewpoint, priority is given to the first two Rs because these prevention approaches resolve the issues regarding waste. Moreover, reduce and reuse methods are energy efficient, save money, reduce land pollution, and assist in protecting ecology (Friedland et al., 2011).

1.6.1 REDUCE

Among the 3Rs, "Reduce" is the first approach because decreasing the resource inputs is the efficient way to attain the optimum decline in solid waste. This is also called as waste prevention and waste minimization as well. There are many ways to minimize the waste generation. One approach is source reduction, which pursues to minimize the waste by reducing in the initial process of manufacture. Thus, source reduction optimizes the energy efficiency as it generates less output materials and avoids long disposal processes. The utilization of fewer resources will also provide economic benefits. This approach is effective at both individual and corporate levels (Diaz-Farina et al., 2020).

For example, in ancient time, computer machine was large in size and occupied many square feet area. Now modern computers are lightweight, small in size, and easy to handle through the use of composite materials. Another example is if a class teacher gives two pages handout material to her students, she reduces 50% of paper use by providing double-sided photocopies to students. The overall energy used by machine to photocopy them over time will probably be reduced. Moreover, if a manager does not hand out any paper sheets but send copies of the document to the employees by email. This is also the good example of source reduction. In the manufacturing process, the application of source reduction will result the reducing the output materials that go for packaging. Hence, new packaging material may provide the same protection to the product with minimum resources.

The second approach is to manufacture the products that are easy to repair, reform, and compost. Third, charge fee to every consumer for the quantity of solid waste they throw away on roads or streets. Provide free service of waste collection for recyclable and reusable materials. Forth, implement cradle-to-grave accountability rules and legislations that require manufacturing factories to get back various used products, such as motor vehicles, and other electronic appliances as some developed countries adopt. Fifth, shift the urban transportation systems from car to mass transit and bicycles (Gichamo and Gökçekuş, 2019).

1.6.2 REUSE

Rather than waste disposal, "Reuse" involves cleaning and using a products that allow an object to cycle within a system for longer time period. This process of waste reduction decreases the pollution, usage of matter, saves money, and improves the economy by creating jobs. No additional energy or resources are required to produce more material.

For example, a used letter envelope can be reused by writing the new information over it. Here, residence times of the envelope in the system are getting increased but the waste generations are reduced. Sometimes reuse involves the repairing an existing material costing money and time, labor, and energy. For example, when we reuse a disposable polystyrene cup more than one time, though reuse involves cleaning the cup and generating some wastewater, also adding some energy cost. Reuse is successful and common practice in many developing countries. In Northeast Thailand, government built 19 Buddhist temples from lots of beer bottles. The colorful used bottles beautifies the temples and allows light penetration into the temple interiors. Bottles caps were also reused to make mosaics artwork (Lew, 2020).

1.6.3 RECYCLING

Recycling is the practice by which waste objects are sorted and converted into input material then used to produce new products. The informal way of waste recycling is a common practice for survival in developing countries. There are push factors that engage poor people into waste picking, fundamentally like economic. Waste pickers are vulnerable groups of our society such as unemployed and disabled people, recent migrants, women, children, and old aged people. They live in an unsafe and filthy environment, and usually work in open dumps and on streets, where they constantly contact with all types of solid waste that poses high risks to their health. In developing countries, only 16% materials are recycled in the waste stream (Lew, 2020).

For example, Lahore roughly recycled almost 27% of waste through the informal way. Currently, there are no operative waste disposal facilities that follow formal recycling systems. This city does not represent the high performance of government in the waste management sector. However, waste pickers collect the used paper and pulp industry recycle it. In Indonesia, recycling reduces approximately 10% of total waste. The scavengers are also playing vital role in decreasing SW in Iran (Dhokhikah and Trihadiningrum, 2012).

1.6.4 ENVIRONMENTAL EDUCATION AND TRAINING

Various projects like environmental training and education have played a significant role in enhancing recycling and reducing waste generation (Han et al., 2018). Increased willingness of the commercial sector to support recycling is an encouraging sign of public support for the environment. For example, Tetra Pak Pakistan is collaborating with various international partners such as World Wilde Fund for nature, waste management companies, pulp industry, and recycling partners to initiate the recycling of used beverage cartons (UBCs) in Pakistan (Tetra Pak, 2017). In 2015, they almost facilitated the recycling of more than 22,000 tonnes of UBCs. Likewise, in Nairobi Kenya, recycling company Taka Taka solutions is working with German organization; Deutsche Gesellschaft für Internationale Zusammenarbeit (GIZ) to recycle an impressive 95% of the 40 tonnes of rubbish generated each day. On the other side, consumers are more likely to buy products that can be easily recycled. Some consumers have purchased small home appliances that reduce the volume of waste and facilitate recycling. Moreover, living habits, traditional and national cultures, and consumption of goods alter the production and composition of domestic waste. The insights into generated waste allow making suggestions for improved education status (GIZ, 2019).

1.7 FUTURE SCENARIO AND SUSTAINABILITY

As the worldwide, volume of solid waste continues to grow, it is clear that the problems related to waste extend beyond the national borders. Solutions to these problems require international coordination among all developing countries. Some international agreements are regional but they have to be international. Like the developed nations such as European Union and United Nations have several regulations about the reductions of solid wastes, developing nations must follow their footprint toward sustainable development. Moreover, steps should be taken to transport of hazardous solid materials from wealthier countries into poor countries because it raises ethical concerns. Interest is gradually increasing in international coordination for combating the solid waste problems that lack environmental regulations.

The population growth and economic development of the developing world have evolved the new challenges in the waste management sector. One challenge is an increasing waste complexity like using nanotechnology adding to the more complexity of products, which requires the more elaborate waste management practices. The second challenge is the data on waste generation are not enough authentic and reliable. So, the understanding of waste characteristics needs to be much improved among the stakeholders who deal the waste management or take decisions for it.

Achieving sustainable development in developing countries involves the planning of the reducing ecological and carbon footprint, and altering the modes of consumption of resources. Lower middle countries are anticipated to mount SWG to 0.79 kg/capita/day by 2050 which is more than the low-income countries (Figure 1.6). The material footprint of developing countries is increased, demonstrating major changes in lifestyle standards. The implementation of solid waste management is constantly lacking in these countries (Olukoju, 2018). International and national nongovernmental organizations can play a vital role in effectively projecting the community's waste generation problems.

We need to take appropriate community steps, to give rights to those in the waste disposal sector, and to find suitable methods that maintain jobs for those who are already working. Regulations given by EPA and international treaties are necessary to protect the most vulnerable countries from exploitation.

1.8 CONCLUSION

SWG is an emerging and problematic issue in developing countries of the world.

As awareness among people toward reducing solid waste increases, it shows a great step toward a sustainable future. The current situation of developing countries is summarized in this chapter. This chapter also presents the ways to reducing the extent of SWG and to avoiding its harmful effects. More research works are required to measure the real potential of hazardous waste releasing from industries and to optimize recycling to attain the waste free developing countries. People training and environmental education are necessary tools required to achieve the goals.

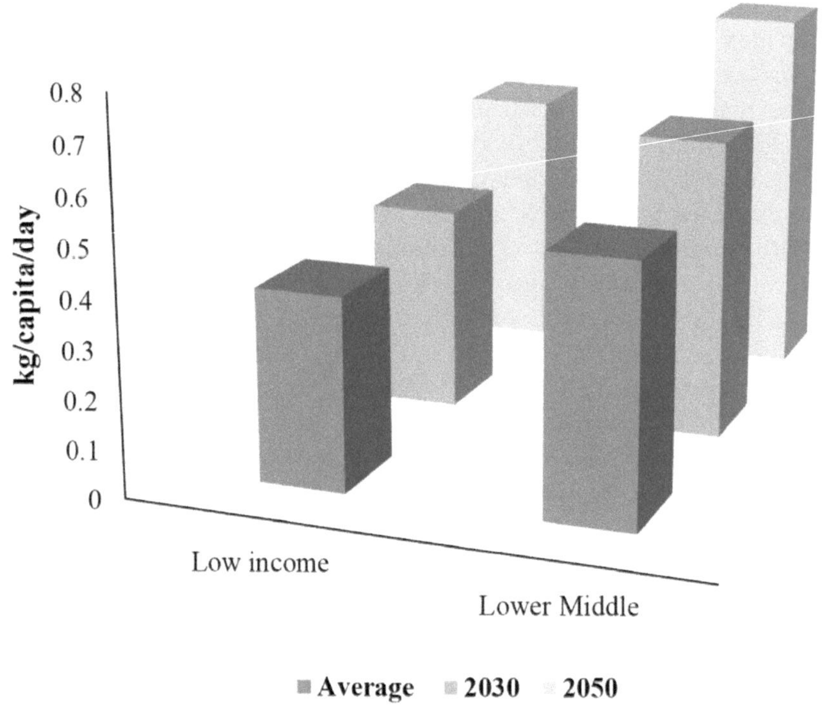

FIGURE 1.6 Future projection of solid waste generation.

ACKNOWLEDGMENT

We are thankful to Mr. Riccardo S. Warunge for helping us in capturing the pictures of waste dumping site in Kibera territory, Nairobi, Kenya.

KEYWORDS

- **solid waste**
- **generation**
- **composition**
- **hazardous waste**
- **developing countries**

REFERENCES

Abdel-Shafy, H. I.; Mansour, M. S. Solid Waste Issue: Sources, Composition, Disposal, Recycling, and Valorization. *Egypt. J. Petrol.* **2018**, *27* (4), 1275–1290.

Alam, O.; Qiao, X. An in-Depth Review on Municipal Solid Waste Management, Treatment and Disposal in Bangladesh. *Sustain. Cities Soc.* **2020**, *52*, 101775.

Aldrin, B. Plaza. *Ditch NIMBY to Fix Philippines' Municipal Solid Waste Problem. Asian Development Blog. Straight Talk from Development Experts.* **2017.** https://blogs.adb.org/blog/ditch-nimby-fix-philippines-municipal-solid-waste-problem (accessed Oct 12, 2020).

Al-Khatib, I. A.: Monou, M.; Zahra, A. S. F. A.; Shaheen, H. Q.; Kassinos, D. Solid Waste Characterization, Quantification and Management Practices in Developing Countries. A Case Study: Nablus District–Palestine. *J. Environ. Manage.* **2010**, *91* (5), 1131–1138.

Azam, M.; Jahromy, S. S.; Raza, W.; Raza, N.; Lee, S. S.; Kim, K. H.; Winter, F. Status, Characterization, and Potential Utilization of Municipal Solid Waste as Renewable Energy Source: Lahore Case Study in Pakistan. *Environ. Int.* **2020**, *134*, 105291.

Baawain, M.; Al-Mamun, A.; Omidvarborna, H.; Al-Amri, W. Ultimate Composition Analysis of Municipal Solid Waste in Muscat. *J. Clean. Product.* **2017**, *148*, 355–362.

Dhokhikah, Y.; Trihadiningrum, Y. Solid Waste Management in Asian Developing Countries: Challenges and Opportunities. *J. Appl. Environ. Biol. Sci.* **2012**, *2* (7), 329–335.

Diaz-Farina, E.; Díaz-Hernández, J. J.; Padrón-Fumero, N. The Contribution of Tourism to Municipal Solid Waste Generation: A Mixed Demand-Supply Approach on the Island of Tenerife. *Waste Manage.* **2020,***102*, 587–597.

Duru, R. U.; Ikpeama, E. E.; Ibekwe, J. A. Challenges and Prospects of Plastic Waste Management in Nigeria. *Waste Disposal Sustain. Energy* **2019,***1* (2), 117–126.

Farmer, B. *Pakistan Becomes Latest Country to Suffer Severe Dengue Outbreak*, **2019**. https://www.telegraph.co.uk/global-health/science-and-disease/pakistan-becomes-latest-country-suffer-severe-dengue-outbreak// (accessed Oct 4, 2020).

Friedland, A.; Relyea, R.; Courard-Hauri, D. *Environmental Science: Foundations and Applications*; W. H. Freeman & Company, 2011; p 717.

Getahun, T.; Mengistie, E.; Haddis, A.; Wasie, F.; Alemayehu, E.; Dadi, D.; Van Gerven, T.; Van der Bruggen, B. Municipal Solid Waste Generation in Growing Urban Areas in Africa: Current Practices and Relation to Socioeconomic Factors in Jimma, Ethiopia. *Environ. Monitor. Assess.* **2012**, *184* (10), 6337–6345.

Gichamo, T.; Gökçekuş, H. Interrelation between Climate Change and Solid Waste. *J. Environ. Pollut. Control* **2019**, *2* (1), 104.

GIZ. *Kenya: Waste Recycling Experts*, 2019. https://www.giz.de/en/mediacenter/78669.html (accessed Aug 23, 2020).

Guan, Y. D. *Pollution Investigation on Abandoned Rural Unsanitary Landfills and Leachate Treatment Using Multi-Media Lavering (MML) System*; Doctor thesis; Zhejiang University, China, 2011; p 17.

Han, Z.; Liu, Y.; Zhong, M.; Shi, G.; Li, Q.; Zeng, D.; Zhang, Y.; Fei, Y.; Xie, Y. Influencing Factors of Domestic Waste Characteristics in Rural Areas of Developing Countries. *Waste Manage.* **2018**, *72*, 45–54.

Ike, C. C.; Ezeibe, C. C.; Anijiofor, S. C.; Daud, N. N. Solid Waste Management in Nigeria: Problems, Prospects, and Policies. *J. Solid Waste Technol. Manage.* **2018**, *44* (2), 163–172.

Iraia, O. G.; Oihane, K. E.; Cristina, M.; Ana, M. M. A.; Ainhoa, A. V. Identification of Municipal Characteristics Regarding Household Waste Generation and Their Forecasting Ability in Biscay. *Waste Manage.* **2015**, *39*, 26–34.

Kawai, K.; Tasaki, T. Revisiting Estimates of Municipal Solid Waste Generation Per Capita and Their Reliability. *J. Mater. Cycl. Waste Manage.* **2016**, *18* (1), 1–13.

Khair, H.; Putri, C. N.; Dalimunthe, R. A.; Matsumoto, T. February. Examining of Solid Waste Generation and Community Awareness between City Center and Suburban Area in Medan City, Indonesia. *Conf. Ser. Mater. Sci. Eng.* **2018,** *309*, 012050.

Khalid, B.; Ghaffar, A. Dengue Transmission Based on Urban Environmental Gradients in Different Cities of Pakistan. *Int. J. Biometeorol* **2015**, *59*, 267–283.

Khan, I. A.; Abbas, F. Managing Dengue Outbreak in Lahore, Pakistan: Efficacy of Government Response and Lessons for the Future. *J. Health Manage.* **2014**, *16* (4), 471–480.

Korai, M. S.; Mahar, R. B.; Uqaili, M. A. The Feasibility of Municipal Solid Waste for Energy Generation and Its Existing Management Practices in Pakistan. *Renew. Sustain. Energy Rev.* **2017,** *72*, 338–353.

Kumar, A.; Samadder, S. R. A Review on Technological Options of Waste to Energy for Effective Management of Municipal Solid Waste. *Waste Manage.* **2017**, *69*, 407–422.

Lahiry, S. *India's Challenges in Waste Management. The Key to Efficient Waste Management Is to Ensure Segregation Source and Resource Recovery.* 2019. https://www.downtoearth.org.in/blog/waste/india-s-challenges-in-waste-management 56753#:~:text=But%20many%20waste%20to%20energy,little%20over%2050%20per%20cent (accessed Sept. 22, 2020).

Lew, R. *Solid Waste Management in Pakistan*, 2020. https://www.bioenergyconsult.com/solid-waste-management-in-pakistan/ (accessed Oct 5, 2020).

Liyala, C. M. *Modernizing Solid Waste Management at Municipal Level: Institutional Arrangements in Urban Centers of East Africa*; PhD Thesis; Environmental Policy Series. Wageningen University, The Netherlands, 2011; p 17.

Miller, G. T.; Spoolman, S. *Living in the Environment: Principles, Connections, and Solutions*, 17th ed.; Nelson Education, 2011; p 804.

Murava, I.; Korobeinykova, Y. The Analysis of the Waste Problem in Tourist Destinations on the Example of Carpathian Region in Ukraine. *J. Ecol. Eng.* **2016**, *17* (2), 43–51.

Nabegu, A. B. An Analysis of Municipal Solid Waste in Kano Metropolis, Nigeria. *J. Human Ecol.* **2017**, *31* (2), 111–119

Olukoju, A. "Filthy Rich" and "Dirt Poor:" Social and Cultural Dimensions of Solid Waste Management (SWM) in Lagos. *Soc. Dynam.* **2018**, *44* (1), 88–106.

Ouma, D. S. The Expectant 'Chocolate City' and the Unfulfilled Attainment of Safe Heavens in Kibera Informal Slums in Nairobi, County. *J. Humanities Soc. Sci. Invention* **2020**, *2* (01), 5–12.

Ramachandra, T. V.; Bharath, H. A.; Kulkarni, G.; Han, S. S. Municipal Solid Waste: Generation, Composition and GHG Emissions in Bangalore, India. *Renew. Sustain. Energy Rev.* **2018**, *82*, 1122–1136.

Romero, P. *Phl Facing Garbage Crisis; 16.6 Million Metric Tons of Waste This Year Can Fill 99 Philippine Arenas*, 2020. https://www.onenews.ph/phl-facing-garbage-crisis-16-6-million-metric-tons-of-waste-this-year-can-fill-99-philippine-arenas (accessed Oct 15, 2020).

Sharma, K. D.; Jain, S. Municipal Solid Waste Generation, Composition, and Management: The Global Scenario. *Soc. Responsibility J.* **2020,** *16* (6), 917–948.

Sridhar, M. K. C.; Hammed, T. B. Turning Waste to Wealth in Nigeria: An Overview. *J. Human Ecol.* **2014**, *46* (2), 195–203.

Supplit, R.; Koch, T.; Schubert, U. Evaluation of the Anti-Corrosive Effect of Acid Pickling and Sol-Gel Coating on Magnesium AZ31 Alloy. *Corros. Sci.* **2007,** *49* (7), 3015–3023.

Swedish cleantech. *Waste Management: The Swedish Success Story*, 2020. https://swedish-cleantech.com/news/recycling-and-waste/waste-management-the-swedish-success-story/ (accessed Aug 9, 2020).

Tetra Pak. *Environmental Activities in Pakistan*, 2017. https://www.tetrapak.com/pk/sustainability/environmental-activities-in-pakistan (accessed Aug 22, 2020).

The New Humanitarian. *Human Waste Woes in Kenya's Slums*, 2012. https://www.thenewhumanitarian.org/feature/2012/12/27/human-waste-woes-kenyas-slums (accessed Oct 2, 2020).

Weltens, R.; Vanermen, G.; Tirez, K.; Robbens, J.; Deprez, K.; Michiels, L. Screening Tests for Hazard Classification of Complex Waste Materials—Selection of Methods. *Waste Manage.* **2012**, *32* (12), 2208–2217.

World Bank. *World Bank indicator: Economy and Growth*, 2020. https://data.worldbank.org/indicator/NY.GNP.PCAP.CD (accessed Oct 2, 2020).

Zhou, H.; Meng, A.; Long, Y.; Li, Q.; Zhang, Y. Classification and Comparison of Municipal Solid Waste Based on Thermochemical Characteristics. *J. Air Waste Manage. Assoc.* **2014,** *64* (5), 597–616.

Zuberi, M. J. S.; Ali, S. F. Greenhouse Effect Reduction by Recovering Energy from Waste Landfills in Pakistan. *Renew. Sustain. Energy Rev.* **2015**, *44*, 117–131.

CHAPTER 2

Mismanagement of Waste in Developing Countries

MUHAMMAD AMEEN[1*], MUHAMMAD ANWAR-UL-HAQ[2], MUHAMMAD IRFAN SOHAIL[2], FATIMA AKMAL[2], and AYESHA SIDDIQUI[3]

[1]*Department of Soil Science, Faculty of Agriculture and Environment, The Islamia University of Bahawalpur, 63100, Pakistan*

[2]*Institute of Soil and Environmental Science, University of Agriculture, Faisalabad.*

[3]*Department of Botany, University of Agriculture, Faisalabad 38040, Pakistan*

**Corresponding author. E-mail: soilchemist.ameen033@gmail.com*

ABSTRACT

Strategies for waste management differs in developed and developing countries. Resultant product of any chemical reaction or process also generate some unwanted or unusable material known as waste. Waste is produced in every industry, named according to their origin. The total amount of waste is generally termed as solid or municipal solid waste. Each type of waste affects the soil, air and water environment. Waste is categorized as municipal, biomedical, hazardous and special hazardous waste. The main industries contributing the waste generation are Agriculture, Steel, Leather, Cement, Textile, Sugar, Plastic, Tyre, Batteries and Automobile. The core techniques for waste management in developing countries are recycling, bioconversion, incineration, however due to multiple factors these techniques are mismanaged. Waste mismanagement includes the open burning and open dumping at

Waste Problems and Management in Developing Countries. Umair Riaz, Shazia Iqbal, & Moazzam Jamil (Eds.)

the end, which affects the environment and human health seriously. Open burning results many toxic aerosols of SO, CO, NO, CO2, PM and other pollutant emissions. Solid waste contaminate water bodies and environment as well. Mismanagement of solid waste or waste is the main concern of social and environmental impacts. Unplanned and uncontrolled disposal of waste is another mismanagement, which severely generate pollution in soil, water and environment. The pollution of surface water is also due to mismanaged dumping and their uncontrolled flow. Lack of appropriate waste management technique is also a main defect in developing nations. The developing nations faces different challenges for waste management including the lack of organization for suitable waste controlling, no regulation prior with these waste fractions. In developing countries, there are a lot of waste material, which are not separated properly and either dumped openly or burned so it leads to either different types of pollutions or their dissemination. There are many factors, which are not working properly in developing countries ultimately causes mismanagement of waste. These includes education, finance, infrastructure and their related resources. The chapter will strengthen the reader knowledge, about the different type of waste, and how to manage efficiently. It will give the concept about impacts of waste on the surrounding, soil, air, water and human.

2.1 INTRODUCTION

Waste is the end outcome of the industry. Waste mismanagement is an environmental issue, which is mainly due to burning and dumping openly. All the strategies of waste management in developing countries are almost same as in developed countries, however in these countries; approach of waste management is not up to the mark. Waste management program is difficult in everywhere in the world, however, in less developed countries these problems are further enhanced (1). Main hurdles in a good waste management program are, insufficient waste management technologies and equipment, shortage of capital for investment and scarcity of protective measures that makes the waste management down the order. This program is deteriorated mainly by ignorance of the citizen, which ultimately deteriorate the environment due to waste dumping, burning and mismanagement, economic interdependence (2). This situation is further enhanced by, type, quality and sources of hazardous waste. Many multinational companies shift their plants in less developed countries, and they have banned the use of technologies in their countries (3). Main sources of waste production after industry are the transporters, disposal facilities and the scavengers. The scavengers use the waste for their livelihood and they claim the waste for recycle after facing a great health risk. Waste also includes:

- Solid waste
- Sewage waste/municipal waste
- Industrial waste
- House hold/kitchen waste
- Hospital/biomedical waste

In developing countries, there are many differences for waste management in villages and cities depending upon the volume and type of waste formed. These differences should be addressed. Soil, water and plants suffers with heavy metal pollution, which is due to the uncontrolled disposal of the waste (4). The mismanagement of waste in any industry is a result of lacking the basic functional element of waste management, which are storage, collection, transportation and treatment. Open burning generates CO, CO2, SO, NO and PM10, which affects the atmosphere and health risks for, peoples working in these areas (5). Waste management is the result of various social and environmental impacts, which needs to improve. Any unwanted or unused material in an industry is known as waste. This is the material, which is discarded after its primary use. We can convert waste to a bi-product by improving their economic value a little or more. The examples are municipal solid waste, water waste etc. The countries, which are low or middle-income countries, are developing countries and around the world, waste generation is very high (3).

Waste production rate is projected to about 20% annually from 2016 level to 3.40 billion tons in 2050 (6). In developing countries more than 90% waste is either dumped or burned off. These practices produce serious health concerns, safety and environmental impacts (6). Mismanaged waste help to increase or produce the diseases and contribute to global climate change through the generation of pollutants (7). Proper management of waste is essential for better environment, also is a great challenge for developing countries. For example proper waste management is too expensive, require 20-50% budget of the organization/factory. Waste mismanagement is aggravated due to large variety of social, cultural, economic and legislative problems (1). There are many mismanaged waste sinks, which can be summarized as below.

1. No proper waste collection.
2. Dumping of waste at an unappropriated place like streets.
3. Managing the mixed waste simultaneously.
4. Open burning.

5. No proper and well-managed sanitary landfills.
6. Less or no awareness of citizens about waste mismanagement and related health problems.

2.2 WASTE PRODUCTION IN DEVELOPING COUNTRIES: CURRENT SCENARIO

At present worldwide municipal solid waste generation is 1.3 BT year^{-1}, which would be 2.2 BT year^{-1} approximately in 2025 (3). Next fifteen year will be very crucial; because of will be increased rate of waste production up to 1.42 kg/person/day/capita from 1.2 kg/person/day/capita. This waste generation is mainly due to rapid urbanization or rapid population increase, industrialization, people habits and climate of the country (8). Waste generation is high when more population shift to urban areas and their living standards become high due to more income (Figure 2.1).

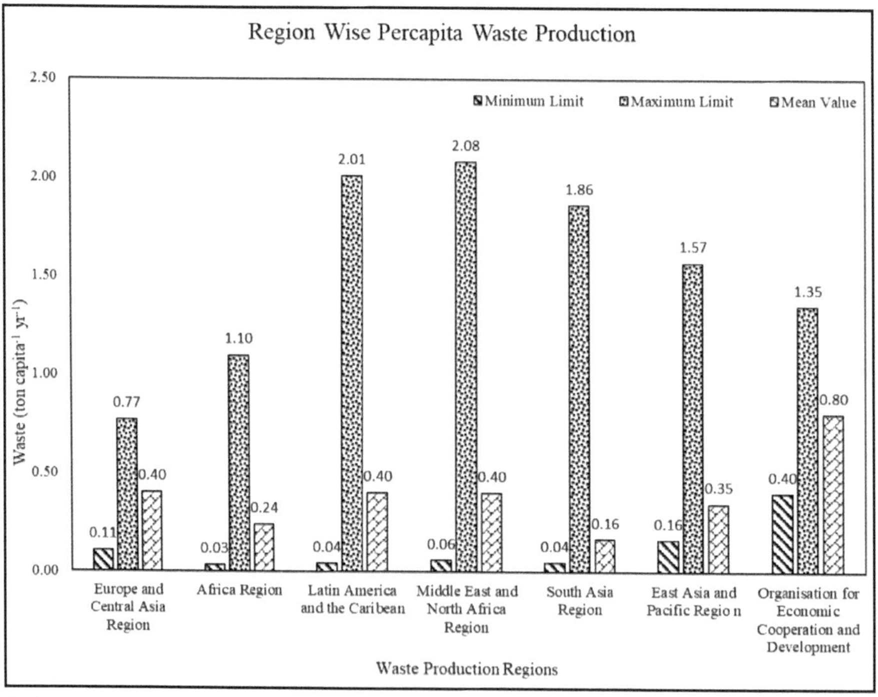

FIGURE 2.1 Region wise per capita waste production

Source: Author, Modified from Ref: (3)

MSW management is the main problem which need to be focused after which the other services should be provided to the citizens, as it requires heavy investment of the budget. Mismanaged waste has many dangerous effects on economy, health and over all climate. Mismanagement of waste increase the cost of efficient waste management. The daily waste production in South Asia (SA), East Asia (EA) and the Pacific collective is nearly 1 MT day^{-1} (3). The low-income nations invest more on collecting the waste rather than disposal while opposite is the caste with high-income countries. Asia, in which waste growth is fastest, contains organics and paper in its waste stream.

Waste production is increasing in the whole world. For example 2.01 billion tons solid waste was generated from the world's cities which was 0.74 kilograms per person per day in 2016 and this annual waste generation would be to 3.40 billion tons in 2050 due to more population increase (3).

Each hospital produce the waste according to its facilities. General District hospital can produce the 1-4 kg of waste/bed/day, whereas 4-9 kg of waste/bed/day is produced by high-income university hospital (9). Current scenario of waste production in middle and low-income country is 0.5-6 kg and 0.5-3 kg respectively (Figure 2.2).

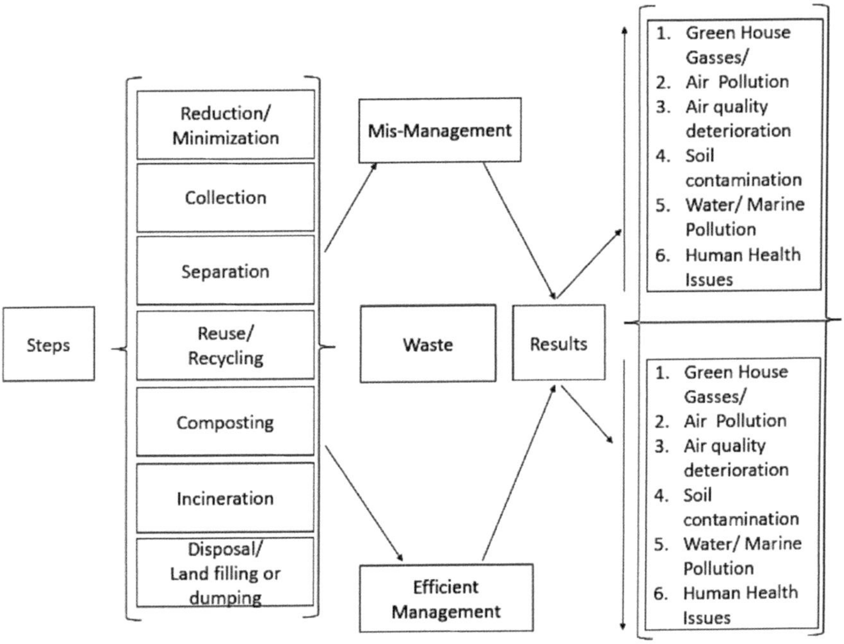

FIGURE 2.2 Current Scenario of waste production, management, mismanagement and results on global environment.

Source: Author

The Fig. 2.2 shows the current scenario, which describes the different steps of waste handling/management. The steps includes reduction, collection, separation, recycling, compositing, incineration and disposal or dumping. The mismanagement at each step results in the increase in greenhouse gasses, air pollution, air quality deterioration, soil contamination, water pollution and human health issues. However if we take efficient measures there should be sudden decrease in the above discussed results.

2.3 TYPES OF WASTE PRODUCED IN DEVELOPING COUNTRIES

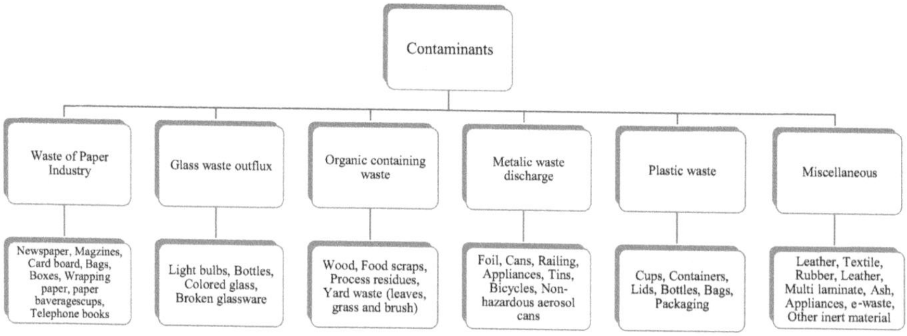

FIGURE 2.3 Type of waste and their sources

Source: Author; Modified from Ref: (3)

In developing nations, generated waste is either burned or disposed of openly, as the main practice of waste treatment, however it involving many environmental and health impacts (10). Adoption of waste treatment technology (burning and dumping) for every fraction of waste pollutes soil, water and air environment. Waste material which are recycled, are picked for recycling as a informal practice decrease the amount of waste for water and open dumps (11) but also a great health and occupational risks (12; 13).

Mostly developed countries produce specific hazardous waste whereas developing countries produce the mixed waste. The informal sector, which is the main problem in developing countries that has small, competitive and labor intensive business, served as the critical source of hazardous solid waste (14).

Mismanagement of hazardous agrochemicals generates more serious conditions in developing countries, especially in Africa and Latin America compare to developed world. In agriculture industry cultivation of crops, rearing of animal and agriculture processing produces the animal and crop residues, which are reused or recycled at a high rate, as soil amendments, animal feed (15). Some of the hazardous waste fractions including fertilizers, pesticides, veterinary products and animal remains are great threat being many impacts on health and environment due to their mismanagement (16). In developing countries, most of the pesticides and fertilizers ultimately add in surface and ground water with huge health impacts.

Modern system of waste management describes the waste into many types, covers almost every type of waste.

1. Municipal waste
   - (1) Household/domestic Waste
   - (2) Construction Waste
   - (3) Demolition waste
2. Hazardous waste
   - (1) Industrial Waste
     - (a) Agriculture industry
     - (b) Cement industry
     - (c) Plastic industry
     - (d) Tyre industry
     - (e) Battery industry
     - (f) Automobile industry
     - (g) Iron/Steel industry
     - (h) Sugar industry
     - (i) Leather industry
     - (j) Textile industry
3. Biomedical waste
   - (1) Clinical waste/Hospital waste
4. Special hazardous waste
   - (1) Radioactive/Nuclear waste
   - (2) Explosive waste,
   - (3) Electronic waste (e-waste)

2.3.1 MUNICIPAL WASTE:

2.3.1.1 HOUSEHOLD WASTE:

The dramatic increase of municipal solid waste in low income nations is mainly due to increase in population and economic growth. This generated MSW is mismanaged and causes serious issues at every step of waste management (17-23). Main problems associated with the MSW is open dumping, improper storage of waste, (18), lack of knowledge and improper placement of collection bins. The municipal waste comprised of household, commercial and demolition waste. In Pakistan, household waste is generated on average basis 0.79 kg/cap/day (24). Due to the household waste mismanagement in low income countries diseases and pollution risks are aggravated. Household waste can be best managed either by safe disposal after incineration or other technique. The effectiveness of the technique depends on its merits and demerits. No single techniques is effective for all type of waste. The household waste can be best managed with training the workers properly, separating the waste in different color of bins after which waste can be disposed of at final and proper location. One of the fraction of municipal solid waste is household which is mismanaged in developing countries (7). Waste produced within homes, under the health-care facilities comes under the household waste. In household waste about 75 -90% is non-hazardous and remaing10-25% is hazardous posses health risks (25). Household waste is mostly dumped in developing countries. Although improve safety management is used at some places. Household mismanagement includes the open dumping as easier option for waste management. In open dumping waste pickers access the waste to collect the material, which are used for recycling but it, transmits the infectious diseases. Another waste mismanagement is the uncontrolled burning, which discharge PCDD like toxins with other pollutants (26). Palestine openly dumped 82.2% household waste and 17.9% hospitals waste (27). Ground water gets contaminated by the leachates of waste disposals flows due to the infiltration process (28). Another example is Bangladesh, where household waste is similarly mismanaged all the items after collection is either burned or disposed of into general sewerage system. This practice leads to many infectious diseases. During rainy season water is polluted with leachate comes from the dumped sites, this water is used for washing, agriculture and household uses (29).

2.3.1.2 CONSTRUCTION AND DEMOLITION WASTE:

Population pressure increases on urban areas, due to more facilities in urban areas as compared to rural areas. Construction of new homes and shelters results in construction and demolition waste, need to be managed properly. Construction and demolition of existing buildings produce the iron, concrete, plastic and ceramics as waste product. Most of them can be recycled or reused especially for iron and plastic containg waste. The waste reached in industry through picking by scavengers. Most of the waste is throne openly in the streets and rest of the waste is dumped at unsuitable places. This mismanagement of waste produces the severe health hazards for humans. Burning of plastic add hazardous gasses/pollutants in the atmosphere (30). During the construction of buildings/homes mismanaged sanitation results the sewage waste, which ultimately pollutes the underground water, sanitary landfills, freshwater (1).

2.3.2 HAZARDOUS WASTE

2.3.2.1 INDUSTRIAL WASTE

The unawareness of the citizens suffers the hazardous waste management system. Mismanagement of waste as dumping suffers the developing countries, which further increases by increasing the type, sources and quantity of waste. Every industry produce the hazardous waste in developing countries as compared to developed countries, where the risk is on low level. Multinational companies shift their plants to the less developed countries where they used the technologies banned in their home countries. Industrial waste is generated in every industry, which is not properly managed and mostly is dumped. Leachate from these dumped sites further pollute the water as surface runoff (31). The chief industries and their waste production can be describe as follows.

1. **Agriculture Industry:** In Developing Countries, agriculture has main share in GDP. In agriculture industry farmers use the pesticides, insecticides, fungicides, weedicides, fertilizers, also rear the animals. From these sources, the chemicals and their related equipment increase the waste. Crop bran, straw, and crop remains, and animal carcasses, animal dung are the main sources of waste need to be managed properly (32). Crop residues are mismanaged as instead

of making compost or green manure they are mostly burned of for energy, which produce the lethal gasses and pollutes the healthy air (33). The animal carcasses are left on the ground for decomposition, which produces the different bacterial, fungal growth, and effect the human health. The heaping of animal dung produce the unpleasant smell, deplete the organic matter in sun. The mismanaged application of fertilizer and pesticides affect the soil, water, air and human health quality (34).

2. **Cement Industry:** In cement industry a lot of waste products are used again but the cement by-pass dust is the main waste product which can be diffused in air and deteriorate the air quality. It enhance the decomposition process and kill the microbes and parasites. CO_2 is another main pollutant or waste comes from cement industry. Calcination and fuel consumption are the main partners of CO_2 generation. Cement industry produces the 5% CO_2 in total (35; 36). Modern cement plant produce the 60 per cent CO_2 by thermal treatment of $CaCO_3$, 30% due to burning of energy sources in the furnace and 10 % from the other plant operations (37).
3. **Plastic Industry:** Plastic is used in everywhere of the world. In plastic industry, the waste is reused or recycle by many times however depending upon the type of plastic and the chemical present in a plastic. The best management of plastic waste is minimization, recycling and the biodegradable plastic usage. However, in developing countries mismanagement of plastic waste occurs in the form of burning, dumping, and open disposal. Mismanagement of plastic waste through burning produces the smoke, which affect the human, wildlife and environment due to carcinogenic effects. Soil is deteriorated with chlorinated plastic, which pollute the water through seepage and ultimately disturb the activity of microorganisms as bacteria (*Flavobecteria* and *Pesdomonas*). Biodegradable plastic some time release the methane, which is powerful greenhouse gas, it increase the global warming (39). Diethylehexyle phthalate pollute the ocean due to its carcinogenic effect, which ultimately contaminate the food chain. Plastic is reused in construction material. Recycling of plastic depends on type of country and polymer type. Plastic containg (polyethylene terephthalate) and polyethylene (high-density) are recycled more as compared to others. Developing countries recycle their plastic waste about 20-40% (38).

4. **Tyre Industry:** In developing countries, waste tyres are not managed, properly due to inefficient management system (40). Most of the waste tyres are disposed and are used in different industries as fuel like in cement industry. The scavengers, scrap dealers and recovery companies (41), mostly use the disposed waste tyres. This mismanagement is a big threat to the environment due to resilient composition and their design (42) and the hazardous contents they contain (43). There is no waste tyre processing and treatment plants in developing countries so waste tyres are either burnt or stockpiled at disposal sites (44). There are numerous approaches for efficient waste tyre control that, when accurately achieved, can be both effective and safe. Incineration for energy production, Landfilling, pyrolysis, small-scale reuse, grinding for several objectives, and retreading are safe and effective management methods of waste tyres. Waste tyres are in bulk volume, so not managed properly decrease the sanitary landfills. Burning of waste tyres pollute the healthy environment enhancing population health risks. Tyre waste in different landfills increase the disease vectors. Dengue is another example of disease spread due to waste tyres. Tyre burning increase the SO_2, CO, poly-aromatic hydrocarbons, NO_2 and volatile organic compounds in the air.

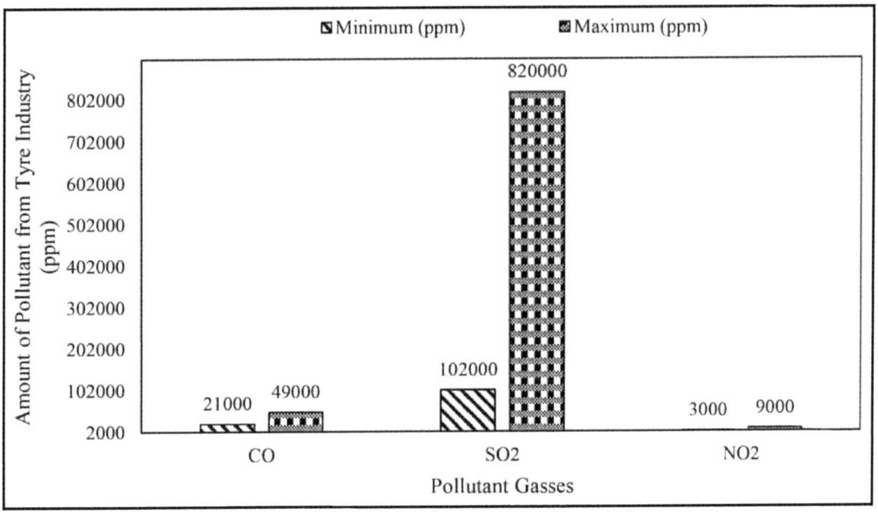

FIG. 2.4 Amount of Pollutant from Tyre Industry Source: Author; Modified from Ref. (1)

5. **Battery industry:** The main mismanagement of used batteries are open dumping and open burning. Due to dumping of batteries, leachates with high heavy metals concentration deteriorate the soil and water environment. Similarly burning of waste batteries produce the PCDD/F, BC and PM. Heavy metals Hg, Pb, Cd and Mn affect the life of all living organisms and humans. For the recovery of different metals, burning is usual and mismanaged practice in developing countries. For example, 10000 tons of batteries were dumped in landfills in Iran after import without separation (45). The dumping of waste batteries in lanfills add high concentration of Cd, Li, Ni, Ar and other toxic metals (45). Mismanagement of waste batteries occurs due to the no proper infrastructure and no special rule and regulations for this waste management (46).
6. **Automobile industry:** Different type of waste like solvents, metals, batteries, plastic and glass are produced in the automobile industry. About 75% of the vehicle weight is recycled. The landfills portion covers the remaining 25% for auto shredding (47). Mainly foams and fluff, plastics, rubbers and metals comes under the auto shredding residue category, however there is no cost effective technology for the recycling of plastics and foam (48).
7. **Iron/Steel industry:** In steel industry coke, coal dust, SMS slag, BF slag, scrape, mill scale, emollient sludge and refractory waste are produced (49). Dumping of these waste become great concerned for environment. In India, 1.2-ton waste is generated per 1 ton of steel production (50). Over all 63% of steel industry waste is dumped which needs to be reused or recycled. In steel industry ferruginous and non-ferruginous waste are generated (51;52). The iron bearing waste is ferruginous waste, which comes from mill scale, flue dust, sludge, steel melting plants; slags either blast kiln and SMS. The second form of steel waste is non-ferruginous or non-metallic iron containg, are broken refractory bricks, lime fines, acetylene plant sludge and broken fire clay bricks.
8. **Sugar industry:** Sugar industry generates the bagasse, press mud, leaf trashes in case of solid waste, however the liquid used in the whole sugar industry from washing to processing is also the liquid waste, which need to be treated before adding in the water reservoirs otherwise it cause the water pollution. High content

of lignin, cellulose and hemicellulose are present in sugarcane bagasse (53). Best management of bagasse includes its usage in textile, pulp and paper, animal feed, fuel, food, chemical, enzymes, polymer and agglomerated board production. China used the bagasse (a solid waste) as fuel in sugar boiler. When the sugar cane juice is in dilute form and unclear it produces the waste known as press mud or sugar cane filter cake, contains nitrogen in sufficient amount (54). The best management of press mud is for fuel utilization, fertilizer and biogas production, wax and soil conditioner production. In countries like china, Brazil and India sugar industry waste is a main problem for Environment (55). waste water also contains the effluent from cane molasses distilleries. It is estimated that one ton of sugar processing requires the 20-30 tons water. The waste from sugar cane industry includes effluent, OM, suspended solid, bagasse, sludge and press mud (56). About 1 ton of sugar cane processing produces the 1000L wastewater in India (57). Mismanagement of sugarcane effluent includes its direct use for irrigation purposes, results in reduction of growth, yield and the soil health (58). The effluent damages the paddy and fish, which is due the presence of harmful substances such as Pb, Zn, Fe, Mn and Cu. Sugar cane leaves are mismanaged as open burning, produces the fly ash. Fly ash severely damage the soil microbial activity and pollutes the air, which triggers the lungs diseases (59; 60).

9. **Leather industry:** At each stage of leather production waste is generated due to the chemical reaction, harmful to the environment (61).Water effluent, which mainly contain chromium, chlorine, cadmium, nickle, zinc and lead, is not managed properly as disposed to the water bodies without treatment. It is well documented that only 75 kg of leather is formed from 500 kg of hides while reaming of these hides known as waste (62). When the waste is dumped of in landfills then it can produce the harmful microbes. Methane is also released from leather industry. Mainly four type of solid waste is produced from tannery, which are as follows. Bronzed collagen, non-collagenous protein waste, untanned collagen and non-proteinious waste. One third of water pollution is due to the water effluent from leather industry.

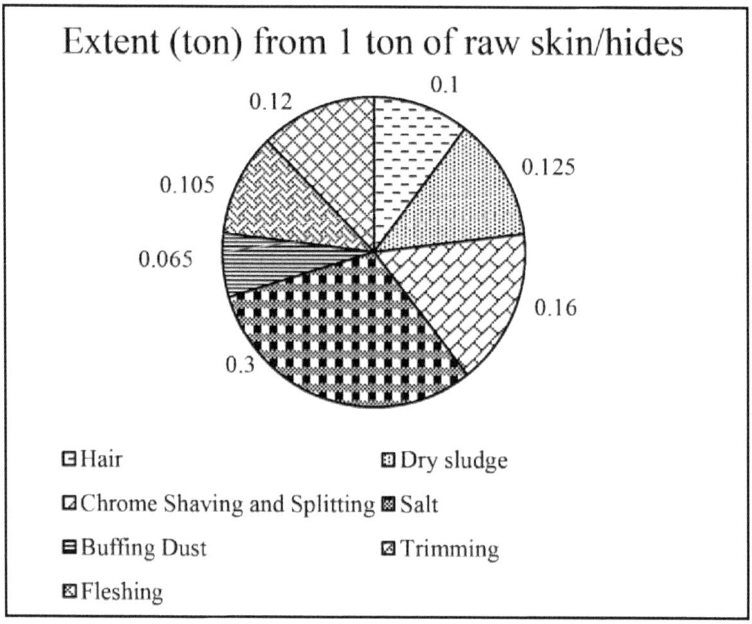

FIGURE 2.5 Forms of waste produced in tons from dyeing of 1 ton of raw skin

Source: Author; Modified from Ref. (62)

10. **Textile industry:** Wastewater is the main component in textile industry known as effluent, comes from washing of fiber to the finished product. For 1 kg of textile production, 200L water is required (63). The untreated effluent cause damage to the environment. Maximum volume of wastewater is produced from the wet processing (64). Low aquatic toxicity is concerned with dyes while detergents, surfactants, emulsified and dispersant are the main contributors of eluent aquatic toxicity. The mixed wastewater is produced from wet finishing treatment, process undergoes in textile bath. Textile bath is discharged into the surrounding, which is main source of pollution (65,66). Coal dust, fiber dust, saw dust and grain dust are the resultant of textile processing, damages the workers after inhalation (67). This dust after mixing with air pollutes the air with suspended solids. Respiratory diseases increases due to the inhalation of air polluted with cotton. The textile industries in developing countries on small scale is still lacking the facilities to avoid such mismanagements.

2.3.3 BIOMEDICAL WASTE

The biomedical waste is produced from different healthcare facilities, has a great health risk for humans (68). There are different categories of biomedical waste including different chemicals, biohazardous waste, pharmaceutical wastes, radioactive substances, pathological wastes and genotoxic wastes, which affect the human health and environment (69). The humans including patients, health care workers and anybody else are being affected about 10-25% from hospital wastes (70). Mostly hospital wastes are hazardous and infectious and their mismanagement increase the different diseases and pollutes the environment. The serious disease like cholera, tuberculosis, typhoid, hepatitis, HIV and AIDS mostly spreads due to hospital waste. Hospital waste is generated 0.5-2.0 kg bed^{-1} day^{-1} globally (71). In Pakistan due to mismanaged disposal of hospital waste posing a serious threats. Pakistan hospital waste ranges from 0.1-0.7 kg (72). Hospital waste has two main components (infectious and non-infectious waste). When the infectious component of waste is mixed with non-infectious portion due to poor disposal then total hospital waste become infectious. Improper disposal of this waste aggravate the risk of diseases of rag pickers, health worker and waste collectors. The disease dissemination is due to the unawareness regarding the waste management, insufficient finance and facilities in Pakistan. Hospital waste is managed properly with incineration technique, which is mostly lacking in developing countries like results in the open dumping in different areas of the country. Hospital waste is generated from nursing homes, clinics, pharmacies, health cares and outpatient surgery centers (73). The mismanagement of biomedical waste serve a great contamination potential for society and environment through water, air contamination, and physical contact of the persons. Maladministration of clinic waste involves a mixture of inappropriate usage of waste through every step of waste management: i.e. generation, collection, storage, transport and treatment (74). Mismanaged handling of biomedical waste also includes, handling the waste without wearing personal protective equipment, poor storage, transporting the waste without proper covering (75). In Pakistan Lahore is the city where only two incinerator are present out of which one is nonfunctional. These practices also attract the rodents and flies. The practice of incineration is also temporary because of environmental issues. The incineration and open burning of medical waste produces the dioxins, greenhouse gasses and other toxic emissions, which can increase the global temperature and cause the cancers as well as respiratory disorders (76).

2.3.4 SPECIAL HAZARDOUS WASTE:

Special hazardous waste deals with three type of waste; explosive, radioactive and electric or electronic waste (1). Each category of special hazardous waste has its own management technologies. The detail of each type of waste is as follows.

2.3.4.1 EXPLOSIVE WASTE:

In the explosive category there waste comes named as propellant, firearm ballistics, Bulk explosives, consumer and commercial explosives, munitions, roadside and signal flares, Marine, Hobby missile propellants and vehicle airbag propellant. A propellant is used to create fluid movement. Propellants are mostly burned, otherwise decomposed. Explosive waste includes munition, which contains propellant, explosive, and fireworks in different forms as bombs, mines, ammunitions and grenades. Fireworks that are damaged, unused or fail to function increase the hazardous waste due to their activity.

2.3.4.2 RADIOACTIVE WASTE:

A very small amount of fuel can generate high energy and with a little amount of waste. However, this produced waste is highly radioactive, need to be managed carefully. Electricity generation process also produce the nuclear waste need to dispose properly. The cost of electricity includes the nuclear waste dispose off cost. The utility of radioactive waste are recorded in agriculture, medicine, research, mineral exploration, non-destructive testing, and manufacturing.

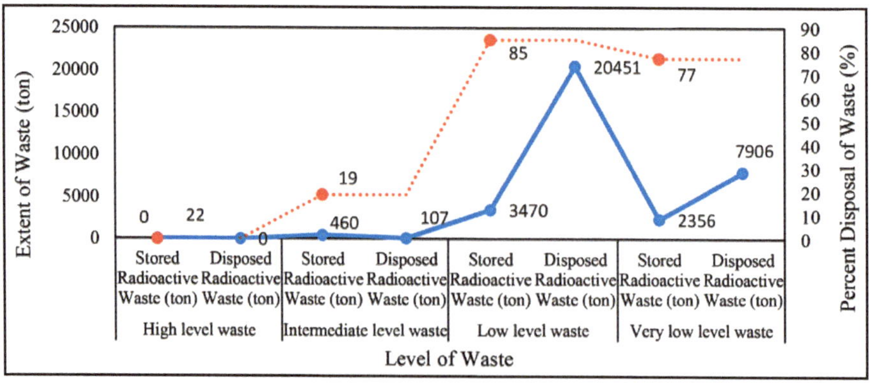

FIGURE 2.6 Level and extent of radioactive waste (Source: Modified from Ref. 151)

Burning of uranium in fuel cycling produces the waste. During the electricity production, high-level waste is produced. Reprocessing of used fuel also generates the radioactive waste. Mostly the waste from decommissioning of nuclear facilities is suppressed and some is reused within the production unit.

2.3.4.3 ELECTRONIC AND ELECTRIC WASTE:

Electronic waste includes waste from computers, internet, print media. Their waste contains the different type of heavy metals cause the air, soil and water pollution. Some parts of electronic are not reused correctly. In USA electronic market is producing, the 20 million computers while 12 million are disposed every year from which remanufactured are only 10% (77). The main reasons of electric or electronic waste are as follows.

- No infrastructure for the management of electronic waste.
- Lack of proper assessment of electrical and electronic waste
- Shifting the electric and electronic waste from developed to developing countries
- Ignorance for the toxicity of electronic waste
- Mixing of electric and electronic waste with municipal solid waste finally dumped openly
- Lack of knowledge about the waste impact on human and environment
- Absence of legislation for the transportation and disposal of waste

Waste in developed countries, which is not recycled, is either dumped, incinerated or exported to developing countries. The US exported 80% E-waste to Asia from there 90% went to China (78). Computer waste contains plastics, iron, aluminum, copper, lead, zinc and tin along with cadmium, mercury, antimony and beryllium. Early PC generation comprised of 4 g of gold but now 1 g is present in PCs (79). Now electronic waste comprised of lead more as compared to other metals.

2.4 WASTE MANAGEMENT PRACTICES IN DEVELOPING COUNTRIES

The incorrect waste management practices results the serious pollution threats for the public mainly arises due to the shortage of control, no proper legislation, different environmental and public health impacts. These causes of poor managements encourage the many federal and national authorities to introduce new infrastructure for the hazardous and unstable waste management. The waste management practices favors the different managemental steps including prevention, reuse of waste, waste recycling and compositing. The selection of waste management practices depends upon the economy and technical aspects of the waste to be handled. The practices mainly adopted in developing countries for waste management are; open burning, open dumping and recycling. All these practices involve unsustainable and unprofessional ways for handling waste, which has worsened the environmental issues i.e. it, covers all sorts of waste (municipal soil waste, electric/electronic wastes, health waste, industrial waste etc.) and unprofessional way of handling waste-by-waste pickers, which transmits severe intimidations to environment and human life. Anyhow, some advance practices are being observed in developing countries, but the least developed countries are still lacking those technologies. The economic growth of country is the biggest indicator to waste management strategies, but the populace behavior is the most significant factor toward efficient waste management. The widely adopted strategies by developing countries are here by discussed.

2.4.1 OPEN DUMPING

Open dumping is a rampant practice in developing countries for waste management, where unregulated sites and uncontrolled disposal exacerbate the situation. This mismanaged disposal of waste would lead to severe health issues, surface and sub-surface water pollution, air contamination and become a hub for vectors of diseases especially in densely populated cities. In the table (1), different countries case studies are reported which practice open dumping for waste management.

TABLE 2.1 Waste management through open dumps/landfill systems in some developing countries

Country/ region	Adopted practice	Description	Reference
Gambia	Open dumping (90% of disposed waste)	Waste generation capacity increased from 204 to 393 tons day^{-1} from 2001 to 2010, yet no significant advancement in management practice has taken	(80)

TABLE 2.1 *(Continued)*

Country/ region	Adopted practice	Description	Reference
Thailand	Open dumping (60% of disposed waste)	Out of 423 disposal sites, 330 are open dumps and the average generation rate of waste per capita is >1 kg/capita/day	(81)
Cambodia	Open/direct dumping (99.5% of disposed waste)	In Phnom Penh city, disposed waste (including all wastes except industrial and medical) dumped into open dumpsite accounting waste from 613.94 Gg to 677.22 Gg in 2014 to 2015 respectively	(82)
Nigeria	Open dumping	In Abuja, four disposal sites were closed due to air pollution and odor. And waste generated at the rate of 250,000 tons/year in 2010	(83)
Mozambique	Open dumping (70% of disposed waste)	Waste of 1.2million people of Maputo City disposed to dumpsite for last 40 years	(84)
Serbia	Open dumping (only practice)	The average waste generation capacity of a person is 0.87kg/day, in Serbia. And only 60% of waste is collected for disposal.	(85)

2.4.2 OPEN BURNING/INCINERATION

Burning waste is another common practice of managing waste in developing countries. It is a direct source of air pollution, which consequently carriages severe threats to environment and increases health risks. As reported by United States Environmental Protection Agency, Incineration of waste caused emission of greenhouse gasses emission into the atmosphere. In addition, burning of different wastes is also a major source of different contaminants release into the environment for example burning of plastic waste would leads to the release of volatile organic compounds and particulate matter (86). Likewise, garbage burning increased the stimulation of particulate chloride up to 60% in Mexico City (87). Globally, 40-50% of municipal solid waste is subjected to unregulated open burning and highest rate of emissions via burning wastes are reported in following countries: China, India, Brazil, Mexico, Pakistan and Turkey (88; 89). The table (2) shows the open burning as a waste management practice by different countries.

TABLE 2.2 The waste management through open burning/ landfill burning and incineration in some developing countries

Country	Description/impacts	Reference
Tanzania	60% of domestic waste is disposed by burning the waste daily	(90)
Lebanon	Based on recent data, there are 941 open dumps in the country, on an average once in a week, more than 150 of these dumps are openly burned	(91)
Mexico	In rural communities, 92% of waste is either disposed by uncontrolled burning in the backyards or dumped unofficially. In general, nearly 24% of municipal solid waste is subjected to open burning	(92)
India	The case study areas (10 metropolitan cities) generated 43% of total solid waste of India. No justifiable data is available regarding open burning of wastes but in case 20% of the MSW is burned, the emission of pollutants by it exceeds daily intake limit.	(93)
Palestine	Out of 133 MSW treatment sites in Westbank region, 116 sites operates through open dumping leading to open air burning	(94)
Ethopia	More than 22% of urban waste is imdiately treated with burning in all public areas	(95)
India	The daily mass of municipal solid waste burned is expected at 90-1170 $kg/km^2 day^{-1}$ and 13-1100 $kg/km^2\ day^{-1}$	(96)
Pakistan	54,850 tons of compact waste is being produced daily in city areas, < 60% of this produced solid waste is being gathered accurately	(97)
Nigeria	The study concluded that about 54.2% solid waste generated is treated through open air burning	(98)
Nepal	Of 319 waste piles studied in the survey and t137 (i.e., 43%) waste heaps were observed to be dynamically burning in diverse ways of the city and sub-city area	(99)
Lebanon	Open burning is predominant outside Beirut and Mount Lebanon	(100)
SriLanka	Study found that, of the 76% of households without collection coverage, 69% burn all their plastic waste	(101)
Uganda	About 74.1% of uncollected waste being burnt openly.	(101)

2.4.3 INFORMAL RECYCLING

In low-income countries, where formal conditioned collection system of waste is not developed yet, the intrusion of informal sector in waste collection could assist the formal waste management program. Because there are certain factors which compel the informal activities i.e. poverty, free waste management, lack of formal waste management system and above all the unemployment. The systematic scheme followed by informal pickers is as: firstly, collect waste from site (household, roads, bins, dump areas etc.), then sold it merchant or to the trading points. The trading points further sell it official or casual recyclers or in a straight line sold it to the manufacturers. This is how the chain is completed and reported in numerous case studies (102). The informal sector activities are in accordance with principles of circular economy, as it tends to decrease the volume of waste being discharge and directly recover the material (transform it into secondary material). In the case studies of many developing countries, informal sector plays key role in waste management as reported in; India (103), Bosnia (104) , Malawi (105), Colombia (106) and many others countries. The informal sector could have become the active member in accompaniment with formal sector to recover the environment sustainability.

The decrease in environmental pollution and waste production-consumption improves the economic growth and sustainable development (11). The waste management aims for worldwide sustainability includes:

- Safe and reasonable price solid waste collection services available to all by 2020
- To discontinue the unrestrained open burning and dumping
- To attain the viable and environmentally safe waste management by 2030

The solid waste management in developing nations as reported by many studies can be improved by waste reshuffling or transferring programs (organic in nature), compost and biogas generation and execution of waste to energy strategies and skill tools. The minimization of waste to energy can be improved by recycling of metals, glass and other materials. So ecological pollution sustained as an alarming issue globally. Whereas normal solutions should be applied after, recognition keeping in mind the suitable solid waste management configuration (107).

2.5 AUGMENTATION OF WASTE MISMANAGEMENT IN DEVELOPING COUNTRIES

More than half of the global population resides in cities. Megacities are the consequences of intense urban developments and expansions carried out world-wide. Therefore, they play primary role in understanding the urban dimensions and planning their respective management strategies (108).

Failure of management of urban dimensions are stimulated by certain driving forces. These mainspring forces are population explosion, rapid urbanization and industrialization, inappropriate attitude of the public and uninterested government particularly in terms of budget allocation. All these driving forces exert pressures on urban development causing social fragmentation and producing unnatural social dynamics. Such rapid urbanization also results in absence of required basic infrastructure and lack of proper administration of such urban dimensions. Because waste mismanagement in megacities is directly linked with the failure of proper administration of these urban developments. (Figure 2.7). Consequently, it is critical to study elements responsible for urban maladministration proportionately responsible for waste mismanagement (109).

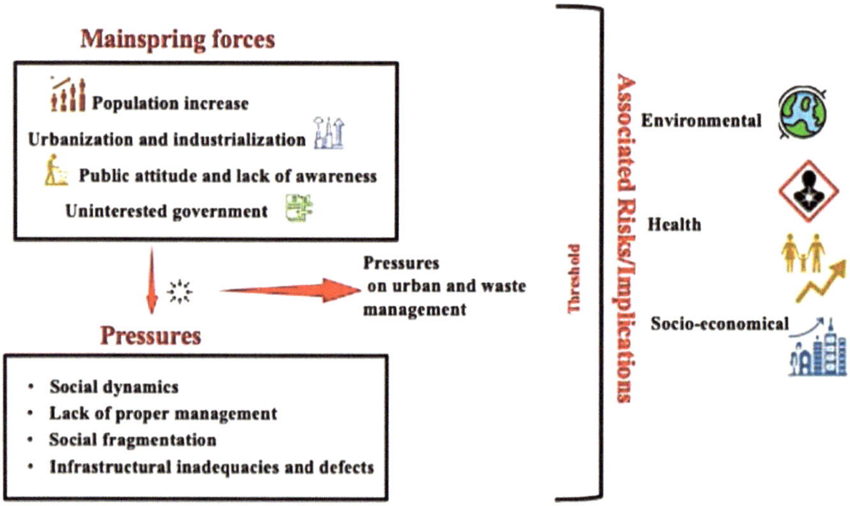

FIGURE 2.7 Driving forces and pressures leading to poor waste management and risks associated, Source: Author

Waste management is a world-wide problem (110). At a larger scale, environmental pollution, social inclusion and lack of economic stability are the direct consequences of this mismanagement. The solution to this global affair obliges integrated and holistic assessments and approaches. However, it is imperative to identify the similarities and dissimilarities between different factors triggering waste mismanagement in developed countries versus under-developed countries. Some causative factors require similar tackling approaches, whereas, others are unique to developing countries and need to be addressed in a related but sustainable way (111; 107). Among several factors that may create hindrance in management of urban and rural wastes are social, financial, economic, technology-related and administration or government related.

2.5.1 FACTORS AFFECTING WASTE MISMANAGEMENT

Waste mismanagement is a nuisance worldwide that requires massive manual efforts. Failure to provide that could produce adverse effects on environment, socio-economic, physical and psychological well-being of the population (Figure 2.7). These impediments include lack of management technology, absence or inadequate government funds, lack of sanitation law implementation, loopholes in policies and strategies, poor attitude of populace and lack of education and awareness (112; 113; 114). Broadly, it can be discussed under following heads (Figure 2.8)

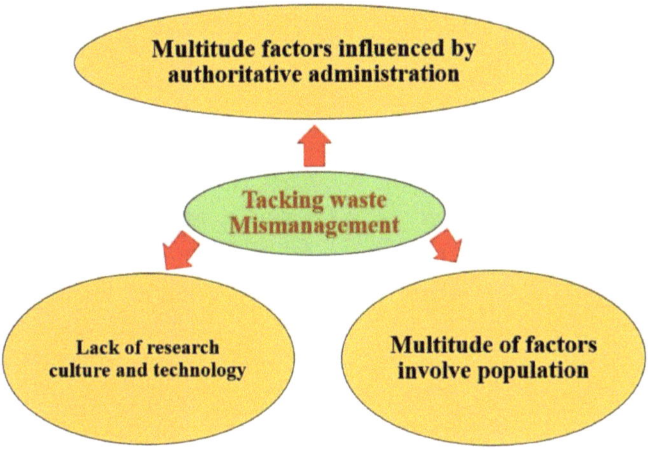

FIGURE 2.8 The key areas to address the mismanagement of waste disposal in developing countries Source: Author

2.5.1.1 MALADMINISTRATION BY GOVERNMENT AND CONCERNED AUTHORITIES

Globally the term integrated solid waste management is used where the objectives of 3R i.e. reduce, reuse, and recycle approach are brought about together with the management of municipal waste. Generally speaking, waste production of a country is directly proportion to its urban development, prosperity and lifestyle (115), however, with the implementation of 3R approaches; *decrease* the size and poisonousness of the discard, reuse the reusable and repairable items e.g. containers, recycle the products, many developed countries such as USA (116) have been able to reduce the burden of this major environmental problem. Likewise, A case study of Adelaide, Australia revealed that despite being one of the largest consuming cities in the globe the generated waste is challenged through an innovate yet effective strategy known as zero waste strategy (ZWS). It results in posing least threat to environmental integrity (117). Nevertheless, successful implementation of such strategies is incomplete without economical and socio-political support provided by the Governments and administrations (118).

On the contrary, Governments in many developing countries such as Asian countries, although have policies and laws enacted to deal with waste management, For instance, National Environmental Policy (NEP) in Pakistan, Local Self Governance Act in Nepal, National Environmental Policy in India and National Strategy for Solid Waste Management (NSSWM) in Sri Lanka deal in waste management in their countries (119). However, loopholes in these policies and strategies and absence of their proper enactment has not only led to wastage of financial and human resources but fail to address the issue. Overall, in low-income countries, governance and administration compromise the social and environmental well-being of the citizens and other indispensable basic needs due to multifold of inevitable factors. Therefore, these cities, particularly, in developing countries are progressively referred to as hotspots of numerous menaces and risks (120;121). One of the major penalties affecting hefty portion of the urban populace is the excessive overloaded and congested living conditions producing infrastructural stress i.e. shortage of basic infrastructure both in terms of numbers and quality (109).

In developing countries, one of the typical examples of waste maladministration is the mismanagement of urban water systems. It is typically assigned to provide safe and clean water supply and safe disposal of wastewater. However, for instance, for urban water systems in Delhi, India, a megacity in developing country; evidently there are inadequate sanitation and wastewater disposal infrastructures (122). These facilities are either missing or challenging to access. Moreover, many times existing infrastructure fails to ensure safe disposal due to ageing of obsolete infrastructure and lack of its proper and repeated maintenance. It results in the formation of unsafe conditions that further leads to social vulnerability

(123). The consequences of these inadequacies of infrastructure, persistent exposure to "unsafe conditions", reduced capacity and potentiality are eventually reflected as health complications and other related troubles. Such impediments cost people high in terms of time, money and burden for the management of their own wastes and for the protection and treatment of themselves against infectious ailments (109; 124). Overall anticipated threats of these harmful conditions are not fully perceived until they assault as major outbursts. However, implications of environmental risks caused by poor management of sewage and wastewater are far more destructive and challenging to assess (125; 126).

Moreover, economic crisis, insufficient and ill-timed discharge of Government funds allocated to the waste management has also been one of the challenges in low income countries (127). Lack of maintenance of outdated waste management tools and shortage of funds for their timely replacement has resulted in failure to collect and dispose of metropolitan waste (128). Additionally, COVID-19 (Coronavirus Disease-19) pandemic has resulted in an unexpected and abrupt disintegration of waste management globally (129). Economic downfall amid pandemic has negatively impacted the world (130). There is a dire need of incorporation of capability of assessment, preparedness and resilience into our waste management system so that it could manage the domestic and medical waste and contain the infectious diseases in case of any such disaster (131).

2.5.1.2 POOR ATTITUDE OF POPULACE

In low-income countries (128), public attitude and awareness played a major role in waste mismanagement. General public do not realize the significance of disposing off waste properly (132). They openly dump the waste in open drains, running water at canals or at other unauthorized places that impose serious environmental and health-related risks (Figure 2.4). It exhibits the failure of sanitation law enforcement agencies on the implementation of the decree (107). Planning of waste management is a subdivision of environment planning. However, in low-income countries like Pakistan (111), and India, 90 percent of collected waste is abandoned in topographic depressions and unoccupied plots. It is estimated that 55000 tons per day waste is generated in Pakistan on daily basis, out of which only 60 percent is hardly collected (133). The rest of the waste is thrown in open sewers, drains, along roads and streets and in open places (Photo-1).

PHOTO-2.1 Picture taken at a Sahiwal metropolitan city located in Punjab Province of Pakistan.

Source: Author

The sanitary workers are unschooled, untrained and poor. The low salaries of these workers thereby, made them loose interest in properly handling the waste. Lack of social acceptability and self-respect is another factor of poor motivation of these workers while performing their duties (134; 135). Moreover, general public fail to cooperate with these workers and their staff in handling the wastes. Lack of awareness of the socio-economic and health risks associated with the waste mismanagement is another factor contributing hugely to this maladministration (133). Education on these factors and associated risks can bring desirable change in public's behavior and can inspire them to consider it obligatory to deal with the waste appropriately (136).

2.5.1.3 LACK OF RESEARCH AND RELATED TECHNOLOGY

In addition to other causative factors; infrastructural stress, obsolete technical tools and lack of innovative research and technology also play integral role in mismanagement of the garbage. Government lacking funds or unable to allocate enough budget in this regard and failure to bring high-end research are some of the factors (137). To sum up, all the causative factors viz., social, economic, technological, political are interrelated and therefore an integrative and holistic approach that involves combined efforts of government and public is required. Moreover, there is a dire need of extensive research and inexpensive innovative technology to achieve sustainable solutions to this issue.

2.6 IMPACTS OF WASTE MISMANAGEMENT

The waste is linked to human since the beginning of human. Waste management had never been a threat to environment and human when the humans were Nordic, soon outburst of population, tackling the waste on this earth became the nerdy challenge. The waste being intruded into environment has dramatically increased with innovation into technology and science. On other hand, science is struggling to coupe this challenge. The nations with strong incomes have been able to somehow, safeguard their demographical environments and communities from hazardous impact of waste. The issue of waste mismanagement is more pronounced in less developed and developing countries mainly due to inadequate waste disposal strategies, infrastructures and burdened economy. Irrespective of geographical boundaries, the meagre and unsustainable management of waste on this biosphere has swiftly commanded to pollution of soil, water and atmosphere with the most damaging impacts on the public health (138). Even the most developed world today i.e. Europe has been constantly challenged by numerous epidemics originated from water contaminations in past. In current century, the impacts of environmental contamination especially to water borne diseases are more pronounce, in developing cities with major impacts on African and Asian regions.

The technological adaptation among peoples is very low, despite of the new technological advancement, better legislation for waste management, enlightened human health supervision and degradation of waste treatment and disposal. The environmental impacts of poor waste management have been summarized here below and in (table 3).

TABLE 2.3 The prominent environmental impacts and pollutant addition through various waste management strategies on ecosystem

	Waste disposal strategies			
	Incineration	**Composting**	**Landfilling**	**Recycling**
Air	SO_2, N_2 oxides, Unstable organic complexes, Methane, Carbon oxides	Bioaerosols, Unstable organic complexes, CH_4, CO_2, dust	Unstable organic complexes, CH_4, CO_2	Dust made of leftovers
Water	Dropped down components of atmospheric pollutants	Percolated pollutants	Lethal organic complexes, Heavy metals	Wastewater
Soil	Ashes of incinerated garbage	Least impact	Heavy metals, Toxic organic compounds	Landfilling of remnants

Source: Author

2.6.1 IMPACT ON TERRESTRIAL ECOSYSTEM

The air, water and soil become more polluted due to high environmental pollution caused by the unscientific and mishandled waste disposal. Waste and wastewater induces numerous intimidations to community health and harshly disturbs fauna, flora, of the surroundings. The uncontrolled release of lethal element to the air and in the soil has already threatened the biosphere. Moreover, the infection of soil and water assets has put the food security and food safety to threat. Life mainly relies on soil, which is composed of minerals and organic components. Without its biota (microorganisms, earthworms, insects and plants remain etc.), soil is lifeless. Disturbance in soil composition and destruction of soil biota cause serious damage to whole biosphere. Poor agriculture practices, sewage water and industrial waste disposal are main causes of soil contamination (139).

The world waste production from urban centers should be from 1300 to 2200 million tons annually upto 2025 (3). This waste production is low in developing countries in which 219 kg/capita/annum waste was produced in 2010. Therefore, this biosphere is receiving such a huge amount of waste every year. From this similar fashion, the waste production will spread to 343 kg/person/annum in 2025. The environmental impacts of waste are directly dependent on composition and characteristics of waste which in turn are subjected by the financial position, technologies being adopted living principles, food lifestyles, customs, learning rate, form of energy foundation, climatic and geographical situations (140).

The concentrated heavy metals are major hazard due to their toxicity, because of impoverished waste management. Heavy metals gets their way in solid waste and liquid industrial effluents due to open dumping and discharging in sewerage system. Air quality of industrial cities in developing countries is already been highlighted among societies. In term of air quality, the mega cities of developing countries are top among the list. The landfills discharge waste gasses including the major component (methane, CO_2 and many trace constituents) have the significant impacts on environment and humans. Methane absorptions can range up to 50% of the configuration of dumping gas at extreme hypoxic decomposition (141). A second difficulty with these smokes is their involvement to the improved greenhouse gas influence and environment change. Among the wastes, the plastic materials are more dangerous to air quality. Plastic and rubber pollute the atmosphere with noxious fumes upon open burning.

2.6.2 IMPACT ON AQUATIC ECOSYSTEM

We agree with the criticism presented by a scientist (142) where the author argues that division of pollution into categories (e.g., air, water, land etc.) is meaningless and highlighted that there is only one pollution' because every pollutant, whether in the air, or on land tends to end up in the ocean. Pollutants of chief apprehensions from underprivileged waste supervision comprise tenacious organic contaminants, radionuclides, nutrients, heavy metals, oils, sediments, litters pathogens etc. the ocean dumping and littering especially of plastic waste is more pronounced in developing countries. The shorelines along the Africa and Asia has been completely polluted (143). The fish and mammals of marine environment has been badly affected (144). Previously 2/3 of marine life is thought out to be a rare species. Because of inappropriately discharged elements and other waste. Plastic waste is found in rivers, lakes, on beaches, in middle of oceans, on the sea floor and even inside the water birds and animals (145). The abundant manifestation of micro plastics has drawn the consideration of Eco-toxicologists on its protection and noxiousness and its influence on biota, a probability for establishing the linkage among the plastics and the ways of intrusive species being a severe ecological threat (146) Global production of plastic is increasing every year. Since, conventional plastic does not degrade soon; it remains in the atmosphere for centuries. This makes plastic accumulate in water bodies, especially oceans, which is a serious environmental problem in all coastal countries (147). Although awareness level of this issue has increased globally, yet there is an immediate need to stop the flow of plastic and other wastes to waters.

2.6.3 IMPACT ON PUBLIC HEALTH, INFECTIONS AND CHRONIC DISEASES

Type of waste, its persistence, population exposure and availability of managemental technologies are the main factors for the variability of waste's impacts on human health. The impacts varies between slight psychosomatic effects to severe injury, and demise. Available scripts on fitness effects of discarded material is still feeble and questionable in most circumstances owing to the problems met in precisely discoverable coverage, monitoring for confounders, recording for interval of exposure and incompetence to monitor those visible to discovered consequences that do not apparent in the short span. While definite fitness influences might be instantaneous, apparent to distinguish and linked with firm waste, extended then problematic to feature the possessions to a precise nature of discarded material. This creates the load of illness attributed to a waste and complete continuum of diseases originating from the contact a difficult task frequently necessitating the huge sample amount and extended times of follow-up (148). Waste mismanagement is one of the major issues of health challenges that include infectious diseases. Cholera, dysentery, malaria, dengue, typhoid and several respiratory complications are common in areas, which lack proper system of waste management. Those are especially at high risks of getting infections who live in neighborhood of open waste dumping sites. In such places, water is contaminated with leachates of waste dumpsites (148). The chief ways of exposure are intake of incinerator and landfills emissions, usage of water polluted with leachate, the food consumption polluted with viruses and bacteria due to the scattered sewage and manure, and persistent organic chemicals discharged from incinerators.

In current decade, the waste tyres have been a major threat to human health. The widespread epidemic of dengue is sourced from waste tyres and plastics. Similarly, the hospital waste in less developed countries is meagrely separated from municipal solid waste or industrial waste. The hospital waste is among the most prominent cause of infection in public. Moreover, the chronic exposure due to open burning and dumping of any type of waste (plastic, hospital, industrial, municipal etc.) is directly linked to chronic disease among public. Between 75% and 90% of biomedical waste is analogous to community solid waste, is "non-hazardous" but left over 10–25% of hospital waste is dangerous and may pose a diversity of ecological and health risk (107). Like the other waste treatments, the open land dumping is also prioritized for hospital waste disposal and therefore, hospital waste transfers contagious pathogenic microbes to the surroundings whichever via straight contact, over inhalation, incorporation, or ancillary contact over the food chain. Burning is marked to decrease the amount of waste and its contagious effect; conversely, abandoned burning triggers the probable basis of lethal discharges. The hazard of contamination as of therapeutic waste is considerable containing HBV, hepatitis C, ebola as well as the most outspread SARS-19 viruses. Another effect of deprived waste controlling is the contamination of water bodies, both the surface and ground water. A significant literature has already been published on water contamination and water-borne diseases. Waste management activities are often low paid.

The workers are mostly unaware of health-related risks and adequate health facilities are not accessible for them (113).

2.6.4 IMPACT ON ECONOMY

Along with ecological damage, waste mismanagement is also associated with economic loss of a country. One of the reasons is that it significantly cuts down the revenue generated by tourism (147). The low-income countries are already going through unplanned urbanization and unsustainable waste management practices; therefore, such countries cannot afford the economic diversion, but provision of quality living standards is the prime objective of every government. Overall, the waste management's issue is more intensified by feeble resident government organizations and the absence of clearness and liability in city supremacy in developing countries. Global Waste Management Outlook (GWMO), and the International Solid Waste Association (ISWA) collectively argues that an vital reply to the world's escalating waste issue is not only a communal fitness and ecological requirement, but also a comprehensive financial investment. Simply, Delay is estimating the all nations five to ten times extra than reserves in appropriate waste supervision.

2.6.5 IMPACT ON LANDSCAPE AND ETHICS

Another significant impact of poor waste management is the deterioration of aesthetic and terrestrial landscapes. Peoples having accommodation near to dumpsites are typically pretentious by stink, the vision of plundering hunting animals and societal disgrace. Distributed solid waste from the prohibited open landfills often slabs the sanitations and gutters as displayed in figure (4). Eventually, these obstructions are producing submerging and unsanitary situations in the city. Another aspect of poor waste management is the odour, aromas are primarily the outcome of the occurrence of slight meditations of fragrant elements (esters, organ sulphurs, hydrogen sulfide, limonene, alkyl benzene and other hydrocarbons) in landfill gas released into the air. The fragrant nature of landfill gas may differ broadly from comparatively sweetened to acidic and pungent dependent on the absorption of the fragrant elements within the gas. These quantities will fluctuate with waste configuration and stage of development, decay period and the rate of gas production, and the type of microbial populaces within the waste, along other causes.

2.7 SUSTAINABLE STRATEGIES FOR WASTE MANAGEMENT IN DEVELOPING COUNTRIES

Uncontrolled dumping in landfill sites is the most common practice used for waste disposal, and many of the working landfills in small-earning countries are providing open dumping sites for the waste. This practice undoubtedly proved serious threats to human health and natural environment but also with time, landfills are filling up very swiftly. Their capacity has been encumbered and, in this land,-scarce situation, building new landfills is becoming too difficult. Therefore, there is need of cost-effective and effective method as an alternative to waste management (107).

Step by step improvement in the efficiency of management system could be introduced. Firstly, there is need to correct the present shortcomings i.e. proper and safe collection mechanism and then prime strategy would be the controlled and regulated dumping, and then gradually moved towards sanitary landfills (with leachate treatment and circulation) and finally recycling programs must be launched to manage the waste disposal. The introduction of 3R practice for waste reduction should be implemented by following the hierarchy of waste management. The concept 'closed loop' must be ensuing to minimize waste and recover resources and then the remaining waste will be dumped safely. Recycling plants introduction in the system would not only reduce environment burden but also create employment and improve economic growth (149; 150). This topic should be thoroughly studied and discussed to come up with an integrated and sustainable approach to improve current waste disposal system. Keeping in mind, all strategies being implemented must be publicly acceptable, cost-effective, eco-friendly and reassure public health. Thus, to improve the waste management system new policies and legislatives must be implemented and introduced in accompaniment with public co-ordination and support.

CONCLUSIONS

Waste is any unwanted material or end remains of the industry, which can be either recycle or reuse efficiently under the roof of waste management. Similar waste management strategies are recommended for developed and developing countries. Waste generation is increasing day by day due to the more export of scrap from developed countries. The waste management differs according to the type of waste, location of the area, education of the peoples and government policies. The waste mismanagement pollutes the soil, air and water, which ultimately affects the human life. The emitted gasses from the potential sources, leachates from landfills and heavy metals are the resultant of solid, industrial, medical and super hazardous waste mismanagement. The waste mismanagement affects the life due to the presence of methane, CO, CO_2, NO, NO_2, SO_2, PM10, and dust. The main waste mismanagements are open burning and open dumping, from which other mismanagement originates. Waste production is increasing at the pace of 1.3 billion tons per year on average. Generally, waste or municipal solid waste comprised of organic, paper, plastic, glass, metal and other sources, some of which are hazardous and some are non-hazardous. Waste management practices includes the prevention, minimization, reuse, recycling, compositing and incineration. All of these technologies depends upon the economy of the country. Informal recycling is another management technology in developing countries. Waste mismanagement are influenced by environment, socio-economic, physical and psychological factors, which are represented by the effects as lack of technologies, inadequate government funds, lack of education, lack of sanitation and poor attitude of the population. All the discussed factors affect the earth environment in many ways. The water born diseases are well known in Africa and Asia. Air is polluted with Sulphur oxide, nitrogen oxide, methane, carbon dioxide due to waste mismanagement in developing countries. When the atmospheric pollutants comes on the earth, they pollutes the water bodies and the soil in each waste management strategy. Industrial effluent, sewerage water, poor agriculture practices are the main contributing factors of soil deterioration. Fish and other aquatic mammals are ultimately affected due to the water pollution arises by the increased amount of plastic, heavy metals, dissolved chemicals in rivers, lakes, and oceans. Human health is affected due to the waste generated diseases as Cholera, dysentery, malaria, dengue, typhoid.

KEYWORDS

- **Biomedical Waste**
- **Chronic diseases**
- **Explosive and Radioactive Waste**
- **Household Waste**
- **Industrial Waste**
- **Municipal Waste**
- **Terrestrial Ecosystem**

REFERENCES

1. Ferronato, N.; Torretta, V. Waste Mismanagement in Developing Countries: A Review of Global Issues. *International journal of environmental Research,* 2019, *16*, 1060.
2. Vaccari, M,; Torretta,V.; Collivignarelli, C. Effect of improving environmental sustainability in developing countries by upgrading solid waste management techniques: A case study. *Sustainability*, 4(11), 2012, 2852-2861.
3. Hoornweg, D. and P. Bhada-Tata. 2012. what a waste: a global review of solid waste management. Urban development series knowledge papers. World Bank, Washington DC, USA.
4. Manaf, L.A.; Samah, M.A.A.; Zukki, N.I.M. Municipal solid waste management in Malaysia: Practices and challenges. *Waste Management.* 2009, *29*, 2902–2906.
5. Wiedinmyer, C.; Yokelson, R.J.; Gullett, B.K. Global emissions of trace gases, particulate matter, and hazardous air pollutants from open burning of domestic waste. *Environmental science and Technology,* 2014, *48*, 9523–9530.
6. World Bank. (1999). *What a waste: Solid waste management in Asia.* Washington, DC: World Bank.
7. Diaz, L.F.; Eggerth, L.L.; Enkhtsetseg, S.; Savage, G.M. Characteristics of healthcare wastes. *Waste Management,* 2008, *28*, 1219–1226.
8. Joseph, K. Perspectives of solid waste management in India. 2002, In *International Symposium on the Technology and Management of the treatment and Reuse of the Municipal Solid Waste*. Shanghai.
9. World Health Organization (WHO), *Management of Solid Healthcare Waste at Primary Healthcare Centre—A DecisionMaking Guide*, World Health Organization (WHO), Geneva, Switzerland, 2005.
10. Ferronato, N.; Bezzi, M.; Zortea, M.; Torretta, V.; Ragazzi, M. An interdisciplinary approach for introducing sustainable integrated solid waste management system in developing countries: the case of La Paz (Bolivia), Procedia Environmental Science, Engineering and Management, 2016, *3*(2), 71-81.
11. Sasaki, S.; Araki, T.; Tambunan, A.H.; Prasadja, H. Household Income, Living and Working Conditions of Dumpsite Waste Pickers in Bantar Gebang: Toward Integrated Waste Management in Indonesia. Resource, Conservation, and Recycling, 2014, *89*, 11-21.
12. Singh, J.; Saxena, R.; Bharti, V.; Singh, A. The Importance of Waste Management to Environmental Sanitation: A Review. Advanced biomedical research, 2018, *9* (2), 202-207.
13. Gutberlet, J. Briefing: social facets of solid waste: insights from the global south. *Waste and Resource Management,*2013, *166*(WR3), 110–113.
14. Vaccari, M.; Perteghella, A. Resource recovery from waste by Roma in the Balkans: A case study from Zavidovici (BiH). *Waste Management. Res.,* 2016, *34*, 866–874.
15. Fontenot J.P.; Baca, R.C.De.; Glimp, H.A. Recycling Animal Wastes by Feeding to Enhance Environmental Quality. The professional animal scientist, 1991, *7*(4):1-8.
16. Katherine, N.P. Hazardous Waste Management. (developing countries). Environment, 1999.
17. Al-Khatib Issam A, Maria, B.M.; Salam, F.A.Z.A; Hafez, Q.S.; Despo, B.K. Solid waste characterization, quantification and management practices in developing countries. A

case study: Nablus District, Palestine, Journal of Environmental Management, 2010, *91*:1131–1138.

18. Batool, S.A.; Chaudhry, M.N. Municipal solid waste management in Lahore city district, Pakistan. Waste Management (Oxford), 2009, *29*:1971–1981.
19. Burnley, S.J. A review of municipal solid waste composition in the United Kingdom. Waste Management, 2007, *27*:1274–1285.
20. Burnley, S.J.; Ellis, J.C.; Flowerdewc,R.; Polld, A.J.; Prosser, H. Assessing the composition of municipal solid waste in Wales. Resources, Conservation and Recycling, 2007, *49*:264–283.
21. Gomez, G.; Meneses, M.; Ballinas, L.; Castells, F. Characterization of urban solid waste in Chihuahua, Mexico. Waste Management (Oxford), 2008, *28*:2465–2471.
22. Ojeda-Benitez, S.; Armijo-de-Vega, C.; Ramırez-Barreto, M.E. Characterization and quantification of household solid wastes in a Mexican city. Resources, Conservation and Recycling, 2003, *39*:211–222.
23. Seo, S.; Aramaki, T.; Hwang, Y.; Hanaki, K. Environmental impact of solid waste treatment methods in Korea. Journal of Environmental Engineering, 2004,*130*:81–89.
24. Jadoon, A.; Batool, S.A.; Chaudhry, M.N. Assessment of factors affecting household solid waste generation and its composition in Gulberg Town, Lahore, Pakistan. Journal of Material Cycle and Waste Management, 2014, *16*:73–81.
25. Tadesse, M.L.; Kumie, A. Healthcare waste generation and management practice in government health centers of Addis Ababa, Ethiopia. Tadesse and Kumie BMC Public Health, 2014, *14*:1221.
26. Hossain, M.S.; Santhanam, A.; Nik Norulaini, N.A.; Omar, A.K.M. Clinical solid waste management practices and its impact on human health and environment—A review. *Waste Management*. 2011, *31*, 754–766.
27. Al-Khatib, I.A.; Sato, C. Solid health care waste management status at health care centers in the West Bank—Palestinian Territory. *Waste Management*. 2009, *29*, 2398–2403.
28. Olomolaiye, P.; Hammond, F.; Booth, C.; Coker, A.; Sridhar, M.; Sangodoyin, A. Medical waste management in Ibadan, Nigeria: Obstacles and prospects. *Waste Management*. 2008, *29*, 804–811.
29. Patwary, M.A.; O-Hare, W.T.; Sarker, M.H. Assessment of occupational and environmental safety associated with medical waste disposal in developing countries: A qualitative approach. *Safety Science*, 2011, *49*, 1200–1207.
30. Verma, R.; Vinoda, K.S.; Papireddy, M.; Gowda, A.N.S. Toxic Pollutants from Plastic Waste- A Review. Procedia Environmental Sciences, 2016, 35, 701 – 708.
31. Zafar, M.; Alappat, B.J. Landfill Surface Runoff and Its Effect on Water Quality on River Yamuna. journal of environmental science and health. 2004, *39*(2), 375–384, 2004.
32. Obil, F.O.; Ugwuishiwu, B.O.; Nwakaire, J.N. Agricultural waste concept, generation, utilization and management. Nigerian Journal of Technology, 2016, Vol. 35, No. 4, October 2016, pp. 957 – 964.
33. Unkovich, M.J.; Pate, J.S. An appraisal of recent field measurements of symbiotic N_2 fixation by annual legumes. Field Crops Research. 2000, *65*, 211–228.
34. Dien, B.V.; Vong, V.D. Analysis of pesticide compound residues in some water sources in the province of Gia Lai and DakLak. Vietnam Food Administrator. 2006.
35. International Energy Agency (IEA) (1999), "The reduction of greenhouse gas emission from the cement industry", IEA Greenhouse Gas R&D Programme Report No. PH3/7, Cheltenham.

36. Madlool, N.A.; Saidura, R.; Hossaina, M.S.; Rahim, N.A. "A critical review on energy use and savings in the cement industries", Renewable and Sustainable Energy Reviews, 2011,15(4), 2042-2060.
37. Bosoaga, A.; Masek, O.; Oakey, J.E. "CO_2 capture technologies for cement industry", Energy Procedia, 2009, 1(1), 133-140.
38. OECD. 2018. Improving Plastics Management: Trends, policy responses, and the role of international co-operation and trade OECD environment policy paper no. 12. Authorized for publication by Rodolfo Lacy, Director, Environment Directorate.
39. Mostafa, N.A.; Awatef, A.; Farag, H.; Aghareed, M.A.; Tayeb, M. Production of biodegradable plastic from agricultural waste. Arabian Journal of Chemistry, 2018, *11*(4), 546-553.
40. Connor, K.; Cortesa, S.; Wattanavit, K.; Issagaliyeva, S.; Meunier,A.; Bijaisoradat,O. Developing a sustainable waste tire management strategy for Thailand: An interactive qualifying project report. B.S. thesis, Faculty of Worcester Polytechnic Institute, Chulalongkorn University, Bangkok, Thailand.2013.
41. Torretta, V.; Rada, E.C.; Ragazzi, M.; Trulli, E.; Istrate, I.A.; Cioca, I.L. Treatment and disposal of tires: Two EU approaches. A review. Waste Management. 2015,*45*, 152–160.
42. U.S. Environmental Protection Agency. 2012. Scrap tires common wastes & materials. http://www.epa.gov/osw/con serve/materials/tires/index.html (accessed March 2016).
43. Junior, A.F.D.A.; Battistelle, R.A.; Bezerra, B.S.; Castro, R.D. Use of scrap tire rubber in place of SBS in modified asphalt as an environmentally correct alternative for Brazil. *J. Cleaner Production,* 2012, *33*:236–238.
44. Grigiante, M.; Ischia, M.; Baratieri, M.; Dal-Maschio, R.D.; Ragazzi, M. Pyrolysis analysis and solid residue stabilization of polymers, waste tyres, spruce sawdust and sewage sludge. Waste Biomass Valorization, 2010, *1*,381–393.
45. Zand, A.D.; Abduli, M.A. Current situation of used household batteries in Iran and appropriate management policies. *Waste Management.* 2008, *28*, 2085–2090.
46. Osibanjo, O.; Nnorom, I.C. The challenge of electronic waste (e-waste) management in developing countries. *Waste Management and Research.* 2007, *25*, 489–501.
47. Sharma, P.; Sharma, A.; Sharma, A.; Srivastava, P. Automobile Waste and Its Management. Research Journal of Chemical and Environmental Science, 2016, *4* (2) April 2016: 01-07.
48. Sivakumar, G.D.; Barnabas S.G.; Anatharam, S. Indian Automobile Material Recycling Management. International Journal of Innovative Research, Science, Engineering and Technology, 2014, *3*(3): 2754-58.
49. Chkravarty, T. K.; Panigrahi, S.K. "Strategies for solid waste management in SAIL steel plants", Proceedings: NS-EWM,1996, 52-62.
50. Sarkar, S.; Mazumder, D. Solid Waste Management in Steel Industry Challenges and Opportunities. World Academy of Science, Engineering and Technology. International Journal of Economics and Management Engineering, 2015, *9*(3).
51. Caijun, S."Steel Slag—Its Production, Processing, Characteristics, and Cementitious Properties", Journal of materials in civil engineering, 2004, 16, 230-236.
52. M.E. Qureshi1, M.K. Wegener1,2 and T. Mallawaarachchi.2001. The economics of sugar mill waste management in the Australian Sugar Industry: Mill mud case study. 45th Annual Conference of the Australian Agricultural and Resource Economics Society, Adelaide, South Australia, 23-25 January 2001.

53. Silva. Properties of Sugar Cane Bagasse to Use in OSB. International Journal of Materials Engineering, 2012, *2*(4): 50-56.
54. Agrawal, K.M.; Barve, B.R.; Khan, S.S. Biogas From Pressmud, IOSR, Journal of Mechanical and Civil Engineering, 2010, 37-41.
55. Wei, Y.A.; Xu, Y.J. Eco-friendly management of sugar industry effluents in Guangxi, China. Sugar Technology, 2004, *6*, 285–290.
56. Muthusamy, P.; Murugan, S.; Smitha, M. Removal of nickel ion from industrial waste water using maize cob. ISCA Journal of Biological Sciences, 2012, *1*, 7–11.
57. Hampannavar, U.S.; Shivayogimath, C.B. Anaerobic treatment of sugar industry wastewater by upflow anaerobic sludge blanket reactor at ambient temperature. International Journal of Environmental Sciences, 2010, *1*, 631–639.
58. Ozoh, P.E.; Oladimeji, A.A. Effects of Nigeria dyestuff effluent on germination latency, growth, and gross growth of *Zea mays*. Bulletin of Environmental Contamination and Toxicology, 1984, *33*, 215–219.
59. Cancado, J.E.D.; Saldiva, P.H.N.; Pereira, L.A.A.; Lara, L.B.L.S.; Artaxo, P. Martinelli, L.A.; Arbex, M.A.; Zanobetti, A.; Braga, A.L.F. The impact of sugar cane burning emissions on the respiratory system of children and the elderly. Environmental Health Perspectives, 2006, *114*, 725–729.
60. Dawson, L.; Boopathy, R. Use of post-harvest sugarcane residue for ethanol production. Bioresource Technology, 2007, *98*, 1695–1699.
61. Dwivedi, A.; Agrawal, A.; Madaan, J. Sustainable manufacturing evaluation model focusing leather industries in India: A TISM approach, Journal of Science and Technology Policy Management, 2019.
62. K. Koppiahraj, S.; Bathrinath, S.; Saravanasankar. Leather Waste Management Scenario in Developed and Developing Nations. International Journal of Engineering and Advanced Technology (IJEAT) , 2019, *9*(1S4), December 2019.
63. Ghaly, A.; Ananthashankar, R.; Alhattab, M.; Ramakrishnan, V. Production, characterization and treatment of textile effluents: a critical review. Journal of Chemical Engineering and Process Technology, 2014, *5*:1–18.
64. Khan, T.I.; Jain, V. Effect of textile industry waste water on the growth and some biochemical parameters on Triticum aestivum. J Environmental Pollution,1995, *2*:47–50.
65. Karthik,T.; Gopalakrishnan, D. Impact of textiles on environmental issues, Part- II, Asian Dyer, 2013, 45–51.
66. Karthik, T.; Gopalakrishnan D. Impact of textiles on environmental issues, Part-IV, Asian Dyer, 2013, 43–46.
67. Karthik, T.; Gopalakrishnan, D. Environmental Analysis of Textile Value Chain: An Overview Springer Science+Business Media Singapore 2014.
68. Muduli, K.; Barve, A. "Barriers to green practices in healthcare waste sector: an Indian perspective," *International Journal of Environmental Science and Development*,2012, *3*(4), 393– 399.
69. Matiko, D. *Managing disposal of unwanted pharmaceuticals at health facilities in Tanzania: a case of Dar es Salaam region public health facilities [MSc. (Pharmaceutical Management) Dissertation]*, Muhimbili University of Health and Allied Sciences, Dares Salaam, Tanzania, 2011.

70. World Health Organization (WHO), *Management of Solid Healthcare Waste at Primary Healthcare Centre—A DecisionMaking Guide*, World Health Organization (WHO), Geneva, Switzerland, 2005.
71. Zhang, H.J.; Zhang, Y.H.; Wang, Y.; Yang, Y.H.; Zhang, J.; Wang, Y.L. Investigation of medical waste management in Gansu Province, China. Waste Management Research, 2013; *31*, 655-699.
72. Hashmi, S.K.; Shahab, S. Hospital and Biomedical waste management. In: Iliyas M, Editor, Community medicine and public health. 4th ed. Karachi: Time Publishers, 2003, pp. 426-437.
73. Omar, D.; Nazli, S.N.; Karuppannan, S. Clinical Waste Management in District Hospitals of Tumpat, Batu Pahat and Taiping, Procedia - Social and Behavioral Sciences, 2012, *68*, 134 – 145.
74. Drabo, H.H.; Ouedraogo, M.K. Healthcare associated infection in Burkina Faso: an assessment in a district hospital, Journal of Public Health in Africa; 2012, *3*, 29.
75. Manyele, S.V. Effects of improper hospital-waste management onoccupational health and safety, African Newsletteer on Occupation, Health and Safety, 2004, *14*, 30-33.
76. Gautam, V.; Thapar, R.; Sharma, M. Biomedical waste management: Incineration vs. environmental safety, Special Article, 2010, *28*(3), 191-192.
77. Ravi, V.; Shanker, R.; Tiwari, M.K. Analyzing alternatives in reverse logistics for end-of-life computers: ANP and balanced scorecard approach. C*omputer and Industrial Engineering,* 2005,*48*: 327–356.
78. Hicks, C.; Dietmar, R.; Eugster, M. The recycling and disposal of electronic waste in China – legislative and market response. *Environmental Impact Assessment Review*,2005, *25,* 459–471.
79. Widmer, R.; Oswald-Krapf, H.; Sinha-Khetriwal, A.; Scnellmann, M.; Boni, H. Global perspectives on the e-waste. *Environmental Impact Assessment Review,2005,* 25, 436–458.
80. Sanneh, E. S.; Hu, A. H.; Chang, Y. M.; Sanyang, E. Introduction of a recycling system for sustainable municipal solid waste management: A case study on the greater Banjul area of the Gambia. *Environment, development and sustainability*, 2011, *13*(6), 1065-1080.
81. Chiemchaisri, C.; Juanga, J. P.; Visvanathan, C. Municipal solid waste management in Thailand and disposal emission inventory. *Environmental monitoring and assessment*, 2007, *135*, 13–20.
82. Seng, B.; Fujiwara, T.; Seng, B. Suitability assessment for handling methods of municipal solid waste. *Global Journal of Environmental Science and Management*, 2018, *4*(2), 113-126.
83. Aderoju, O. M.; Dias, G. A.; Gonçalves, A. J. A GIS-based analysis for sanitary landfill sites in Abuja, Nigeria. *Environment, Development and Sustainability*, 2020, *22*(1), 551-574.
84. dos Muchangos, L. S.; Tokai, A.; Hanashima, A. Analyzing the structure of barriers to municipal solid waste management policy planning in Maputo city, Mozambique. *Environmental Development*, 2015, *16*, 76–89.
85. Prokic, D.; Mihajlov, A. Contaminated sites. Practice of solid waste management in a developing country (Serbia). *Environment Protection Engineering*, 2012, *38*(1), 81-90.
86. Barabad, M.L.M.; Jung, W.; Versoza, M.E.; Kim, M.; Ko, S.; Park, D.; Lee, K. Emission Characteristics of Particulate Matter, Volatile Organic Compounds, and Trace Elements

from the Combustion of Coals in Mongolia. *International Journal Environmental Research and Public Health,* 2018, *15*, 1706.
87. Li, G.; Lei,W.; Bei, N.; Molina, L.T. Contribution of garbage burning to chloride and PM2.5 in Mexico City. *Atmospheric Chemistry and Physics,* 2012, *12*, 8751-8761.
88. Christian, T.J.; Yokelson, R.J.; Cardenas, B.; Molina, L.T.; Engling, G.; Hsu, S.C. Trace gas and particle emissions from domestic and industrial biofuel use and garbage burning in central Mexico. *Atmospheric Chemistry and Physics*, 2010, *10*(2), 565-584.
89. Wiedinmyer, C.; Yokelson, R.J.; Gullett, B.K. Global emissions of trace gases, particulate matter, and hazardous air pollutants from open burning of domestic waste. *Environmental science & technology*, 2014, *48*(16), 9523-9530.
90. Kassim, S. M. *Sustainability of private sector in solid waste collection-a case of Dar es Salaam Tanzania.*"Doctoral dissertation; Loughborough University, England; 2006.
91. Human RightsWatch: "As if You're Inhaling Your Death". The Health Risks of BurningWaste in Lebanon. 2017.
92. Reyna-Bensusan, N.; Wilson, D. C.; Smith, S. R. Uncontrolled burning of solid waste by households in Mexico is a significant contributor to climate change in the country. *Environmental research*, 2018, *163*, 280-288.
93. Kumara, K..; Kumar, S.; Rajagopal, V.; Khare, A.; Kumar, R. Emission from open burning of municipal solid waste in India. *Environmental technology*, 2019, *40*(17), 2201-2214.
94. Getahun, T., Mengistie, E., Haddis, A., Wasie, F., Alemayehu, E., Dadi, D., Van Gerven, T. and Van der Bruggen, B., 2012. Municipal solid waste generation in growing urban areas in Africa: current practices and relation to socioeconomic factors in Jimma, Ethiopia. Environmental monitoring and assessment, 184(10), pp.6337-6345.
95. Nagpure, A.S., Ramaswami, A. and Russell, A., 2015. Characterizing the spatial and temporal patterns of open burning of municipal solid waste (MSW) in Indian cities. Environmental science & technology, 49(21), pp.12904-12912.
96. Ejaz, N., Akhtar, N., Hashmi, H. and Naeem, U.A., 2010. Environmental impacts of improper solid waste management in developing countries: A case study of Rawalpindi city. The sustainable world, 142, pp.379-387.
97. Babayemi, J.O. and Dauda, K.T., 2009. Evaluation of solid waste generation, categories and disposal options in developing countries: a case study of Nigeria. Journal of Applied Sciences and Environmental Management, 13(3).
98. Das, B., Bhave, P.V., Sapkota, A. and Byanju, R.M., 2018. Estimating emissions from open burning of municipal solid waste in municipalities of Nepal. Waste management, 79, pp.481-490.
99. Abbas, I., Chaaban, J.K., Al-Rabaa, A.R. and Shaar, A.A., 2017. Solid Waste Management in Lebanon: Challenges and Recommendations. Journal of Environment and Waste Management, 4(3), pp.235-243.
100. Cogut, A., 2016. Open burning of waste: a global health disaster. R20 Regions of climate action.
101. Steuer, B.; Ramusch, R.; Salhofer, S.P. Can Beijing's informal waste recycling sector survive amidst worsening circumstances? *Resources, Conservation and Recycling*, 2018, *128*, 59-68.
102. Kumar, A.; Samadder, S.R.; Kumar, N.; Singh, C. Estimation of the generation rate of different types of plastic wastes and possible revenue recovery from informal recycling. *Waste Management*, 2018, *79*, 781-790.

103. Vaccari, M.; Perteghella, A. Resource recovery from waste by Roma in the Balkans: A case study from Zavidovici (BiH). *Waste Management & Research*, 2016, *34*(9), 866-874.
104. Kasinja, C.; Tilley, E. Formalization of informal waste pickers' cooperatives in Blantyre, Malawi: A feasibility assessment. *Sustainability*, 2018, *10*(4), 1149.
105. Vergara, S.E.; Damgaard, A.; Gomez, D. The Efficiency of Informality: Quantifying Greenhouse Gas Reductions from Informal Recycling in Bogota, Colombia. *Journal of Industrial Ecology*, 2016, *20*(1), 107-119.
106. Ferronato, N.; Torretta, V. Waste mismanagement in developing countries: A review of global issues. International journal of environmental research and public health, 2019,*16* (6), 1060.
107. Kötter, T. Risks and opportunities of urbanisation and megacities. Proceedings of the FIG Working Week, Athens, Greece.2004.
108. Singh, R. Wastewater Related Risks and Social Vulnerability: A Case Study of Delhi. Resilience and Social Vulnerability, 2008, 121.
109. Matter, A.; Ahsan, M.; Marbach, M.; Zurbrügg, C. Impacts of policy and market incentives for solid waste recycling in Dhaka, Bangladesh. Waste Management, 2015,*39*, 321-328.
110. Hasan, S.A.; Subhani, M.I.; Osman, M. Waste Management in the Various Municipalities of Various Socio-Economic Conditions (An Empirical Evidence from Pakistan).2011.
111. Triassi, M.; Alfano, R.; Illario, M.; Nardone, A.; Caporale, O.; Montuori, P. Environmental pollution from illegal waste disposal and health effects: A review on the "Triangle of Death". International Journal of Environmental research and public health, 2015,*12*(2), 1216-1236.
112. Ferronato, N.; Torretta, V.; Ragazzi, M.; Rada, E. C. Waste mismanagement in developing countries: A case study of environmental contamination. UPB Sci. Bull,2017, *79*(2), 185-196.
113. Ramachandra, T.V.; Bharath, H.A.; Kulkarni,G.; Han, S.S. Municipal solid waste: Generation, composition and GHG emissions in Bangalore, India. Renewable and Sustainable Energy Reviews, 2018, *82*, 1122-1136.
114. Factbook, O.E.C.D. Economic, Environmental and Social Statistics. Organisation for Economic Cooperation and Development Publishing.2013,
115. Kollikkathara, N.; Feng, H. Stern, E. A purview of waste management evolution: Special emphasis on USA. Waste management, 2009, *29*(2), 974-985.
116. Zaman, A.U. Measuring waste management performance using the 'Zero Waste Index': the case of Adelaide, Australia. Journal of Cleaner Production, 2014,*66*, 407-419.
117. Psomopoulos, C.S.; Bourka, A.; Themelis, N.J. Waste-to-energy: A review of the status and benefits in USA. Waste management, 2009, *29*(5), 1718-1724.
118. Visvanathan, C.; Glawe, U. Domestic solid waste management in South Asian countries–a comparative analysis. Promoting Reduce, Reuse, and Recycle in South Asia, 2006, 27.
119. Parkinson, J.; Tayler, K. Decentralized wastewater management in peri-urban areas in low-income countries. Environment and Urbanization, 2003,*15*(1), 75-90.
120. Murtaza, G. Zia, M.H. Wastewater production, treatment and use in Pakistan. In Second regional workshop of the project 'Safe use of wastewater in agriculture (pp. 16-18).2012.
121. Karpouzoglou, T.; Zimmer, A. Ways of knowing the wastewater scape: Urban political ecology and the politics of wastewater in Delhi, India. Habitat International, 2016, 54, pp.150-160.

122. Agarwal, A.; Singhmar, A.; Kulshrestha, M.; Mittal, A.K. Municipal solid waste recycling and associated markets in Delhi, India. Resources, Conservation and Recycling, 2005, *44*(1), pp.73-90.
123. Biswas, A. K.; Tortajada, C. (2016, February). Delhi: megacity, megatraffic, and megapollution. In Policy Forum.
124. Asano, T.; Cotruvo, J.A. Groundwater recharge with reclaimed municipal wastewater: health and regulatory considerations. Water research, 2004, *38*(8), 1941-1951.
125. Bos, R.; Carr, R.; Keraita, B. Assessing and mitigating wastewater-related health risks in low-income countries: An introduction. Wastewater irrigation and health: Assessing and mitigating risk in low-income countries, 2010, pp.29-47.
126. Osei-Mensah, P.; Adjaottor, A.A.; Owusu-Boateng, G. Characterization of solid waste in the Atwima-Nwabiagya District of the Ashanti Region, Kumasi-Ghana.2008.
127. Fagariba, C.J.; Song, S. Assessment of Impediments and Factors Affecting Waste Management: A Case of Accra Metropolis.2016.
128. You, S.; Sonne, C.; Ok, Y.S. COVID-19's unsustainable waste waste dump sites and selection of sanitary landfill sites in the West Bank, Palestinian territory. Environment,2020,
129. Baldwin, R.; Mauro, B.W.D. (2020). Economics in the Time of COVID-19.
130. Windfeld, E.S.; Brooks, M.S.L. Medical waste management–A review. Journal of environmental management, 2015,163, 98-108.
131. Hasan, S.E. Public awareness is key to successful waste management. Journal of Environmental Science and Health, Part A, 2004, *39*(2), 483-492.
132. Batool, S.A.; Chuadhry, M.N. The impact of municipal solid waste treatment methods on greenhouse gas emissions in Lahore, Pakistan. Waste management, 2009, *29*(1), 63-69.
133. Tiwari, R.R. Occupational health hazards in sewage and sanitary workers. Indian journal of occupational and environmental medicine, 2008, *12*(3), 112.
134. Kumar, R.; Samrongthong, R.; Shaikh, B.T. Knowledge, attitude and practices of health staff regarding infectious waste handling of tertiary care health facilities at metropolitan city of Pakistan. Journal of Ayub Medical College Abbottabad,2013, *25*(1-2), 109-112.
135. Desa, A., Kadir, N.B.A.; Yusooff, F. Waste education and awareness strategy: towards solid waste management (SWM) program at UKM. Procedia-Social and Behavioral Sciences, 2012, *59*, 47-50.
136. Pillai, R.; Shah, R. Municipal solid waste management: current practices and futuristic approach. SCMS Journal of Indian Management, 2014,*11*(4), 72.
137. Giusti, L. A review of waste management practices and their impact on human health. Waste management, 2009, *29*(8), 2227-2239.
138. Lenart-Boron, A.; Boron, P. 2014The effect of industrial heavy metal pollution on microbial abundance and diversity in soils—a review. In Environmental, risk assessment of soil contamination. IntechOpen.
139. Jin, J.; Wang, Z.; Ran, S. Solid waste management in Macao: practices and challenges. Waste Manageement, 2006, *26*, 1045–1051.
140. Alam, P.; Ahmade, K. Impact of solid waste on health and the environment. International Journal of Sustainable Development and Green Economics, 2013, *2*(1), pp.165-168.
141. Williams, C. Combatting marine pollution from land-based activities: Australian initiatives. Ocean & coastal management, 1996,*33* (1-3), pp.87-112.

142. Islam, M.S.; Tanaka, M. Impacts of pollution on coastal and marine ecosystems including coastal and marine fisheries and approach for management: a review and synthesis. Marine pollution bulletin, 2004, *48*(7-8), 624-649.
143. Amoatey, P.; Baawain, M.S.. Effects of pollution on freshwater aquatic organisms. Water Environment Research, 2019, *91*(10), 1272-1287.
144. Geyer, R.; Jambeck, J.R.; Law, K.L. Production, use, and fate of all plastics ever made. Science advances, 2017, *3*(7), 1700782.
145. Anbumani, S.; Kakkar, P. Ecotoxicological effects of microplastics on biota: a review. Environmental Science and Pollution Research, 2018, *25*(15), 14373-14396.
146. Jambeck, J.; Hardesty, B.D.; Brooks, A.L.; Friend, T.; Teleki, K.; Fabres, J.; Beaudoin, Y.; Bamba, A.; Francis, J.; Ribbink, A.J.; Baleta, T. Challenges and emerging solutions to the land-based plastic waste issue in Africa. Marine Policy, 2018, *96*, 256-263.
147. Ziraba, A.K.; Haregu, T.N.; Mberu, B. A review and framework for understanding the potential impact of poor solid waste management on health in developing countries. Archives of Public Health, 2016, *74*(1), 1-11.
148. Sanneh, E. S.; Hu, A. H.; Chang, Y. M.; Sanyang, E. Introduction of a recycling system for sustainable municipal solid waste management: A case study on the greater Banjul area of the Gambia. *Environment, development and sustainability*, 2011, *13*(6), 1065-1080.
149. Al-Khatib, I. A.; Abu Hammad, A.; Sharkas, O. A.; Sato, C. Public concerns about and perceptions of solid waste dumpsites and selection of sanitary landfill sites in the West Bank, Palestinian territory. *Environmental monitoring and Assessment*, 2015, *187*(4), 186.
150. IAEA. 2018. Status and trends in spent fuel and radioactive waste management. International Atomic Energy Agency, Vienna.

CHAPTER 3

Sustainable Management of Waste in Developing Countries: Insight into Sustainability and Waste Management: Why It Is Needed?

LAILA SHAHZAD*, ASMA YASIN, FAIZA SHARIF, and MUHAMMAD UMER HAYYAT

Sustainable Development Study Center, GCU Katchery Rd, Lahore 54000, Pakistan

**Corresponding author. E-mail: lailashahzad@gcu.edu.pk*

ABSTRACT

Humans have been responsible for resource extraction, consumption, and producing enough waste in environment, which cannot be reclaimed naturally. In the 21st century, sustainable waste management can be one of the principal challenges to find a sustainable solution. The idea of sustainability originated from 1987, "our common future report" which described utilizing your resources sustainably by leaving them for future generation therefore waste generation need to be managed by reducing input by mankind. This chapter covers the idea of sustainability incorporated into the management of waste, especially for developing countries. One can think of the need of sustainability in waste management; however, it provides an opportunity of handling the waste before its production. Minimizing or changing the consumption patterns of waste can lead to having sustainable long-term benefits to resource management of developing countries. Solutions for waste management can lie in collection, segregation, recycling, & reusing,

Waste Problems and Management in Developing Countries. Umair Riaz, Shazia Iqbal, & Moazzam Jamil (Eds.)

and creating secondary use of collected waste. In addition to this, handling waste properly has an opportunity to provide employment of a large scale and will defiantly generate revenue. Waste has to be seen as an opportunity rather as a burden!

3.1 INTRODUCTION

Since the beginning of the 20th century, industrialization has boomed economic development resulting in rapid economic growth, and significant improvements in the living standards of the global population are witnessed due to suburbanization (Jagdeep et al., 2014). Improved economic conditions with advanced living conditions and rapidly growing population have also accelerated the generation rate of waste production around the world which has become a serious threat and a worldwide challenge (Seo et al., 2004). Waste generation has increased in recent times because people have increased their consumption of the resources that are highly material intensive which causes the release of huge concentrations of waste contaminating the surrounding environment (Blanchard, 1992; Gerbens-Leenes et al., 2010; Wenheng and Shuwen, 2008). American National Academy of Science records state that approximately 95% of all the material substances that are taken out from the natural resources for use become the part of waste streams within a period of 30 days (Bylinsky, 1995).

Management of municipal solid waste is an enormously neglected regime majorly in the urban municipalities in developing and emerging nations (Asad et al., 2020; Batool and Ch, 2009; Chung and Charlos, 2008; Ahmeda and Alib, 2004). Due to improper handling of this waste can have huge impacts on the soil, air, and public health therefore its hazards need urgent handling and proper addressing to the public (Batool and Ch, 2009; Sharholy et al., 2008). Due to the lacking capacity for effectively adopting the technologies implemented expressively by the developed nations is the major factor behind the failure of the councils in providing the adequate solid waste management (SWM) services (Rotich et al., 2006; Almendro-Candel et al., 2019). Management has worsened due to a lot of barriers faced by the developing nations like lack of sufficient resources, restricted financial supremacies, and meager governance (Manga et al., 2008; Oosterveer and Van Vliet, 2010). It has been observed that on an average scale about 50% of the community residing in developing or underdeveloped countries are deprived of the basic waste collection systems and other management services (Parizeau et al., 2006). The development of a waste management system that is sustainable has limited implication at the moment as the government has a limited budget and waste collection is not considered as an important aspect in sustainability is therefore overlooked and the total management cost comprises of appropriate solid waste disposal (McBean et al., 2005).

Environment when comes in contact with human activities, waste production is the ultimate result Giusti, 2009). Therefore, waste management must be made sustainable and effective in its nature. Sustainability literally means using the present resources in such a way that next generation does not have to suffer for their share and their needs are not sacrificed. "Our Common Future" first coined the term sustainable development from which it became prominent globally (Mebrstu, 1998). In Agenda 21, Rio de Janerio stated an action plan which focused on the sustainable waste management. Integrated and sustainable SWM is a systematic and holistic approach that has been addressed in this conference along with the full-fledge interpretation of environment, economic, and social pillars with special emphasis on management and treatment practices to attain sustainability (Diaz et al., 1996).

The notion of sustainable development came to global prominence by way of the report Our Common Future, published by the World Commission on Environment and Development in 1987, but it consolidated on and interpreted precursor versions (Mebrstu, 1998). Agenda 21—the action plan on sustainable development agreed at the United Nations Conference on Environment and Development in Rio de Janeiro in 1992—has been influential in SWM. Diaz et al. (1996) argued for a holistic and systems approach to "integrated and sustainable SWM," addressing all three aspects of triple bottom line interpretation of sustainability (environmental, social, and economic) with a particular emphasis on spatially combining SWM, wastewater treatment, energy production, and food production facilities.

Annual solid waste generation rate has been estimated to be approx. 2.01 billion tones and in the next 30 years, it is expected to grow to 3.4 billion tones on annual basis (Kaza et al., 2018). Socioeconomic state of a nation can predict its generation rate and the waste composition as countries like USA, Europe, and other developed nations are responsible for producing great concentrations of plastic, paper, and other nonorganic matter while the developing countries are majorly agricultural-based countries, therefore waste composition is more like organic in nature as higher fractions of such wastes are found in waste streams indicating this phenomenon (Das et al., 2019; ATSDR, 1995). Developed countries like USA and UK use high-end technologies for the appropriate treatment of solid wastes. Municipal solid wastes (MSWs) are mainly treated via sanitary landfills, treatment using heat such as incineration procedures, gasification, and pyrolysis methods. While some of the biological methods are also made into practice for the proper treatment of certain types of wastes, which involves the composting method and the technique of anaerobic digestion (AD). Although high level of financial investment and skilled professionals is demanded for keeping such advanced technologies in operation still they cannot completely neutralize the harmful impacts of solid wastes. Underdeveloped countries lack such technologies and proper management methods therefore they adopt the simplest and cost-effective methods for waste disposal such as waste burning, open dumping, and landfilling being completely ignorant about the leachate and gas it produces, which have even more hazardous impacts on human health and overall environment. Physical and mental health is deteriorated due to the noxious effect of the harmful gases produced from such waste dumping and it results in serious respiratory and neurological diseases (Kaza et al., 2018; Asad et al., 2020).

Integrated systems of waste management have been shown in Figure 3.1, which have been used all around the world for treating the waste that is being produced heftily. The sustainable management of the solid waste being produced involved a special focus in the collection phase and the phenomenon of 3Rs is made into practice.

This chapter explores the possibility of using sustainable waste management schemes in developing nations. Since every step involved in waste management is important therefore public participation and their effective involvement along with the government policies are necessary to attain this system in the proximate future.

3.2 WASTE CLASSIFICATION

Classification of waste is not an easy task since its composition is heterogeneous in nature with the great level of variation but their managing processes are similar. The type of waste can be best classified depending upon its source of generation and the varied waste characteristic.

Following are the major classes of wastes (Goren and Ozdemer, 2015):

- Domestic or household waste.
- Industrially produced waste.
- Medical waste.
- Hazardous waste.

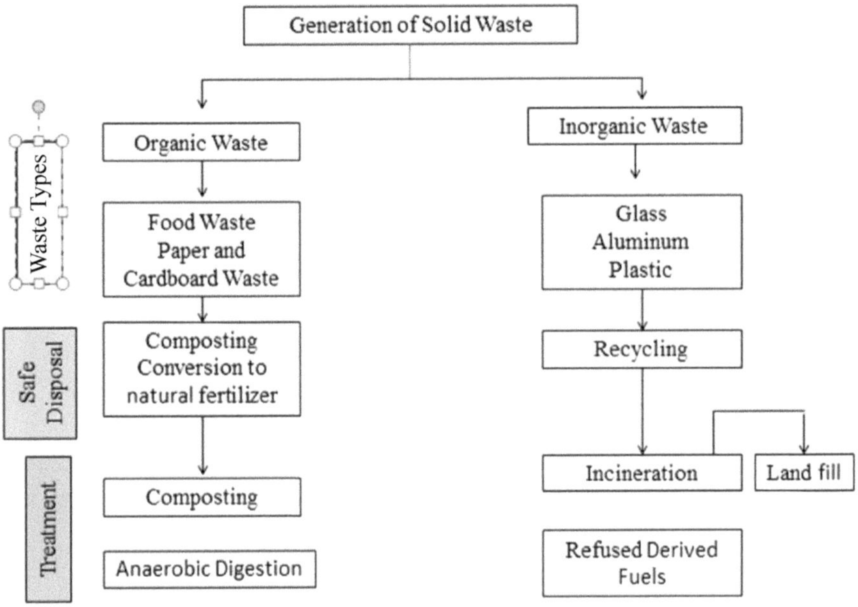

FIGURE 3.1 Framework of integrated waste management system.

3.2.1 DOMESTIC WASTE (MUNICIPAL SOLID WASTE, MSW)

Everyday items usually known as garbage or trash are included in domestic waste. It further includes food waste, garden waste (grass, yard waste, etc.), durable as well as nondurable goods namely old furniture, containers, plastic one-time used bottles, paper/newspaper, inorganic wastes, and other appliances no longer in use by the owner. Such wastes are basically household waste since it comes from people's homes but it is not a hard and fast rule. Commercial, institutions might also produce waste with similar composition. Due to this, it is acknowledged as municipal solid waste that excludes wastewater treatment plant effluents, etc. (Chalmin and Gaillochet, 2009).

3.2.2 INDUSTRIAL WASTE

Industrial waste as the name indicates is the type of waste obtained after a manufacturing procedure including wastewater sludge and other factory waste material. It is difficult to predict accurately about the type of waste being produced as it is mostly handled by the factory itself of some private disposal

TABLE 3.1 Types and Sources of Wastes.

Sources of waste	Typical waste generators	Types of solid waste
Household waste/ residential waste	Basically produced by families living in colonies	Plastic waste busters, backyard waste, textile (clothing) waste, kitchen waste (including food waste), leather wasted products, broken glass, metal cans, wasted paper, cardboard, wood ashes, and other special waste including items in bulks, electronic waste, car tires, batteries, etc. Plus other hazardous waste products
Waste from commercial buildings	Major commercial buildings involve hotels, cafes, restaurants, office buildings, and other goods markets	Plastic waste, paper and cardboard goods, kitchen waste, wood wastes (old furniture), glass and metal waste items, hazardous commercial waste also including special waste types
Institutional wastes	Waste generators include hospitals, clinics, nursing institutes, jails, prisons, government concentration camps, and schools	Plastic waste, paper and cardboard goods, kitchen waste, wood wastes (old furniture), glass and metal waste items, hazardous commercial waste also including special waste types
Waste from industries	Manufacturing industries (whether heavy or light), construction areas, chemical plants, assembly lines, and power generating companies	Food and packaging waste, special waste, hazardous waste items, waste from housekeeping activities, ashes wastes, and finally construction, and demolition wastes
Construction and demolition waste	Wastes from the newly formed construction sites, waste produced while repairing of roads, renovation sites, and waste from demolishing a building	Dirt, concrete from the walls of buildings, steel and discarded wood, etc.
Municipal and metropolitan services	Waste generated after street cleaning and landscaping, waste like grass, soil waste from parks, waste collected from beaches, sludge waste from wastewater treatment plants, and other wastes from recreational sites	Street sweeping dust, waste leaves and tree trimmings (dead branches and leaves), sludge waste, other wastes from recreational zones like that of beaches, parks, amusement parks, etc.
Processing waste	Waste generated from chemical industries, power plants, manufacturing refineries, and other waste produced during the mineral extraction processes	Usually includes the process waste from industries, scrap waste material, tailings and slags, and products which are off-specification.
Agricultural waste/field waste	Major waste generators are crops, dairy and livestock farms, orchards, and vineyards	Basically food waste more appropriately spoiled food, hazardous farm waste (fertilizers, pesticides, etc.), and finally agricultural waste

companies (Chalmin and Gaillochet, 2009). If some wastes from industries are hazardous in nature then they must be treated with special care prior to their disposal. The industrial waste can also include E-waste (electronic waste). If the industries produce certain harmful chemicals then they might be categorized as special waste (Goren and Ozdemer, 2015; Klundert, 1999).

3.2.3 HAZARDOUS WASTE

The waste which is dangerous to life in any aspect, difficult to treat or dispose due to its toxic, corrosive of reactive nature, can directly or indirectly pose threat to the environment and human health. In European nations, manufacturing industries are responsible for producing extremely hazardous waste in significantly greater concentrations (European Commission, 2003). Improperly disposed chemicals may enter the waste system as it has the potential to pollute our food sources (Ulas, 2014).

3.2.4 MEDICAL WASTE

Clinical or medical waste is generally discarded products from healthcare centers like hospitals, laboratories, nursing centers, etc. It also includes infectious medical equipment, pathological chemicals, or diagnostic samples, needles, syringes, and other miscellaneous radioactive materials. All the medical wastes cannot be recycled therefore they are separated at source so that they could be incinerated but this process is extremely expensive but other methods are not viable. The majority of the infectious waste is autoclaved or incinerated and their residues are deposited in landfills later on to prevent transference of any infection (Tchobanoglous, 2002). Management of clinical waste is directly linked to public health and welfare due to the probability of any disease spread. Precaution is financially effective than any kind of therapy for its treatment.

3.3 WASTE GENERATION AND MANAGEMENT

Production of waste varies in every country depending upon the uses and demands of their population. Therefore, the generated waste is treated accordingly to the best of its interest. Table 3.2 shows the management practices employed in low-income, middle-income, and high-revenue-based nations.

TABLE 3.2 Comparison of Waste Management Practices in Low-, Medium-, and High-Revenue Generating Countries.

Management practices	Low-revenue generating countries	Medium-revenue generating countries	High-revenue generating countries
Reduction at source	Low-income countries lack fully organized waste management programs but have the concept of reusing and the waste generation rates are low per capita	On the track toward source reduction but still very rare implementation of properly organized such programs	Source reduction and reuse of material have been supported by proper and organized educational programs which form the basis of sustainable waste management
Waste collection	Low-income countries have an erratic and insufficient waste collection system as the service is limited to businesses and wealthy areas that are willing to pay for this service	Considerably upgraded service and residential areas receive better facility as larger vehicles are used and more mechanized working have been established	Highly mechanized waste collection vehicles are common which increase the collection rate higher than 90%
Waste recycling	Recycling occurs majorly by the informal sector and waste collecting crew. Recycling is actually salvaging which is limited to the material of products that were imported and other local market items	Middle-income states have employed some high-end sorting machines to upgrade their processing facilities. Some informal sectors are also involved	High-end technology for material sorting, waste collection service for the collection of recyclable materials, and other processing facilities are available readily. Goals being developed for long term markets
Composting	Not done at commercial level although waste contain greater organic matter percentage	Composting plants constructed at greater scale were futile as small scale projects are considered more sustainable	Backyard and large-scale composting facilities are on the move. Waste streams have fewer portions of compostables as compared to low or middle-income countries
Incinerations	Incineration has a high capital cost (operational cost) therefore it is neither common nor successful in low-income states	Although this facility is used occasionally but due to its high operating and commercial cost its fulltime implementation is difficult financially	Common in high-income states as its prevalent in high land cost areas. Majority of them are environmentally safe having some energy control mechanisms and controls

TABLE 3.2 *(Continued)*

Management practices	Low-revenue generating countries	Medium-revenue generating countries	High-revenue generating countries
Landfilling	Open dumping is common as landfilling is not that common	Although many controlled mechanisms with environmental controls and sanitary landfills are there still open dumping is communal	Proper sanitary landfills with the facility of leak detection and leachate collection systems, liners in the landfills with gas collection and treatment systems available
Total cost	Approximately 80–90% of total MSWM budget is expended on the waste collection. Fee collection system is not efficient as most waste fees are regulated by the local government	In total, 50–80% of the total budget of MSWM represents the collection cost of waste. Waste fee is regulated by national as well as local government sectors	High-income countries have spent approximately less than 10% on its waste collection systems. Greater budgets are allocated for other intermediate waste treatment facilities. Active community participation has made it possible to reduce the cost and enhance the options present for waste planners

Source: Adapted with permission from World Bank World Bank Report (1999).

3.4 TRADITIONAL WASTE TREATMENT SYSTEMS

Conventional waste management system is an integration of waste collection, transformation, and ways of waste processing to its phase of recycling or disposal in incinerators, resource recovery facilities, and in other sanitary landfills. The collection of solid waste is typically the most challenging task since it needs to be collected from point to point which is a costly process and approximately 80% of the total community budget is consumed by this management system alone (Ulas, 2014). The most common treatment methods used for waste treatment are:

- Composting/Biodegradation
- Incineration
- Recycling
- Sanitary landfilling
- Mechanical biological treatment (MBT)
- Anaerobic digestion.

In developing nations, an integrated waste management system was adopted that differs from the traditional systems of managing waste as it involves stakeholders for resource recovery and aid in the interaction of various other systems to form a bond between them (Klundert, 1999). Figure 3.2 shows the waste hierarchy in an inverted pyramid form to show the most to the least preferred method.

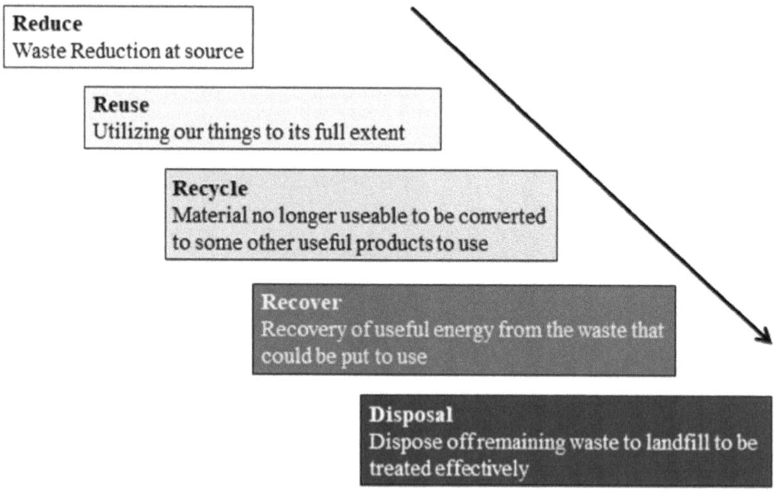

FIGURE 3.2 Waste hierarchy.

3.4.1 COMPOSTING AND BIODEGRADATION

Yard and food wastes are decomposed biologically since they are organic in nature. Such form of recycling of nutrients back to its origin is known as composting/biodegradation. Microbes inside the organic matter start their cyclic process due to the presence of aerobic conditions. Composting is the natural process of degrading waste which is helpful for many applications and acts as natural fertilizer (Ulas, 2014). This method of soil enrichment and organic waste treatment is highly appreciated in the developing countries since most of the nations are agriculture-based. Developed countries like that of European Union use this method in gardening and sustainable agriculture. Compost-enriched soil is commercially effective in suppressing pest attack and any plant disease. Composting is a growing interest area for many sectors in developing nations as they could follow the examples of developed nations since they discover such things to be more environmental friendly than other mechanically used methods for waste treatment. Figure 3.3 delivers a graphical representation of a list of developed nations along with their preferable usage of the type of waste management/treatment system (Daniel and Perinaz, 2012; Zorpa and Lasaridi, 2013).

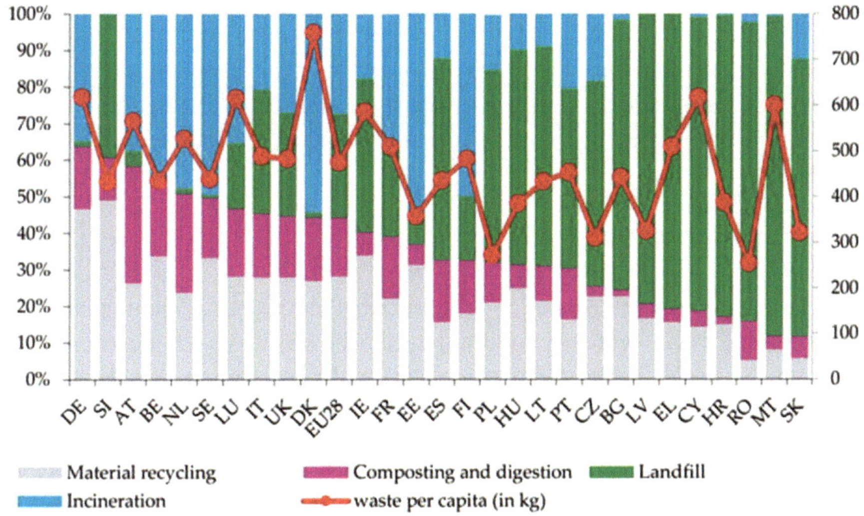

FIGURE 3.3 Waste generation and management in European Union Countries.

Source: Reprinted from Almendro-Candel et al. (2019). https://creativecommons.org/licenses/by/3.0/

3.4.2 INCINERATIONS

Controlled burning procedure for significantly reducing the total volume of waste is incineration process which is used by the government as well as local operators. If this combustion process is handled with accuracy and extreme care, the effluent water can be converted to the stream which would be helpful in generating electricity and can also act as fuel for industrial of commercial heating systems. Various types of wastes can be burnt in incinerators such as MSW, tire scraps, or even sewage sludge from wastewater treatment plants (Williams, 2005). The incineration process is widely used around the world and is considered as the second largest waste handling process in the whole world. But the main issue related to its use is air pollution due to the release of ash residues in air. Although scrubbers can be accompanied to prevent this spreading. Burning of waste is done at extremely high temperature, which has the capability to eliminate hazardous chemical components and other harmful microbes present in waste (Goren and Ozdemer, 2015). In developed countries like Europe, traditional or old incinerators are banned from being used due to their air polluting factor but previously this technology was widely utilized as it can reduce 90% of waste volume and approximately 75% by weight is reduced. Incineration rates for Japan, Swiss, and Switzerland are 60, 80, and 70%, respectively (Klundert, 1999). Developing countries like India, China, Bangladesh, Pakistan use incineration for reducing the volume of extremely harmful waste like chemicals, hospital waste, etc., wide applicability is hindered due to its high operating cost and high initial investment. But if the waste-to-energy concept is followed then this high cost can be compensated to some extent.

3.4.3 SANITARY LANDFILLING

In developing countries, major proportion of the waste is dumped in landfills. Landfills are not just a dumping site but its well-engineered locality formulated under proper regulations. The only problem is that the growing population generates more waste which makes it difficult for landfills to accommodate the amount due to insufficient and limited resources and the waste management staff is also no trained professionally (Fell et al., 2010). Sanitary landfills are not dangerous for people as it has soil covering over it and the gas being produced from the landfill should be effectively removed to prevent any explosion. Every country owns a certain number of landfills to allow its citizens safe waste disposal (Gottberg and Longhurst, 2010). If prepared according to the engineered technology, landfills can also provide solutions for waste-to-energy production but many countries adopt approaches that are unsustainable and focused on its technical facet only (Klundert, 1999; Wilson et al., 2012). Developing countries are usually short in their budget to formulate a fully functioning sanitary landfill site. As its design demands proper planning for effective and safe waste disposal facilities for various types of solid wastes. Developed nations using landfills for waste disposal have been depicted by the graphical representation in Figure 3.3 (Sharp et al., 2010; Zorpas and Lasaridi, 2013).

3.4.4 MECHANICAL BIOLOGICAL TREATMENT

MBT is basically an augmented progression to recover constituents from the waste materials and can be used for different purposes (Eunomia, 2015). As the name indicates, it involves both mechanical and biological treatments for its effective functioning (Eunomia, 2015). It is advantageous to use as it demands lesser space but provides valuable material and good recovered energy in return (Halkos and Petrou, 2016). This process follows two ideas:

- Firstly, collected waste is segregated and then treated accordingly.
- Secondly, the wastes are treated first and then separated (Eunomia, 2015).

Through this process, an aerobic biological unit is used to stabilize the organic content of the waste that reduces its biodegradability level so that a greater level of methane can be generated (Halkos and Petrou, 2016). Then an anaerobic unit is used to convert the organic components of the MSWs to methane. This process is exceptionally useful as it has contributed in the waste recycling process since recyclable materials have lower market value. Other separated organic compounds obtained during the recovery process can be used to produce compost (Halkos and Petrou, 2016). Additionally, other energy recovery entities (plastics, cardboard, and paper) can be extracted during waste sorting process which will be effectively be utilized in the production of refuse-derived fuel.

3.4.5 ANAEROBIC DIGESTION

Organic biomass is degraded content under anaerobic conditions using bacterial species to extract useful methane gas that is termed as AD (Eunomia, 2015). Following are the major types of AD:

- Mesophilic AD.
- Thermophilic AD.

The major distinction between these types is temperature, which is being employed in varying (Anjum et al., 2017). Mesophilic AD takes place at 30–40°C while that of thermophilic process is conducted at 60°C (WRAP, 2016). This process is much more flexible in its range of treating various types of waste than any other method ranging from clean organic waste content to grey waste or waste with moisture content reaching up to 60–90% (Halkos and Petrou, 2016). Therefore, this technology would be effectively used for kitchen wastes that are mostly with greater moisture content (Anjum et al., 2017). It is a great source for the production of renewable energy as the organic content of waste is broken to harvest more methane. The biogas produced from is sustainable in its functioning as it can be used in energy production. Similarly, its by-products are a rich source of nutrients and can be used effectively to enhance soil fertility and is an essential element in plant growth. It can also function best as a biofertilizer (WRAP, 2016).

3.4.6 *RECYCLING*

Firstly, population much refrains from producing excess waste. For this purpose, people should first reduce or effectively consume their products allowing little waste, secondly items such as shoes, clothing, and other things can be reused again and again. Recycling is the last resort in the 3R cycle which allows the utilization of a similar item in a different form (Wilts, 2012). Source reduction is actually super effective as it involves throwing away less waste and is the most preferred method in MSW. These methods ensure environmental protection in the longer run. The concept of 3R is depicted in Figure 3.4.

3.5 PRODUCTION OF MSW IN PAKISTAN

Any approach used for waste management in the developing countries is usually formulated with the help of data of waste generated per capita. These data are collected by surveying the selected regions (Dhokhikah and Trihadiningrum, 2012). According to the case study conducted in Peshawar, Pakistan total waste production was estimated depending upon the data collected from various towns and colonies. Daily waste of the city was calculated to be 650.8 tons in Peshawar. Waste from all the towns selected was predominantly organically rich as most of them were food waste which makes up about 14.5–15.2% of the overall waste production of the area. Similarly, plastic waste is among the next largest among the waste stream followed by paper waste and broken/discarded glass waste. The collected data indicated a clear graph where food waste is maximum and other plastic, wood, rubber, paper, and glass waste are also contributing to it. Total MSW generation rate was 0.4 kg/day. A similar case has been reported in India where the small and big or major cities have a waste production of about 0.1–0.5 kg/capita (Calabrò and Komilis, 2019). However, in heavily populated major cities developing countries have a high-income rate as well might have waste above 1 kg/capita/day (National Research Council, 1991).

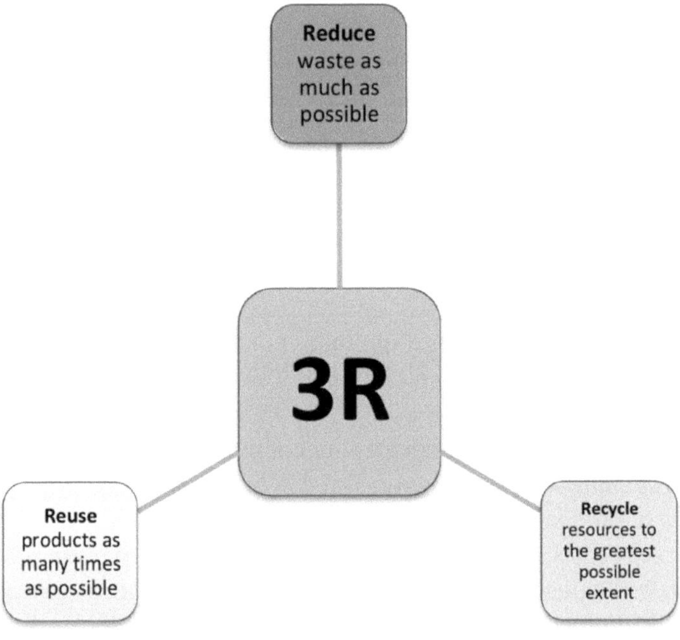

FIGURE 3.4 The concept of 3Rs.

World Bank report states that the waste production per day in urban regions of Asian republics was projected to be exceeding 760,000 tons/day but it has been predicted to correspondingly enhance to almost 1.8 million tons/day till 2025 than that in the period of 1998–1999 (Dhokhikah and Trihadiningrum, 2012). These reports were formulated by estimated calculation and the real-world data might differ in a sense that it could be reduced or doubled in the years to come. As a result, we might have to face huge pressures on the sanitary landfills as the waste production might increase continuously which demands proper management approach to reuse and recycle waste materials.

3.6 SUSTAINABLE WASTE MANAGEMENT

People in every community are responsible for producing domestic waste including animal manure, fire ash, rubbish, food waste, old clothes, etc. Waste reabsorbs into the nature entering the natural cyclic mineral process in an agricultural community. Livestock and other domesticated farm animals rely on the food remains, which makes little food waste in dustbins. But ever since industrialization bloomed and urban sprawl further increased in the 18th century, majority of the solid waste collected was not used in other forms, reusing of such waste was gradually diminishing. In order to plan for future management of wastes, the projected increase should be recorded with time. Having sufficient knowledge about disposal options to choose the best among them is also important to predict which waste belongs where (Calabrò and Komilis, 2019; Yadav and Karmakar, 2020).

Now, people value leisure therefore the demand for packaged food has been increased significantly which made the waste production in low revenue-generating countries is approximately to be 0.4–0.6 kg/capita per day. And for developing and high-income countries or developed nations, the waste production was approximately 0.5–0.9 and 0.7–1.8 kg/capita per day, respectively, clearly defining those high-income countries producing more MSW than that of the developing countries (De et al., 2019). The type of waste suitable for recycling is defined by its quality, moisture content, and densities (Goren and Ozdemer, 2015).

The concept of sustainability consists of waste prevention which is basically reduction at the source which is the most successful method of waste reduction. It is environmentally beneficial for reducing the pollution levels, energy, and resource conservation which ultimately reduces the need of large combustors and landfill sites. The size of the population along with the quantity of waste being produced and proper waste characterization can help in choosing which treatment method would be best suited (Jaunich et al., 2016). Sustainable management of waste is largely based upon the policies being formulated by the national government and firmly implemented on its people for proper segregation of their individual waste produced which means proper three-bin system should be implemented where recyclables are separated from food scrap and other nonrecyclable types of waste. These wastes must be collected separately as plastics and broken glass can be recycled while food waste is majorly organic in nature and can be used as a natural fertilizer of compost. Similarly, collection and disposal would be much easier and viable for the collectors as they can properly dispose them. Sustainable management demands the effective involvement of every citizen so that they could provide a better environment for the future generation and not a world full of garbage and waste present at every inch of their walking distance.

Figure 3.5 shows the collection systems at the source.

FIGURE 3.5 Various waste collection systems at source in practice. (A) A 3–5 bin collection system practiced at source and later transferred to larger bins is shown in an office setup in Korea. (B) A single bin system with a label of only paper and cardboard was captured from an institute in Lahore, Pakistan. (C) A single bin collection method without any label on it has also been used in various institutions in Pakistan.

3.6.1 IMPORTANCE OF SOCIOECONOMIC FACTORS IN MSWM IMPLEMENTATION

Waste segregation needs strong stakeholder/customer commitment along with their general knowledge for appropriate waste separation (Calabrò and Komilis, 2019; Ozonoff et al., 1987). Without their adequate interest, municipal solid waste management (MSWM) system can become extremely complex in terms of:

- varied nature of waste fractions (like organic waste, recyclable paper, plastic, glass, etc.),
- proper system for the implementation of the rules present for waste collection and disposal.

Customer commitment has a positive impact on social norms, public education, and further improves the quality of local education regarding waste management (Calabrò and Komilis, 2019; Agovino et al., 2019; Kattoua et al., 2019).

Studies have stated that people in developing countries like China, Iran, etc., have perceived the suitability of recycling but are still in the developing phase to advance their management schemes for solid waste (Calabrò and Komilis, 2019).

3.6.2 ENVIRONMENTAL AND HEALTH IMPACTS CAUSED BY MUNICIPAL SOLID WASTE

Open dumping and flare burning of MSW tend to release harmful gaseous emissions including volatile organic carbons (VOCs) and other particulate matter in the air which is also responsible for soil and groundwater contamination through the seepage process (ATSDR, 1995; Vrijheid, 2000). Huge concentrations of GHG emissions are subjected to release due to the organic waste decomposition that causes a change in the overall global climatic conditions. Significant enhancement has been recorded in the growth rate of various pathogens due to the greater presence of organic nutrients available from the wastes being dumped openly. This increase in microbial pathogens has seriously contributed in the chronic and viral infections in waste collectors and other people living in close vicinity of the areas where landfill is near (Katsouyanni et al., 1997). Serious health concerns are connected with waste as many people working in waste collection system or its management process comes in direct contact with waste and with due time they are afflicted with respiratory diseases due to its unpleasant odor and other problems, allergies of eyes, nose, and skin have been noticed because of MSW burning or pathogenic microbes (National Research Council, 1991; Ozonoff et al., 1987). It was validated through a study conducted in India that approximately 71% and 25% of the workers and waste collectors are found to be afflicted with various respiratory diseases in their two major cities (Hu and Shy, 2001). Hazardous or E-waste, if not properly incinerated, might release smokes that have the ability to cause chemical poisoning and many other neurological disorders, especially in children and aged people (Dockery and Pope III, 1994; Katsouyanni et al., 1997). Mixed wastes (including household waste, biomedical waste, food waste, and waste containing healthcare products combined with domestic waste) when dumped openly without any level of treatment have increased the risk of HIV and hepatitis B (Zanobetti et al., 2000). All these problems have been faced in developing and under-developed nations since their economic conditions are poor and proper profound treatment is not possible for them to provide for the huge amount of wastes increasing incredibly.

3.7 ASPECTS TO ASSESS THE WASTE MANAGEMENT SYSTEM

MSWM systems could be assessed in terms of sustainability through various aspects. Major aspects are social, economic, and environmental aspects of waste management. Although a lot of aspects are there to consider while considering management activities but only these three are major among them (Allesch and Brunner, 2014).

3.7.1 ECONOMIC ASPECTS

Economic aspects are majorly the most important thing to consider when any management system is considered as it is the most important factor because financial stability is needed to adopt any technology which could contribute to any failure faced and total process efficiency. Since the economic aspect is so important, it is likely to be considered at both micro and macro levels (Spoann et al., 2018). At the macro scale of waste management, the total cost of the waste in a region or a whole country is usually comprehensively calculated and evaluated to formulate an average GDP percentage, whereas we only consider the investment and operational cost at micro level (Allesch and Brunner, 2014).

3.7.2 SOCIAL ASPECTS

It incorporates three different perspectives;

- *Social acceptability* where people accept the MSW management system presented.
- *Social equity* where the benefits attained from the waste management system needs to be shared mutually among all the residents.
- *Social functioning* refers to the social assistance that comes in package with the good aspects of administration systems.

Social aspect is also linked to the country's educational system, its governance, employment rate, market, and security. Other key factors include public health and social safety, which have a close coordination with the economy and environment (Seo et al., 2004).

3.7.3 ENVIRONMENTAL ASPECTS

Among all the aspects, environmental aspect is the essential among them as it assesses the effects on natural resources, water, soil, and air due to waste management techniques (Waqas et al., 2019). Moreover, it is especially necessary as it protects nature as well as biodiversity from the harmful consequences of waste management processes (Nizami et al., 2017). Such environmental aspects as basically evaluated through the life cycle assessment (LCA) method where several categories as included like global warming potential, depletion of ozone, ecotoxicity, human toxicity, and acidification.

3.8 CONCLUSION

Developing countries are faced with a lot of MSW management problems such as waste collection efficiency is considerably very low, awareness level is mediocre, and knowledge is limited regarding management strategies and the concept of sustainable waste management and efforts regarding it. Wastes are openly dumped and even landfill capacities have exceeded which demands proper legislation. Due to a lack of efficient management policies, wastes are dumped which leads to many health and environmental problems. Many treatments are used for sustainable MSW management including, MBT, composting, incineration, and recycling of waste. But such technologies demand effective strategies for their working which include that people should be aware of waste characterization, LCA for the suitable strategy to adopt, and other researches are required to highlight the problems areas and those gaps could be highlighted. Any plan is sustainable that considers all the perspectives of its social acceptability, economic, and most importantly environmental aspect of any waste management process.

3.9 SUMMARY

As the saying goes "One man's waste is another man's treasure" is actually true in its literal meaning. But since the waste utilization is not efficient in underdeveloped countries so the waste generation is greater than it could actually put to some useful purpose. Greater global population is responsible for the generation of the huge sum of waste. The waste type depends upon its source as residential waste is produced from human localities, commercial waste from commercial buildings, agriculture and construction waste from agricultural fields, and other building construction, respectively. Majority of the low and middle revenue-generating countries lack greater technology and efficient waste management techniques to manage their huge waste produced as utilized by the developed nations. Waste is usually managed by the traditional methods of waste collection, landfilling, open-dumping, incineration, etc. These approaches are not sustainable as waste is just still cause various harm in some other manner therefore management of waste which is also sustainable in its action is in dire need so we can preserve our nature for future generations. For this to be put into practice, collection phase needs to be done properly with separations as degradable must be separated from nondegradable. Similarly, recyclable waste must be separated from nonrecyclables at the household level. By doing this, collection of waste would be easier, and as major segregations have already been made so the one needs to recycle can be sent for the purpose. This method is easier to sustain than the segregation of a huge pile of waste. 3R formula can be applied for attaining sustainable waste management. This is not only useful but also helpful in reducing many of the health diseases related to waste collection and open dumping. Sustainable waste management is applied to all overdeveloped nations since they found it to be effective and developed nations need to follow their perspective for creating a sustainable future with good management practices.

KEYWORDS

- **municipal solid waste**
- **recycling**
- **waste collection systems**
- **sustainability**
- **reuse**
- **incineration**
- **bioeconomy**
- **waste production**
- **waste management**
- **policy**
- **strategy**

REFERENCES

Ahmeda, S. A.; Alib, M. Partnerships for Solid Waste Management in Developing Countries: Linking Theories to Realities. *J. Habitat Int.* **2004,** *28*, 467–479.

Allesch, A.; Brunner, P. H. Assessment Methods for Solid Waste Management: A Literature Review. *Waste Manage. Res.* **2014,** 32(6), 461–473.

Almendro-Candel, M. B.; Navarro-Pedreño, J.; Lucas, I. G.; Zorpas, A. A.; Voukkali, I.; Loizia, P. The Use of Composted Municipal Solid Waste under the Concept of Circular Economy and as a Source of Plant Nutrients and Pollutants, Municipal Solid Waste Management, Hosam El-Din Mostafa Saleh, IntechOpen. *Waste Manage.* **2019,** 42, 233–432. DOI: 10.5772/intechopen.83386. https://www.intechopen.com/books/municipal-solid-waste-management/the-use-of-composted-municipal-solid-waste-under-the-concept-of-circular-economy-and-as-a-source-of-

Agency for Toxic Substances and Disease Register. Hazardous Substances Emergency Events Surveillance. Annual report; US Department of Health and Human Services, Public Health Service Atlanta, 1995.

Agovino, M.; Cerciello, M.; Musella, G. The Good and the Bad: Identifying Homogeneous Groups of Municipalities in Terms of Separate Waste Collection Determinants in Italy. *Ecol. Indicat.* **2019,** *98*, 297–309. https://doi.org/10.1016/j.ecolind.2018.11.003.

Anjum, M.; Khalid, A.; Qadeer, S.; Miandad, R. Synergistic Effect of Co-Digestion to Enhance Anaerobic Degradation of Catering Waste and Orange Peel for Biogas Production. *Waste Manage. Res.* **2017,** *35* (9), 967–977.

Asad, I.; Xiaoming, L.; Guang-Hao, Chen. Municipal Solid Waste: Review of Best Practices in Application of Life Cycle Assessment and Sustainable Management Techniques. *Sci. Total Environ.* **2020,** *729*, 2–14.

Batool, S. A.; Ch, M. N. Municipal Solid Waste Management in Lahore City District, Pakistan. *Waste Manage.* **2009,** *29*, 1971–1981.

Blanchard, O. Energy Consumption and Modes of Industrialization: Four Developing Countries. *Energy Policy*, **1992,** *20*, 1174–1185.

Bylinsky, G. Manufacturing for Reuse. *Fortune* **1995,** *131*, 102–103.

Calabrò, P. S.; Satira, A. Recent Advancements toward Resilient and Sustainable Municipal Solid Waste Collection Systems. *Curr. Opin. Green Sustain. Chem.* **2020,** *26*, 2452–2236. https://doi.org/10.1016/j.cogsc.2020.100375.

Chalmin, P.; Gaillochet, C. An Abstract of World Waste Survey 2009. *Waste Manage. Resour.* **2009,** *32* (9), 800–812. http://www.rebnews.com/pdfs/news/from_waste_to_ resource_ report_veolia.pdf (accessed Dec 12, 2013).

Chung, S. S.; Carlos Lo, W. H. Local Waste Management Constraints and Waste Administrators in China. *Waste Manage.* **2008,** *28* (2), 272–281.

Daniel, H.; Perinaz, B. T. What a Waste: A Global Review of Solid Waste Management. Urban Development Series; Knowledge Papers No. 15; World Bank: Washington, DC, 2012; pp 345–768. https://openknowledge.worldbank.org/handle/10986/17388 License: CC BY 3.0 IGO.

Das, S.; Lee, S. H.; Kumar, P.; Kim, K. H.; Lee, S. S.; Bhattacharya, S. S. Solid Waste Management: Scope and the Challenge of Sustainability. *J. Clean. Prod.* **2019,** *228*, 658–678. https://doi.org/10.1016/j.jclepro.2019.04.323.

De, F. G.; Ferrara, C.; Iannone, V.; Parente, P. Improving The Efficacy of Municipal Solid Waste Collection With a Communicative Approach Based on Easily Understandable Indicators. *Sci. Total Environ.* **2019,** *651*, 2380–2390. https://doi.org/10.1016/ j.scitotenv.2018.10.161.

Defra. Mechanical Biological Treatment of Municipal Solid Waste. *Depart. Environ. Food Rural Affairs* **2013,** *13*, 3–57 This document is available on the Defra website: www.defra.gov.uk

Dhokhikah, Y.; Trihadiningrum, Y. Solid Waste Management in Asian Developing Countries: Challenges And Opportunities. *J. Appl. Environ. Biol. Sci.* **2012,** *2* (7), 329–335.

Diaz, L. F.; Savage, G. M.; Golueke, C. G. Sustainable Community Systems: The Role of Integrated Solid Waste Management. *Proc. 19th Int. Madison Waste Conf.*, Madison, WI, USA, **1996,** *19*, 280–291.

Dockery, D. W.; Pope III, C. A. Acute Respiratory Effects of Particulate Air Pollution. *Annu. Rev. Public Health* **1994,** *15*, 107–32.

Eunomia. Economic Analysis of Options for Managing Biodegradable Municipal Waste, Final Report to the European Commission **2015,** *17*, 20–205. http://ec.europa.eu/environment/ waste/ compost/pdf/econanalysis_appendices.pdf

European Commission. Waste Generated and Treated in Europe, Luxembourg, European Commission, Office for Official Publications of the European Communities, 2003.

Fell, D.; Cox, J.; Wilson, D. C. Future Waste Growth, Modelling and Decoupling. *Waste Manage. Res.* **2010,** *28*, 281–286.

Gerbens-Leenes, P. W.; Nonhebel, S.; Krol, M. S. Food Consumption Patterns and Economic Growth. *Increas. Affluence Use Natural Resour.* **2010,** *55*, 597–608.

Goren, S; Ozdemer, F. Regulation of Waste and Waste Management in Turkey. *Waste Manage. Res.* **2015,** *29* (4), 433–441.

Gottberg, A.; Longhurst, P. J.; Cook, M. B. Exploring The Potential of Product Service Systems to Achieve Household Waste Prevention on New Housing Developments in the UK. *Waste Manage. Res.* **2010,** *28*, 228–235.

Giusti, L. A Review of Waste Management Practices and Their Impact on Human Health. *Waste Manage.* **2009,** *29* (8), 2227–2239. https://doi.org/10.1016/j.wasman.2009.03.028
Halkos, G.; Petrou, K. N. Efficient Waste Management Practices: A Review. University Library of Munich, Germany. *MPRA Paper* **2016,** *3* (1), 595–602.
Hu, S.; Shy, C. M. Health Effects of Waste Incineration: A Review of Epidemiologic Studies. *J. Air Waste Manage. Assoc.* **2001,** *51*, 1100–9.
Jagdeep, S.; Rafael, L.; Rajib, S.; Björn, F. Progress and Challenges to the Global Waste Management System. *Waste Manage. Res.* **2014,** *32* (9), 2–14.
Jaunich, M. K.; Levis, J. W.; DeCarolis, J. F,; Gaston, E. V.; Barlaz, M. A.; Bartelt-Hunt, S. L.; Jones, E. G.; Hauser, L.; Jaikumar, R. Characterization Of Municipal Solid Waste Collection Operations. *Resour. Conserv. Recycl.* **2016,** *114*, 92–102. https://doi.org/10.1016/j.resconrec.2016.07.012.
Kaseva, M. E.; Mbuligwe, S. E. Appraisal of Solid Waste Collection Following Private Sector Involvement in Dar e Salaam. *Habitat Int.* **2005,** *29*, 353–366.
Katsouyanni, K.; Touloumi, G.; Spix, C.; Schwartz, J.; Balducci, F.; Medina, S. Short Term Effects of Ambient Sulphur Dioxide and Particulate Matter on Mortality in 12 European Cities: Results from Time Series Data from the APHEA Project. *BMJ* **1997,** *314*, 1658–1663.
Kattoua, M. G.; Al-Khatib, I. A.; Kontogianni, S. Barriers on the Propagation of Household Solid Waste Recycling Practices in Developing Countries: State of Palestine Example. *J. Mater. Cycl. Waste Manage.* **2019,** *21*, 774–785. https://doi.org/10.1007/ s10163-019-00833-5.
Kaza, S.; Yao, L. C.; Bhada-Tata, P.; Van Woerden, F. What a Waste 2.0: A Global Snapshot of Solid Waste Management to 2050. *Urban Dev.* **2018,** *4*, 2–38. https://openknowledge.worldbank.org/handle/10986/30317 License: CC BY 3.0 IGO.
Klundert, A. V. Integrated Sustainable Waste Management, CEDARE/IETC Inter-Regional Workshop on Technologies for Sustainable Waste Management, Held July 13–15, 1999 in Alexandria, Egypt, 1999.
Manga, V. E.; Forton, O. T.; Read, A. D. Waste Management in Cameroon: A New Policy Perspective? *Resour. Conserv. Recycl.* **2008,** *52*, 592–600.
Mebrstu, D. Sustainability and Sustainable Development: Historical and Conceptual Review. *Environ. Impact Assess. Rev.* **1998,** *18* (6), 493–520.
McBean, E. A.; Rosso, E.; Rovers, F. A. Improvement in Financing for Sustainability in Solid Waste Management. *J. Resour. Conserv. Recycl.* **2005,** *43*, 391–401.
National Research Council. *Environmental Epidemiology: Public Health and Hazardous Wastes*, Vol. 1; National Academy Press: Washington, DC, 1991.
Nizami, A. S.; Rehan, M.; Waqas, M.; Naqvi, M.; Ouda, O. K. M.; Shahzad, K.; Pant, D. Waste Biorefineries: Enabling Circular Economies in Developing Countries. *Bioresour. Technol.* **2017,** *241*, 1101–1117.
Oosterveer, P.; Van Vliet, B. Environmental Systems and Local Actors: Decentralizing Environmental Management in Uganda. *Environ. Manage.* **2010,** *45*, 284–295.
Ozonoff, D.; Colten, M. E.; Cupples, A.; Heeren, T.; Schatzin, A.; Mangione, T. Health Problems Reported by Residents of A Neighborhood Contaminated by A Hazardous Waste Facility. *Am. J. Ind. Med.* **1987,** *11*, 581–97.
Parizeau, K.; Virginia, M.; Chanthy, L. Waste Characterization as an Element of Waste Management Planning: Lessons Learned from A Study in Siem Reap, Cambodia. *Resour. Conserv. Recycl.* **2006,** *49* (2), 110– 128.
Rotich, H. K.; Yongsheng, Z.; Jun, D. Municipal Solid Waste Management Challenges In Developing Countries: Kenyan Case Study. *Waste Manage.* **2006,** *26* (1), 92–100.

Sharholy, M. K.; Ahmad, G.; Mahmood, R. C.; Trivedi. Municipal Solid Waste Management in Indian Cities—A Review. *Waste Manage.* **2008,** *28* (2), 459–467.

Sharp, V.; Giorgi, S.; Wilson, D. C. Methods to Monitor and Evaluate Household Waste Prevention. *Waste Manage. Res.* **2010,** *28*, 269–280.

Seo, S.; Aramaki, T.; Hwang, Y.; Hanaki, K. Environmental Impact of Solid Waste Treatment Methods in Korea. *J. Environ. Eng. Div.* **2004,** *130* (1), 81–89.

Slorach, P. C.; Jeswani, H. K.; Cuéllar-Franca, R.; Azapagic, A. Environmental and Economic Implications of Recovering Resources from Food Waste in a Circular Economy. *Sci. Total Environ.* **2019,** *693*, 133–516. https://doi.org/10.1016/j.scitotenv.2019.07.322.

Spoann, V.; Fujiwara, T.; Seng, B.; Lay, C. Municipal Solid Waste Management: Constraints and Opportunities to Improve Capacity of Local Government Authorities of Phnom Penh Capital. *Waste Manage. Res.* **2018,** *36* (10), 985–992.

Tchobanoglous, G. Chapter 5—Solid Waste Management, Edited by Salvato J. A., Nemerow, N., Agardy, F. Environmental Engineering. *Waste Manage.* **2002,** 755–888.

The World Bank. How We Classify Country? **2010,** *4* (2), 10–297. http://data.worldbank.org/about/countryclassification (accessed Nov 2011).

Tsalis, T.; Amarantidou, S.; Calabró, P.; Nikolaou, I.; Komilis, D. Door-to-Door Recyclables Collection Programmes: Willingness to Participate and Influential Factors with A Case Study in the City of Xanthi (Greece). *Waste Manage. Res.* **2018,** *36*, 760–766. https://doi.org/10.1177/0734242X18764291.

Ulas, A., Eds. *Handbook of Research on Developing Sustainable Value in Economics, Finance, and Marketing*, 4th ed.; Business Science Reference (an imprint of IGI Global), 2014; p 535. DOI: 10.4018/978-1-4666-6635-1.

UNEP. Global Waste Management Outlook. United Nations Environment Programme. **2015,** *4*, 20–79. http://www.unep.org/ietc/ Portals/136/Publications/Waste%20Management/GWMO%20report/GWMO%20full%20re port.pdf (accessed Feb 3, 2016).

Upton, A. C. Public Health Aspects of Toxic Chemical Disposal Sites. *Annu. Rev. Public Health* **1989,** *10*, 1–25.

Vrijheid, M. Health Effects of Residence Near Hazardous Waste Landfill Sites: A Review of Epidemiologic Literature. *Environ. Health Perspect.* **2000,** *108*, 101–112.

Waqas, M.; Nizami, A. S.; Aburiazaiza, A. S.; Barakat, M. A.; Asam, Z. Z.; Khattak, B.; Rashid, M. I. Untapped Potential of Zeolites in Optimization of Food Waste Composting. *J. Environ. Manage.* **2019,** *241*, 99–112.

Wenheng, W.; Shuwen, N. Impact Study on Human Activity to the Resource-Environment Based on the Consumption Level Difference of China's Provinces or Autonomous Regions. *China Population Resour. Environ.* **2008,** *18*, 121–127.

Williams, P. T. *Waste Treatment and Disposal*, 2nd ed.; John Wiley & Sons, Ltd.: West Sussex, 2005.

Wilson, D. C.; Parker, D.; Cox, J. Business Waste Prevention: A Review of the Evidence. *Waste Manage. Res.* **2012,** *30*, 17–28.

Wilts, H. National Waste Prevention Programs: Indicators on Progress and Barriers. *Waste Manage. Res.* **2012,** *30*, 29–35.

WRAP. Anaerobic Digestion. *Waste Resour. Action* **2016,** *82*, 608–990. Programme. http://www.wrap.org.uk/content/anaerobic-digestion-1 (accessed Mar 5, 2016).

World Bank, World Development Report 1999/2000. *Entering the 21st Century*; Oxford University Press: New York, 1999. https://openknowledge.worldbank.org/handle/10986/5982 License: CC BY 3.0 IGO.

Yadav, V.; Karmakar, S. Sustainable Collection and Transportation of Municipal Solid Waste in Urban Centers. *Sustain. Cities Soc.* **2020,** *53*, 101–137. https://doi.org/10.1016/j.scs.2019.101937.

Zanobetti, A.; Schwartz, J.; Gold, D. Are There Sensitive Subgroups for the Effects of Airborne Particles? *Environ. Health Perspect.* **2000,** *108*, 841–855.

Zhen-shan, L.; Lei, Y.; Xiao-Yan, Q.; Yu-mei, S. Municipal Solid Waste Management in Beijing City. *Waste Manage.* **2009,** *29* (10), 2618–2624.

Zorpas, A. A.; Lasaridi, K. Measuring Waste Prevention. *Waste Manage.* **2013,** *33*, 1047–1056.

CHAPTER 4

Problems and Challenges Associated with Waste: Waste Prevention Techniques

TARIQ MEHMOOD[1*], SAIRA BIBI[2], AFZAL AHMED DAR[3], MUHAMMAD AAMMAR TUFAIL[4], MUHAMMAD SOHAIB[5], UMAIR RIAZ[6], GHULAM RASOOL[7], ANAM ASHRAF[8], AWAIS SHAKOOR[9], and MUKKARAM EJAZ[10]

[1]*College of Environment, Hohai University Nanjing, 210098, China*

[2]*Pak-Austria Fachhochschule, Institute of Applied Science and Technology, Mang, Haripur, Khyber Pakhtunkhwa, 24421, Pakistan*

[3]*School of Environment Science and Engineering, Shaanxi University of Science and Technology, Xian, 710000, China*

[4]*Department of Civil, Environmental and Mechanical Engineering, University of Trento, Trento, 38123, Italy*

[5]*College of Food and Agricultural Sciences King Saud University, Riyadh 11451, KSA*

[6]*Department of Soil & Environmental Sciences, MNS University of Agriculture Multan, Pakistan*

[7]*College of Hydrology and Water Resources, Hohai University, Nanjing 210098*

[8]*School of Environment, Tsinghua University, Beijing 100084, China*

[9]*Department of Environment and Soil Science, University of Lleida, Avinguda Alcalde Rovira Roure 191, 25198, Lleida Spain*

[10]*School of Environmental and Municipal Engineering, Lanzhou Jiaotong University, Lanzhou, 730070, Gansu, PR China*

**Corresponding author. E-mail: tariqslamat@hhu.edu.cn*

Waste Problems and Management in Developing Countries. Umair Riaz, Shazia Iqbal, & Moazzam Jamil (Eds.)

ABSTRACT

Wastes are such undesirable and unusable materials disposed of by the general public's various sectors. It may be biodegradable or nonbiodegradable and recyclable or nonrecyclable. Due to development and industrialization, more products are being introduced in the market. Various packaging materials on the other hand are also being thrown in the waste bin. Increasing the population demands more food and utilization of other commodities. All these factors are producing waste that needs to be managed properly. Several managing practices are proposed and implemented around the world. Landfilling pyrolysis, incineration, and gasification are major strategies for the safe disposal of waste. Different countries' economic classification reviled different waste profiles in these countries, depending on countries' economic status. For instance, the waste of developing societies mostly includes food material (organic waste), but developed countries produce various types of waste with a bigger proportion of paper and plastic. Moreover, the lack of technical expertise, deficiency of advanced equipment, shortage of resources, and poor legislation make waste management a big challenge in many countries. Due to such discrepancies, a single management strategy has not potential to implement in all scenarios. Hence only the waste prevention strategy promises exceptional solutions of waste in multiple challenging conditions. Waste prevention is a top priority in Waste Hierarchy due to its environmentally friendly nature, sustainability, economic advantages, and time-saving attributes. As environmental pollution is directly related to waste production and the major constraint for the affective management system is the deficiency of a righteous worker, so to achieve the desired outcomes public participation is mandatory. Individual behavior (manufacturing, consumptions, storage, reuse, and recycling) and government actions such as subsidies, promotions, and facilitation to promote waste production have appeared to be significant factors that determine waste prevention success. There are several challenges in adopting waste prevention, including lack of ecological safety awareness, climate destruction, wildlife demolition, human health concerns, economic costs etc. Regardless of these challenges waste prevention has been successfully implemented and resulted in desirable results in different countries. This chapter concludes that green life has proved to be the most sustainable and advantageous in favor of planet earth.

4.1 INTRODUCTION TO WASTE POLLUTION WASTE PREVENTION

Development and modernization have taken its number of drawback shares, and one of the major facets of fear is the waste pollution as it is causing devastating impacts on the earth, including land, water, and air (Lestari and Trihadiningrum, 2019). Bulk waste generation as another rising problem attributed to population growth has appeared as a major concern worldwide. Wastes are such undesirable and unusable materials that are disposed of by the various sectors of the general public, including homes, hospitals, mines, industries, schools, etc. The nature of waste differs significantly from region to region and deviates with time. Waste may be composed of solid, liquid, hazardous, organic, inorganic, recyclable, and/or not nonrecyclable. Waste may include cupboards, food grain, glass, leather, metal, paper, plastic, textiles, and wood. In short, any material that is disposed of after its basic use, or is valueless, faulty, and of no more usage, is categorized as waste (Kumar, 2016). Handling waste appropriately is mandatory for edifice and civilized cities, but its remains a challenge for a number of developing countries. Yet, either because of inefficient infrastructures or resource crunches, not all such wastes get collected and transported to the final dump locations. If at this phase the disposal and management is inadequately done, it can cause severe influences on health and complications to the surrounding atmosphere (Palmiotto et al., 2014). Effective waste management is costly and often consumes 20–50% of municipal budgets of the developing nations. Resultantly, this waste is thrown into the municipal waste streams from where the area municipalities collect it and further throw in the dumps and landfills (Hung et al., 2012).

However, dumping of waste is not a desirable solution, and we cannot evacuate this problem for long time. Rising demand for foods and other needs are creating a rise in the amount of waste that every household generates every day. In 2016, the solid waste generated by cities throughout the world was 2.01 billion tonnes, amounting to a footprint of 0.74 kg/person/day (Figure 4.1). It is an estimation that, with current statistics of development, population growth, and urbanization, annual waste production will be augmented by 70% to about 2.01–3.40 billion tonnes in 2050 (The World Bank, 2019).

The condition is equally shocking for rural areas where there are absence/limited waste-handling strategies, particularly in the situation of under-developed, and developing societies. Nevertheless, in any case dumping and landfills are not promising solutions; researchers and policy makers are considering other sustainable management strategies to overcome this overwhelming global problem.

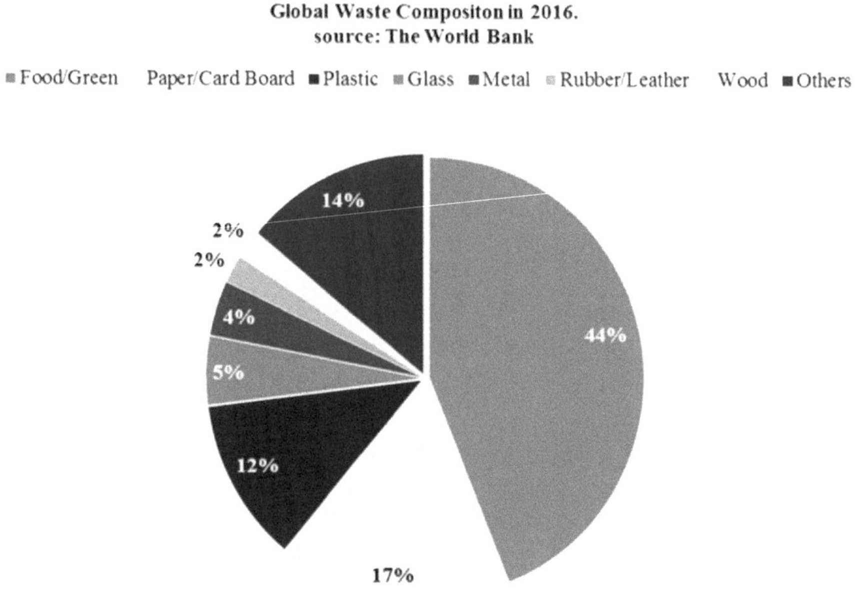

FIGURE 4.1 Global waste composition in 2016 (The World Bank).

Many measures have been engaged for promoting waste management, including reuses, recycling, incarnation biological treatment. However, these all required time, labor, cost and also affect our planet somehow. Moreover, discrepancies in waste profile, concentration, and available resources between developed, developing, and under developing countries are big hurdles in the implementation of proper management strategies worldwide. In this particular scenario, reduction, preventing, or avoidance of waste generation has become a top priority and most desirable in waste management hierarchy. Article 4 of Waste Framework Directive explained this as "decreasing the proportion of waste produced at source and dropping the perilous content of that waste is considered the uppermost priority in Waste management Hierarchy" (Cecere et al., 2014). More concisely, prevention of waste is a combination of actions taken before a material, product, or a substance become waste for reducing: (1) the adversarial influences of the waste on human health and environment; (2) the harmful content in material products; (3) the waste extent, through the products recycling or the addition of the product's life period (Cristóbal, 2018).

Waste prevention is linked closely with enlightening the manufacturing approaches and guiding people to demand less packaging and green products

(Zorpas et al. 2015). Hence, educating people to avoid waste production by their behavioral change is one of the key factors for achieving the optimum goals of waste prevention strategies. Besides, several government incentives, promotion, and encouraging polices could also boost this whole approach. A wide-ranging standpoint on the region of waste is obligatory to achieve viable waste management. Several measures that decrease the waste volumes and control streams of wastes consistent with the hierarchy for diverse treatment methods are required. The key is to raise the recovery of material waste (Chalak et al., 2016). By recovering resources, manufacturers may boost manufacturing efficiency while also protecting the environment.

Acknowledging the waste prevention as a top priority strategy, the present chapter covers major aspects of waste prevention strategy. Particularly, characteristics of waste in different parts of the world based on economic status, impact of waste on lifestyle, challenges, and waste prevention strategies, avoiding the waste by behavioral changes, optimizing national and global waste prevention strategies, and finally problems and limitation in waste prevention are major aspects of this chapter, which are discussed in detail. It is expected that an overview of the aforementioned segments of the present chapter will be helpful in the decision-making process of sustainable waste management and could suggest more viable strategies worldwide.

4.2 WASTE PRODUCTION AND ITS MANAGEMENT IN DEVELOPING AND DEVELOPED COUNTRIES

Generally, nature of municipal solid-waste (MSW) differs from society to society and is also influenced by geographical position, industries, energy resources, and living standards of public. Waste of developing countries mainly comprises organic materials, while developed countries produce different nature of wastes with plastic and paper as its major proportion. The waste composed of organic materials called as organic waste and it includes mostly food, green manure, farm yard manure, and paper waste. In developing countries (i.e., Indonesia), 60–70% of total waste produced is considered organic waste. But, in developed countries including US, EU, and Japan, the proportion of organic waste in wastes differs moderately as 24%, 34%, and 40%, respectively. In 2013, the environmental protection agency estimated the amount of municipal solid waste generation in the USA with 0.254 billion tons and a typical profile of waste produce in US is presented in Figure 4.2 (EPA, 2013).

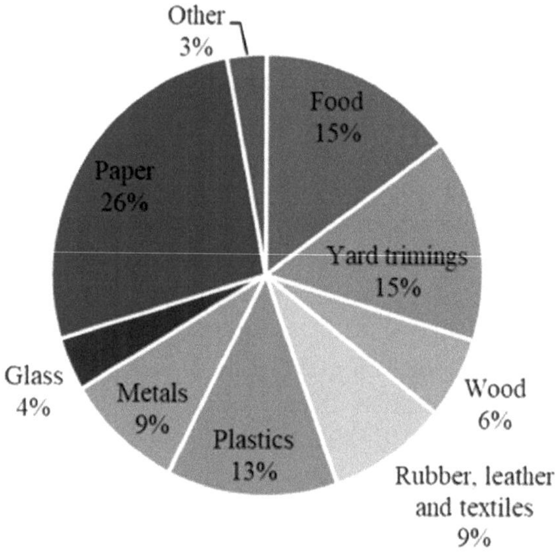

FIGURE 4.2 Waste constituents and major elements of municipal solid waste by the US (U.S. Environmental Protection Agency 2013).

On the basis of the developmental status, the countries are grouped into lesser developed, developing, and developed countries. This developmental classification is based upon per capita gross domestic product, socioeconomic development, HDI, and GNI. Lesser developed or poor countries exhibit the lowest socioeconomic indicators with the lowest HDI ratings and GNI of US$1025 or less. But, the GNI values for the developing and developed counties are US$1026–12,475 and US$12,476 or more, respectively (Folayan et al., 2020).

In perspective of solid waste there are different definitions of SW among countries. Developing countries such as India, Indonesia, Philippines, and Vietnam define the SW as all the waste produced from metropolitan zones and handled by the metropolises. The countries such as Myanmar define the SW as waste produced in municipal areas, by industrial and agricultural sectors, while developed countries such as USA define the SW as waste that is handled historically in the area under municipal administration.

However, there is difference about scope and definition of SW across all republics: underdeveloped, developed, and developing countries. While designing a waste management strategy among countries of all categories, the primary focus should be on characteristics of waste including the amount of waste produced, type of waste, moisture contents, and calorific value of each type of waste. Hence, this difference in waste definitions and waste characteristics among the countries is the big challenge for data compilation and comparison among the countries (Folayan et al., 2020; Mmereki et al., 2016).

Indeed, it is estimated that in underdeveloped nations such as Bangladesh and Pakistan, the organic waste constituents are more than 65%, whereas in developing countries such as Japan and the Korea this portion is less than 30%. Previous researches revealed that organic fraction is directly associated with the density and moisture contents in waste channels; this should be kept in mind while collecting, transporting, and treatment of waste. Economic progress is also directly associated with the per unit rate of waste production, and with ingesting of viable goods, and trends of development. Previous research articles reported that more affluent population generate more residual waster (Mmereki et al., 2016).

Lesser developed and developing countries are producing lesser per person waste as compared with developed countries. Rate of waste generation in developed countries is about 1 kg/person on a daily basis, whereas under-developed and developing countries show waste generation rate less than 1 kg. As reported, the waste production patterns estimated in lesser developed and developing countries are 0.33 and 0.73 kg/person/day, respectively.

Solid-waste mismanagement results in environmental contamination, so it is a global issue. Public involvement and economic sustainability are required for integrated assessments and universal approaches for solving issues regarding SW mismanagement. In developing/lesser developed countries unsustainable waste management is common, besides there are also differences between big cities and rural areas within any country. So, these differences should be emphasized, where management issues are different, specifically regarding the quantity of waste produced and the WM facilities available. However, both developing and lesser developed countries suffer negative economic legislatives, political issues, technical and operational limitations.

Compared with residents of developed and developing countries, particularly the poor are more vulnerable to unsuitable waste management. The poor cities are more severely impacted by unsustainably managed waste. In poor countries, about 90% either is disposed of in open dumps or is burned openly. These practices carry out severe health, safety, and environmental concerns. Mismanaged disposal acts as a generation of disease, source of GHGs emission, and promoter for urban violence (Ferronato and Torretta, 2019).

The inappropriate administration of WM is due to the series of factors which limits affective WM; the major factors are lack of teamwork among stakeholders, weak governing body, organizational framework, and managerial capacity, along with limiting factors of poor WM. There are some difficulties for implementing WM strategies, reported by several researchers, such as lack of human resources, insufficient public awareness about affective WM, the absence of formal training projects about recycling of waste at the source point (household level), and the lack of financial resources mainly in developing societies. The other main concern for affective waste management is insufficient collaboration between public and private authorities.

In the presence of the above-mentioned problems, it is hard for lesser developed and developing countries to handle emerging challenges regarding waste management of diversified characteristics. These challenges may cause destructive impacts on public health, safety, and the environment. However, developed countries have proper control on possible issues regarding waste management; this is only possible on behalf of implementing efficient functional frameworks/policy, and application of modern cost-effective waste management techniques.

Commonly, land filling is practiced worldwide for disposing of waste because of its ease of implementation. There is an amended form of landfilling, called "sanitary landfilling," which is practiced mostly in developed countries. A sanitary landfill is a modern landfilling technique that certifies nominal pollution hazards with maximum consumption of solid wastes. This modern technology duly followed the USEPA standard protocols. However, a communal landfill is a profound reservoir where solid waste is dumped and shielded by soil. Sanitary landfills are generally safer for environment as compared with common landfills; these are constructed with some technical upgradations in joint landfills. A salubrious landfill characteristically encompasses four layers (Christensen, 2012):

First layer: this bottom layer is made up of thick clay lining with a sheet of HDPE. The covering of polythene sheet on clay inhibits the seepage of toxins.

Second layer: this drainage layer includes drainage channels to direct the flow of inbound toxins out of the dump toward treatment facility.

Third layer: this collection layer is to trap the gases of landfill and direct them for treatment and usage.

Fourth layer: this top-most layer is the huge basin or reservoir to accommodate the solid waste.

After the completion of all four layers of sanitary landfill, the mouth of landfill concealed with soil for controlling the scent, and possible release of infective microbes.

4.3 IMPLICATION OF WASTE

The waste generated by humans has been damaging the ecosystem for a long time until today. Humans are producing huge garbage and cannot manage it in a maintainable way. Waste, which is not biodegradable and cannot be appropriately recycled, is filling many landfills and vast oceans (Tabasová et al., 2012). Let us take an instance of plastic waste; the latest study reported the amount of plastic waste of about 6.3 billion metric tons had been generated from which only 9% had been recycled. In 2017, the approximations of Environmental Protection Agency stated that the overall public solid leftover of 0.267 billion tons was formed in the USA. In comparison with the levels of 2015, it has been increased by 5.7 million tons (Rim-Rukeh, 2014). As a whole, the proportion of generated waste affects the ecosystem in several ways, such as its involvement in the worsening crisis of climate, its adverse impression on the wildlife, and its detriments to every individual public health (Butnariu et al., 2018).

4.3.1 ENVIRONMENTAL IMPLICATION

Environment pollution is directly associated with waste production and its management strategies. The content of landfill's leachate highly depends upon the type of landfills, which may be (a) closed, (b) load of both MSW and incinerated bottom ash, and (c) only direct load of MSW. Previous literature revealed that the landfills with only direct load of MSW resulted in meaningfully high leakage of heavy metals such as lead (Pb), chromium (Cr), and cadmium (Cd) than the other two types (a and b).

Managing stocks of unwanted ozone-depleting substances (ODS) is highly linked with greenhouse gases (GHGs). Bestowing to World Meterological Organization (World Meterological Organization, 2019), MSW and landfills are the third major anthropogenic reason of global carbon dioxide (CO_2), methane (CH_4), and nitrous oxide (N_2O) emission. Apart from CH_4, CO_2 is the single most important anthropogenic GHG in the atmosphere. SW emit approximately 800 tons of CO_2 to the air yearly. In preindustrial era the atmospheric CH_4 level was approximately 722 ppb, and now it increased

257% due to amplified emissions from anthropogenic activities. About 60% of CH_4 and 40% of N_2O in air originated from anthropogenetic sources (e.g., biomass burning, fossil, fuel exploitation, landfills, ruminants, and rice agriculture).

4.3.2 CLIMATE DESTRUCTION

The methane gas is released from dumped trash in different landfills. The step taken further to represent landfills reported about 91% of all landfills emit methane worldwide (Alam and Ahmade, 2013). The burning of huge, open heaps of trash in several regions of the world emits hazardous levels of a greenhouse gas carbon dioxide which is warming the planet Earth. Studies have estimated that about 40% of the biosphere's trash is burned in the same way, imposing massive-scale risks to both environment and the people living closed to such burning spots (Khan and Ghouri, 2011).

4.3.3 WILDLIFE DEMOLITION

One of the most enormous penalties of global waste issue demonstrates itself concerning waterways and marine life. Basically, it affects the individuals who are dependent upon the sea for their means of support. They cannot differentiate between foodstuffs and pollutants. They throw the trash that results in death as the marine animals are not able to digest it (Lemly and Skorupa, 2012). It affects seals, fish, whales, turtles, and several other sea animals, as experts have also reported various plastic fragments in more than thousand aquatic species' bodies. Owing to ingestion of plastics or trash, malnourishment is commonly the next step as few species do not have higher levels of acids in stomach for breaking the ingested objects (Kühn et al., 2015).

4.3.4 HUMAN HEALTH CONCERNS

Human health is at risk through our procrastination because of generating huge proportion of trash and not disposing of appropriately that ultimately causes downfall of the wildlife and ecosystems with human share (Kanagaraj et al., 2015). It has become challenging to promote or prevent endurance with how humans treat the Earth. The more emissions of trash by human activities affect them on a more comprehensive scale. One can develop diseases, for example, asthma, cancer, birth defects, cardiovascular diseases, childhood cancer, low birth weights, infectious diseases, and preterm deliveries. Bacteria, insects, and pests can also be additional to the human health challenges caused by trash (Wilcox et al., 2016).

4.3.5 ECONOMIC COSTS

There are higher economic costs of waste management and costs are mostly paid by municipal administrations; money can be saved with more proficiently manufactured collection ways, developed vehicles, and public awareness. Waste reclamation by recycling can control economic expenses as it avoids exploring raw materials and mostly cuts costs of transportation. The waste treatment location and disposing facilities often decrease values of material goods because of noise, unsightliness, negative stigma, and pollution (Verma et al., 2016).

4.4 WASTE PREVENTION THROUGH BEHAVIORAL CHANGE

4.4.1 INDIVIDUAL EFFORTS TO PREVENT WASTE PRODUCTION AT HOME, SCHOOL, AND WORKPLACE

Source reduction is one of the main approaches to reduce waste.

This should be possible by:

1. Utilizing less material when making an item
2. Reuse of items on location
3. Designing items or bundling to reduce their amount.

On an individual level we can lessen the utilization of pointless things while shopping, avoid purchasing things with insignificant packaging, avoid purchasing expendable things, and furthermore avoid requesting plastic convey bags (Bortoleto, 2014).

This classification of source reduction strategies identifies with adjustment of techniques or administrative and organized aspects of a manufacturing operation. Waste generation is decreased through better plan of the executives or housekeeping practices. Great housekeeping practices are made out of the following elements:

- Training of workers
- Practices of management

- Inventory management control
- Segregation of waste-stream
- Efficient material handling
- Enhancements planning
- Avoidance of spill and leak
- Regular maintenance
- Process recording

Housekeeping procedures are moderately easy and inexpensive to update and thus include most of the methods of source reduction employed to industry to date. For instance, simple procedures to train plant workers to properly use equipment or abstain from combining hazardous and nonhazardous materials will go far to avoid the generation of hazardous waste. For example, properly qualified and effectively monitored employees can produce fewer products that do not comply with requirements that would conceivably have to be disposed. Suitably monitored and processed inventories of raw materials or products prevent contamination or termination of materials or the procurement of unwanted materials (Bortoleto et al., 2012; Bortoleto, 2014).

The act of separating waste has proven to be helpful in the development of solvent recyclability. For instance, issuing right collection tanks and drums for each sort of spent solvent results in more affordable recovery since a reprocessing step might be allotted with. Similarly, in cases where a combination of such wastes is considered dangerous, the separation of hazardous waste sources from nonhazardous waste streams leads to volume reductions in hazardous waste. The measurement of hazardous waste generated by avoiding leaks, spills, and/or pollution can also be minimized by better treatment of raw materials.

A material-processing scenario with improvements can be referred to from a minor modification in metal-plating processes.

In this case waste was prevented by bringing an intermission into the apparatus that transfers parts across tanks; this permitted the drag out solution to drip back into the apparatus tank instead of polluting the rinsing tank. An effective waste reduction management aid is the establishment of an Executive Management Level (EML) program designed to study workshop or office managers accountable for regularly addressing the amount of waste to their agencies or departments (Liu et al., 2010). Encouraging force arrangements may also be established to prevent the volume of such unwanted waste. Certifiably, this approach is not a practice of waste elimination deprived of anyone else, but added in cost of disposal of hazardous waste, and eventually brings workouts that allow the amount of waste generated to be decreased (Bortoleto et al., 2012; Bortoleto, 2014).

Such approaches reveal that, by looking at plant operations, there are easy, cost-effective chances of reducing waste. By arranging excellent housekeeping activities, only sections of the waste stream are always minimized, but implementation is typically fast and instantly inexpensive.

4.4.1.1 WASTE PREVENTION AT HOME

With the exception of a zero-waste lifestyle, there is still a chance to get better at dealing with the world. While the concept of reducing one's waste can be overwhelming, inside the comfort of your own home is a great place to start.

For instance, shifting to cloth bags instead of plastic bags or making a concerted effort to reuse or recycle cans and bottles can have a huge impact on your waste production (Bortoleto et al., 2012; Oke, 2015).

Below, you can make nine simple improvements to minimize waste in your house.

1. Get familiar with recycling or reusing values.
2. Abandon the sacks (bags) of plastic.
3. Create a menu for dinner.
4. Start with reusable containers.
5. Begin to fertilize the soil.
6. Learn to restore rather than dispose of.
7. Delete mail that is redundant.
8. Avoid using plates that are disposable.
9. And in any case, if it is not too much hassle, stop buying plastic drinking water bottles.

4.4.1.2 WASTE PREVENTION AT SCHOOL

There are a variety of ways in which educational institutions can be effective in their programs for reprocessing, and waste management. As of now, learners are reusing at home, so it is critical that individuals corresponding reprocessing and reusing openings are also made useful at institute. School money, resources, and environmental capital can be set aside by preventing waste and recycling. Moreover, it is a perfect way to train students about how to make an alternative to their classroom, society, and the environment to eliminate, recycle, and reuse waste (Boschini et al., 2020; De Menna et al., 2020).

1. School materials or appliances that are not desirable to discarding them to a limited agency. Act with parents and teachers to make "Back to School" greener as well.
2. Initiate soil management subject to decrease food waste in the school cafeteria.
3. Placement of stainless/hardened steel waste compactors in kitchen.
4. Bring together a department supervising recycling and waste. The recycling community of your school could consist of a pupil leader(s), caretaker, instructor, and management.
5. Instruct students and workers about the initiative for reuse, recycling, and waste reduction. Irrespective of how little the recycling program is, teaching "how" and "why" to the students and employees is essential to their stability and growth.
6. Think without paper. In contrast to copying hard paper versions (copies), there is a lot that should be possible by email, electronic documents, and online. Workers and organizations should be taught how it can have a huge impact on reducing paper waste and increasing productivity to make little changes to make their education program paperless.

4.4.1.3 WASTE PREVENTION AT WORK

We acknowledge that reducing waste is an important aspect of maintaining and protecting the resources of the planet for a long time. Fortunately, by reusing, returning bottles, using ceramic dishes over paper plates, etc., many of us are mindful of the effects and aim to reduce waste at home.

In any case, in the work environment, shouldn't anything be said about decreasing waste?

Companies may not invest time, resources, or energy into reducing waste in the work environment in the case that they feel it is too poorly structured to even consider setting up a plan, or that it is insignificant because it does not generate revenue to the company. Therefore, while waste elimination may not generate income in the conventional sense, it will save your company's money directly.

There are specific approaches to minimizing waste that are easy but difficult to implement, benefiting nature and saving you money over the long term, regardless of whether the organization is not prepared to set up a dedicated community to help take greening measures (Bortoleto, 2014; Wilson et al., 2012):

1. Adopt paper-free computing business solutions.
2. Hold the paper reuse bin within the span of the shoulder.
3. Brighter printing.
4. Provide silverware and quality dishes.
5. Free yourself from the K-Cup machine.
6. Mass purchase.
7. Reuse fasteners and paper folders.
8. Establish a priority on recycling/reusing.
9. Provide water that is clean and filtered.
10. Offer staff a water bottle that is reusable.

4.4.2 GLOBAL/NATIONAL STRATEGIES TO FACILITATING WASTE PREVENTION

4.4.2.1 WASTE PREVENTION

Waste prevention covers a range of policy choices and has a wide variety of advantages. Focusing on waste production inventory, it decreases the volume and toxicity of waste until recycling, energy retrieval, composting, and landfilling become a secondary option. The prevention of waste also requires steps to decrease the undesirable effects on the health, and the environment of the waste produced. The avoidance of waste can be accomplished by decreasing the amount of material involved in the manufacturing of goods and by increasing the quality with which commodities are used after they have been produced. Preventing waste by off-putting excessive utilization and designing and employing fewer waste-generating goods is a form of strict waste avoidance. Extending the lifespan of a commodity or exploring alternatives such as reuse are ways of mitigation by waste flow diversion (Oke, 2015; Šomplák et al., 2019).

4.4.2.2 QUALITATIVE PREVENTION

Instead of affecting the total volume of waste, minimizing the hazardous fraction of waste is known as qualitative waste prevention and helps in limiting environmental and human exposure to hazardous waste.

4.4.2.3 SCOPE OF A COMPREHENSIVE WASTE PREVENTION PROGRAMME

The programme dealing with waste prevention has its roots in the waste administration sector, but its scope encompasses the entire economy, all the streams of materials and the goods used by a country, from their corresponding cradles to their disposal. Thus, not only in the waste-handling sector, but also in the mining, fecund sectors, public and private users should be concerned with a robust waste reduction programme. Accordingly, all economic sectors can be allies in the waste management and preparation programme (Cecere et al., 2014).

4.4.2.4 STRATEGIES

The strategies for waste reduction available to Member States fall into three broad categories, suggesting different levels of public authorities' involvement: informative/intelligence, promotion, and regulation.

Informative approaches aimed at improving behavior and making better decisions include:

- Sensitivity campaigns
- Details on strategies for waste management
- Systems of instruction for responsible officials
- Ecolabeling.

Promotional approaches, the promotion of behavioral improvement, and the provision of financial and logistical support for effective interventions include:

- Favor for voluntary contracts
- Reuse and repair promotion
- Support of environmental management systems
- Renewable rewards for consumption
- Support of research and development.

Regulatory policies, the implementation of waste generation quotas, the extension of environmental obligations, and the imposition of environmental requirements on public contracts include:

- Planning steps
- Taxes and benefits, such as pay as you launch schemes
- Extended policies of producer responsibility

- Green public procurement policies
- Eco-design standards

These initiatives are complementary and may be incorporated into other related established policy areas, such as sustainable consumption and development policies, environmental or waste management policies, or may constitute an independent national programme for the prevention of waste. Economic instruments can participate very effectively in the prevention of waste and should be taken into account, if well planned and supported by complementary steps. Waste management can be addressed in several different ways and specific strategies for targeting key partner and key waste sources (Cecere et al., 2014; Šomplák et al., 2019).

However, individuals and the need for improvements in behavior are the key to waste management, and insight into customer and company behavior can increase the effectiveness of selected initiatives. To significantly address the waste issue and modify the way resources are handled, an integrated combination of interventions is ultimately needed.

4.5 WASTE PREVENTION AS WHOLE

4.5.1 WASTE LEFT OVER PREVENTION IN THE WASTE CONTEXT ORDER

Waste prevention can be demarcated as "the measures before object, a product that minimizes residual amounts, and harms human health" (Based on the waste system directive (WFD-2008/98/98/EC)). Preventing waste involves reducing waste production, including by-products breaching environmental effects. Waste management requires stringent waste generation reduction (both quantitative and qualitative reductions) at source and product reuse. This may not require resource recycling, and a separate collection of waste (Elbeshbishy et al., 2012). Over the past 10 years, a range of national and international attempts have been made to identify waste reduction, coupled with regulatory goals and recommendations to ensure effectiveness, and conclusive waste prevention (Salhofer et al., 2008).

4.5.2 HOUSEHOLD WASTE PREVENTION DEFINITION

Unstructured interviews deliberately minimize waste to include recycling (RECAP, 2008). However, this opinion can cause residents to conclude that they already "do their best" to avoid waste, and may also restrict their further participation or intervention. In this sense, the WFD notes that avoidance requires taking steps before a substance, resource, or commodity is converted into waste and thus reduces: In particular in Article 3, clauses 12–13: (1) Total amount of waste, including by reuse of products or by extending the product life cycle; (2) adverse environmental and human health consequences of waste created; or (3) In fabrics and goods, considering the quality of hazardous substances. Reuse is hereby characterized as any process in which non-waste goods or parts are used again for the same purpose for which they are designed (Jacobsen et al., 2002; Wilson, 2005). The author provides a strong distinction between waste management, and recycling with a visual depiction. However, house composting is included in waste management according to Wilson (2004) since it stops waste from inflowing to waste stream. The EEA (2002) generally uses the word "waste minimization," although there is no specific definition and it can be difficult to distinguish between preventative intervention and minimization (Jacobsen et al., 2002). The concept of waste minimization was accepted in a meeting held in Berlin (1996). As this explanation implies, waste avoidance is a wider concept than waste prevention. Waste avoidance encompasses "avoidance," "source reduction," and "product reuse." However, the elimination of waste also requires "quality management" (such as threat reduction) and "recycling" of disposal steps.

4.5.3 METHODS FOR THE REDUCTION OF POLLUTION

Several approaches for calculating waste avoidance have been used and can be outlined as follows:

a) Direct source decreases quantification, referred to by volume or weight on recorded changes in waste stream amounts. This approach involves direct programs of inspection by case studies, surveys, audits, and trials of waste.
b) Source cost-efficiency analyses that typically combine two financial factors: an efficiency in source cost and saving in buying and disposing of costs together to measure the overall cost of the exertion. The fundamental measures comprise the detection of source lessening, and the costs to be calculated before and after the reduction of the source (e.g., buying, disposal, manpower, or other related factors).

c) Another calculating approach is the use of metrics to assess both the specific waste management policy capacity and the performance of the policy following implementation (whether on an economic, resource, and waste basis). Such metrics might include waste production per person.

d) The waste efficiency ratios of resources are basic measures of a commodity/product/service separated into resources sufficient to manufacture the product/commodity/service. The efficiency in which services are used is calculated in each ratio (Table 4.1).

TABLE 4.1 Measurement of Hazardous Waste Processes and Approaches.

Sr. no.	Processes	Approaches
1	Automation, detection, or monitoring	Act on waste management for volunteer families. Households are measuring, tracking, or auditing their waste and recording it with diaries or input sheets
2	Usage of round data collection for a precise calculation of waste	A mixture of monitoring strategies is utilized to determine the effect of campaigns, e.g., measuring waste rises through waste tonnage/collection round results.
3	Groups of pilots and power	Charge in order in an area where there is no intervention in a comparable field
4	Surveys, indicators, interviews, perceptions, and focus groups.	At times, diaries also document behaviors and response actions. Surveys are commonly used to estimate how many individuals perform a certain action; focus groups are being used to figure out why and/or how they respond to campaign content
5	Surveys of the participation (or reporting of the participation) including lines of assistance requests, Site statistics, number of registrants, written articles, etc.	To determine the complexity of the proposed program. They also track rewards, such as red packet, coupons, sales of home compost bins, and email choice service registrations.
6	Analysis of composition	To consider the influence of the program on multiple waste materials
7	Conversion, assessment, and simulation considerations	Conversion variables, proxies, and ratio model with data on the use and output of waste available in depth.
8	P.O.S (Point of sales data)	Estimate commodity waste avoidance and research the applicability of waste management methodology. When goods are scanned and bought at stores, POS data are generated, providing detailed sales data by product.
9	Hybrid—a mixture of one or both of the above	Using a combination of approaches for control and assessment

Waste reduction is a policy domain that is cross-referential and specifically applicable both within the environmental fields (e.g., environmental management systems) and within particular nonenvironmental (e.g., innovation, management, and policy) that hold a strong capacity to minimize the quantity consequences of the produced waste. The general goals of infringement of the connexion amid economic development, and the environmental consequences of producing leftover/residual, and working in the direction of a zero-waste economy should be taken into account in waste reduction programs. In this sense, quantitative priorities and specific timelines help mobilize a change toward behaviors, and activities to avoid waste or toward more effective inventory handling. (i) A geographical size, (ii) contextual goals, (iii) compilation of data, (iv) voluntary/obligatory periods, as deadlines can be shared through voluntary commitments, the following measures can help decide the required objective (EU, 2012) (Table 4.2).

4.5.4 INCENTIVES AND BARRIERS FOR WASTE PREVENTION

It is a reality that all humans need a reason to respond. Even small- to average-sized enterprises (SMEs) are responsible for the atmosphere and minimizing their waste. With 76% claiming sustainable business practices, 85% of companies in the EU pointed to economic gains and convictions as the justification for environmental work (FSB, 2006; Zorpas, 2010). Some of them emphasized a determination to reduce their effect on the atmosphere and a quarter acknowledged the consequence of environmental duties in public relations (Zorpas, 2010). From the householder's point of view, Wilson (2005) said that no single waste management practice was involved and not just one but a lot of behavior (Wilson, 2004, 2005). These behavioral variables are somewhat different from the recorded surveys; up to 60% of the population at least one, at least part of them, is calculated by one source. The reasons behind waste reduction are various and different: They span many facets of philosophy and often concentrate on those practices (e.g., food, home composting, reuse) (Table 4.3).

4.6 WASTE PREVENTION TECHNIQUES

The model alternate of waste management is to avert production of residual waste in the preliminary phase. Therefore, waste prevention is an elementary objective of all the waste management approaches. Several technologies can

TABLE 4.2 Monitoring and Assessment Methods Have Their Strengths and Limitations.

Sr. no.	Methodology	Advantages (strengths)	Disadvantages (weaknesses)
1	Self-weighing, monitoring, or reporting	1. Quantitative data on reduction is observable, direct, and measurable 2. Places participants in communication with their waste—the effect on visibility. 3. Provides participants with motivating feedback. 4. Monitoring may provide alternatives to weighting	1. Due to multiple start-and-stop hours, new entrants, insufficient diaries, and lack of participant buy-in for calculation for waste translation, contradictory results can be collected 2. Factors of conversion are necessary 3. The project advances at high drop-out rates 4. There are dangers of screening yourself
2	Usage of round data collection for a precise calculation of waste	1. Makes exact amounts and comparisons of improvements in the rise in waste 2. Can have adequate time and preparation to perform a variety of separate tests 3. Large sample sizes can provide more statistically robust representative populations 4. Can help reduce distortions by preselecting target groups	1. And in a particular geographical place will waste be tracked 2. These populations/collection schemes require thorough and thoughtful preparation 3. Quality and accurate data for waste collection are critical but difficult to obtain 4. This strategy cannot be used where populations have not been identified geographically
3	Researches of speech and actions including measurements, interviews, and focus groups (focused results)	1. Forms an outline of changes 2. Provides objective and qualitative data and analysis for assessment 3. Provides useful feedback for campaign growth 4. Permits to perform large-scale surveys 5. Focus groups offer insight into habits and beliefs that can be structured reasonably quickly and cost-effectively	1. Small samples or poor response rates can be satisfactory or robust 2. Careful survey design is required to provide a comparative analysis of waste data 3. The survey may be skewed by citizens' jury 4. Self-fulfillment surveys will lead to bias 5. Concentrate classes are not sufficient for weight data collection

TABLE 4.2 *(Continued)*

Sr. no.	Methodology	Advantages (strengths)	Disadvantages (weaknesses)
4	P.O.S (Point of Sales Data)	POS data for estimating product waste reduction and the applicability of waste prevention methodology	Providing comprehensive retail distribution data
5	Hybrid – a mixture of one or both of the above	1. Gives the background of the interim self-weighing / observation monitoring surveys before and after 2. Enables the use of short- and long-term tracking and assessment of mixed approaches 3. Results from one approach can be used as a control of another method (e.g., survey community focus groups)	Can the data surveillance and surveys and the self-weighing and observer appraisal be complicated and effort consuming, and it requires meticulous initial preparation, which can make small-scale projects more challenging.

be engaged all over the industrialization, use, or post-use shares of product life cycles to eradicate waste and, sequentially, prevent or decrease pollution (Chalak et al., 2016). Few characteristic approaches include environmentally friendly manufacturing systems that integrate less harmful or hazardous resources, the practices of modern detection technologies for material storing, innovative techniques of chemical neutralization to decrease reactivity, and water-conserving approaches that reduce the necessity for fresh water contributions (Kumar and Chandra, 2020).

TABLE 4.3 Barriers to Pollution Management.

Sr. no.	Barrier category	Barrier	Description
1	Hard Barriers	High Financial Costs	Capital spending and current implementation and organizational waste management initiatives
		Availability of Personnel Resources	Ability to provide staff with information required to develop and implement waste management initiatives
		Legal Restrictions	Present legislation, policies, and regulations that contradict lasting thought
2	Soft Barriers	Low Acceptance	Unwillingness to engage in waste management
		Measure Unknown	Waste reduction initiatives not (yet) recognized or undefined as waste prevention measures
		Measure ineffective	The inefficiency of waste management programs, real or projected
3	Hybrid	Lack of Potential	True waste prevention capacity missing or not noticed/underestimated
4	Other	Other reasons	

4.6.1 BIOREMEDIATION

It is a process of transforming, prosperous, cost-efficient, and environmental-friendly strategy for waste management for the cleansing of wastewater and polluted soil containing poisonous and noxious waste products (Dubchak and Bondar, 2019). It is a natural exercise that utilizes the metabolic potentials of living organisms (principally microorganisms) for the degradation and reduction of perilous pollutants by transforming or mineralizing them into less injurious constituents under in situ or ex situ settings. It is a huge term that is generally related to several other eco-friendly approaches, for example, phytoremediation, bioleaching, land farming, bioventing, composting, rhizo filtration, bio-augmentation, and bio-stimulation (Lourenço et al., 2019). Microorganisms such as fungi and bacteria capable of growing and persisting in polluted environments can be affected remarkably through several biotic and abiotic factors. These factors may affect indirectly or directly, so long as the primary information about bioremediation assists in processing it efficiently (Rhodes, 2013).

4.6.2 BIODEGRADABLE PRODUCTS

Biodegradable waste is also found in municipal waste including plastics, food waste, green waste, and paper waste. Other biodegradable wastes comprise human waste, sewage, manure, slaughterhouse waste, and sewage sludge (Okan et al., 2019). In the lack of oxygen, much of this waste decomposes to methane through anaerobic digestions. In several regions of the developed countries, biodegradable waste is separated from the remaining waste streams, either by separating curb collection or by sorting of waste after collection. At the terminal of collection such wastes is defined as green wastes (Avérous and Pollet, 2012).

4.6.3 WASTE MINIMIZATION

Waste minimization is a collection of practices and processes planned to decrease the amount of produced waste. By eliminating or reducing the production of persistent and harmful wastes, waste minimizations support efforts for promoting a more sustained society (Schott and Andersson, 2015). It comprises redesigning processes and products and changing social patterns of consumptions and productions. The most ecologically resourceful, financially effective, and cost-efficient way of waste management often is to not have to discourse the problems at the starting phase. Administrators see waste minimization as a principal focus for most waste management approaches (Iwata, 2015).

4.6.4 REUSE AND RECYCLING

Recycling is defined as the recovery of valuable materials, for example, glass, plastics, paper, metals, and wood from the waste collection so they might be incorporated into the manufacturing of different products. With larger incorporations of recycled ingredients, the obligatory use of raw materials for same applications is compact (Sandin and Peters, 2018). Recycling decreases the necessity of natural resource exploitations for raw elements; however, it also permits waste materials to be recovered and consumed as valued resource elements. Waste recycling conserves directly natural resources, decreases energy consumptions and emissions generated by abstraction of materials and their succeeding manufacture into new products, decreases greenhouse gas emissions and inclusive energy consumption which participate in the worldwide climate revolution, and decreases the landfilling of the recycled materials (Walser et al., 2012).

4.6.5 ENGINEERING CONTROL OF WASTE PRODUCTION

Present habits of use, lack of facts, knowledge, and scepticism related to the efficacy of interventions are among the common challenges to elucidate literature-based waste deterrence. Also, the long-term effects of waste management are considered financially, but interventions will produce high costs. Actions thus often focus instead of preventive action on the source of waste management (Cecere et al., 2014). There are consistently expressed potential negative implications for the industry (Melanen et al., 2002; Salhofer et al., 2008).

The understanding of waste reduction and its separation from recycling is also another important aspect to remember. Waste reduction is often misunderstood or perceived as the same reduction that avoids incineration or depletion of waste and thereby "losses" the economy. Through this concept, the material becomes part of the economic system. Recycling is also referred to as a form of prevention. Today's efforts to go zero waste are in line with this because zero waste makes it easier to get rid of waste at places that meet these definitions (Cole et al., 2014; Zaman and Lehmann, 2013). Often practical recycling can also minimize the drive for waste prevention (Wilts, 2012). These misunderstandings, knowledge shortages, and a general loss of consciousness about the world contribute to indifferent behavior. A variety of techniques have been introduced to improve that. Long-term improvements in behavior will result in higher costs of products and waste disposal, although education and conscientious initiatives can resolve this concern for centuries (Polanec et al., 2013; Tasaki and Yamakawa, 2011). The first stage in addressing them is the recognition and review of obstacles to the inclusion in waste management work at a group level. We measure the significance of current hurdles using a quantitative approach, while qualitative tools allow us to explore the causes in detail. High prices, urban life, and consumer ethos as well as the lack of education, and understanding are well-known hurdles. This collection of barriers was addressed, consolidated, and expanded in collaboration with the project steering board of local governments. By following this, mainly seven objects have been listed as possible obstacles. These can be graded as tough and gentle obstacles. The "other factors" column was used to balance this list.

4.7 PROBLEM ASSOCIATED WITH WASTE PREVENTION

4.7.1 GROWING GLOBAL WASTES

Growing proportion of waste across the globe economies exhibits a complex relation between waste amounts and economic development. In low to middle class, waste production percentages per capita are comparatively down compared to high class income. But, they are growing with higher frequency (Papargyropoulou et al., 2014). In higher revenue states, the rates of per capita waste production are already very high and still growing with economic development. It is because of the way capitals are being "consumed" in our social structure. A consumption of linear classification of reserve is principally leading in the parsimonies of entire world. Several resources are transformed into diverse products in a primarily production system worldwide. The commodities are dispersed to consumers on the international marketplace, and wilds nursed into the waste management system (Rajmohan et al., 2019).

4.7.2 GROWING PRODUCT COMPLEXITY AND DIVERSITY IN PRODUCTION SYSTEMS

The natural resources experience many transformations in the course of their life-cycle phases. High-standard natural resources generated throughout the extraction procedures are used to utilize various products in the industrial systems by merging several natural resources. So, the industrialized sectors act as "resource diluting arrangements" with respect to the mass of a specific natural resource. The products that enter the consuming system leave in the form of numerous waste elements and result in resources dilution in time and space. Furthermore, these conversions also cause changes in the chemical, biological, and physical characteristics of the natural resources (Tian et al., 2013).

4.7.3 LACK OF ECOLOGICAL SAFETY AWARENESS

Ecological cognizance in the society plays a significant role in sustained consumption and dumping of material substances. Public participation in waste prevention has an imperative potential in the whole operational efficacies of any management system. Deprived behavior of waste categorization among people directly hinders the activities of recycling. Recyclable elements mixed with other types of wastes cause both environmentally and economically ineffective recycling processes, subsequently making other treatment possibilities more attractive for example landfill and incineration. There is a deficiency of knowledge regarding how resource problems are associated with worldwide matters such as climate change (Wilcox et al., 2015).

4.7.4 BARRIERS TO PRACTICAL IMPLEMENTATION OF WASTE PREVENTION APPROACHES

There are several elements responsible for practical performance and implementation of multiple innovative technologies in waste prevention management. Subsequently, policy decisions encounter failures throughout the implementation phases, such as due to fiscal causes or rivalry with ongoing methods. Therefore, the policy interventions flop to achieve the planned objectives. It also involved political, social, cultural, and economic influences on the way of implementation of these interventions (Jiang et al., 2013).

4.8 LEADING PROJECTS AND SUCCESSFUL STORIES ON WASTE PREVENTION WORLDWIDE

The growing complexity and volume of waste is imposing different severe risks to human health and ecosystems. Each year, an approximated 11.2 billion tons of solid waste is released globally and the organic proportion decaying is contributing around 5% of worldwide emissions of greenhouse gases (Lestari and Trihadiningrum, 2019). In the last few years, there are several waste prevention projects being launched recently across the globe. Some of the most prominent and leading projects have been described as follows (Figure 4.3):

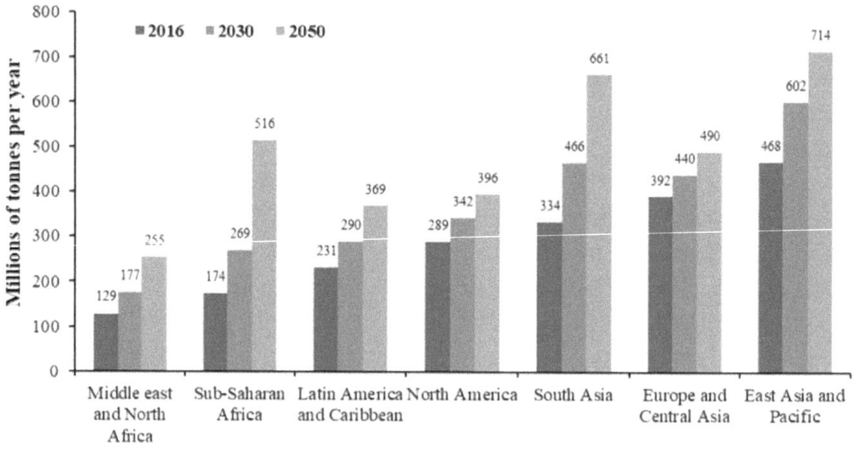

FIGURE 4.3 Projected global waste generation by 2050 (Kaza et al. 2018).

4.8.1 RENEW GANGA—RENEW OCEANS; ALLIANCE TO END PLASTIC WASTE

The Indian Ganga river is about 2600 km in length which starts from the Himalayas, and sinuous through the Indian and Bangladeshi plains till ultimately being emptied into the Indian Ocean. But it is also accounted as one of the rivers worldwide to carry 90% of land-side plastic waste into the sea with about 544,000 tons accumulation of plastic left-over yearly. This is the reason why Alliance is participating in funding, logistics, materials, competencies, and technical expertise for waste prevention in India. The Renew Ganga project aimed at developing a circular national economy through collecting low-value waste of plastic and transforming it into fuels for local corporations like diesel for vehicles. In 2019, around 50-ton plastic waste was collected via this project, and it is projected to be double in 2020 ("Renew Ganga—Renew Oceans" 2019) (Play, 2020).

4.8.2 END PLASTIC WASTE INNOVATION PLATFORM—PLUG AND PLAY

It concentrates on subsidizing startups throughout the world for innovation in the plastic waste management. The platform is secured directly to the innovation pillar of Alliance, progressing, and intended to design novel technologies to minimize waste, make recycle, and recover plastics easily, and create worth from used plastics. Three global hubs including six accelerator programs are working in Silicon Valley, Singapore, and Paris. In Paris, 11 projects have been preferred from around Europe, with a variety of solutions, for example, The Great Bubble Barrier that is intended to clear plastic pollution in seawaters using artificial intelligence to sort and prevent waste ("End Plastic Waste Innovation Platform - Plug and Play - End Plastic Waste" 2020) (Play, 2020).

4.8.3 PROJECT STOP JEMBRANA

The province Balinese of Jembrana is the island of approximately 150,000 populaces having a limited waste prevention system. It leads to a higher risk of plastic waste leakage into the local atmosphere and adjacent Bali Sea. The Ijo Gading River that flows over Jembrana is the largest plastic contributor in the ocean across the island, and accounts for 12% of the leakage into the ocean. The project has been selected for supporting a new waste prevention system in the island. The service is projected to be completely functional by the end of 2020 ("Jembrana-Project STOP" 2020) (Stop, 2020).

4.8.4 MINISTRY OF ENVIRONMENT AND WATER BULGARIA

It is a project for recycling waste electronic and electrical tools launched in Bulgaria in 2010. Participation in embellishment of legislative alterations and deliberations in Bulgarian recycling chamber, trader's assembly along with scrap metals were the millstones of this system ("Ministry of Environment and Water | MOEW" 2020) (MOEW, 2020).

4.8.5 HYDEA/IADB BELIZE

The master plan of solid waste prevention in Brazil was launched in developing tourism regions in Southern and North Belize corridors; preparation of minimum budget analysis, estimated balance sheet, financial models, income, and cash flow statements. Strong emphasis has been made on cohesive interaction between rural and urban certainty of Belize ("IDB, Government and Private Sector Discuss Ways to Ensure Belize's Sustainable Development | IADB" 2017) (IDB, 2017).

4.8.6 UNITED NATIONS INTEGRATED WASTE MANAGEMENT FOR IMPROVED LIVELIHOODS (INWAMI) PROJECT

This activity aims to establish an integrated waste management system to improve livelihoods in Kisumu County, Kenya. Improved waste management can create jobs and income opportunities for poor youth and women living within informal settlements, thus helping vulnerable populations adapt to and mitigate climate change. Reducing waste and managing it properly prevents leachate from contaminating soil and water. Likewise, this averts the open burning of waste that causes air pollution, including greenhouse gases. Recycling materials from waste prevents depletion of virgin raw materials, which minimizes resource extraction in Kisumu and in Kenya at large ("Integrated Waste Management for Improved Livelihoods (INWAMI) Project | UNFCCC" 2020) (De Menna et al., 2020).

4.9 HOW GREEN LIFE/GREEN ENERGY PRODUCTION CONTRIBUTE TO WASTE PREVENTION

The concept of lean production underlines the value of resource conservation by retaining equivalent outputs with reduced inputs, including waste reduction (Dora et al., 2014). As such, it is an invitation to research potential synergetic results through the relation between lean output and environmental costs minimized by increasing resource quality. The advantages of combining both methods were consistently highlighted (Corbett and Klassen, 2006; Zokaei et al., 2013). The synergistic impact of green and lean food waste is motivated not only by high investment in infrastructure, but also due to its low profitability margins (Gustavsson et al., 2011). Companies are now responding to the growing focus and emphasize on healthier, environmental goods. Recent food waste research tends to concentrate more on life cycle ending and introduction of technologies, instead of exploring the capacity for waste reduction in former stages before the commodity is reached by the customer (Schneider, 2011). The root factors need to be identified and removed to avoid food waste proactively. However, developing a more stable environment of less pollution, because of the highly dynamic and expanded types of the food and drinking industries, is not a simple task, but rather an incredibly complex operation. Such a broad scope of reasons, such as management decisions at the start of production, technological design of production lines, break-up or failed production lines, a lack of climate control, or many other fields, contribute also to a single production line investigation. Also, there could be several contributing factors and no single cause. Such a study, therefore, needs significant skill to guarantee good consistency and secure results (Gray, 2009; Schneider and Obersteiner, 2007; WRAP, 2009). In other areas, lean-philosophy is implemented, for example by reducing wastes through stable processes, which has been able to increase resource efficiency. It will be comparable in findings to apply this approach to the food industry (Folinas et al., 2013). The software engineering tool will in this sense compliment the lean approach, as it has proved highly productive to demonstrate the overshadowed complexities of ordinary processes and disclose the reasons behind variations that generate waste and inefficiency such as food waste.

The goals of these two approaches, the lean approach and the six sigma approach, are to minimize variations while growing income, lowering costs and eradicating errors and waste, taking care of efficiency considerations (Kovach and Cho, 2011). The findings of lean Six Sigma are the synthesis of six Sigma root causes research and the perception of lean thought waste reduction. This dramatically increases the aptitude to recognize the primary reason for waste. Though lean practices rely to a large extent on their background, this is true of the lean six sigma system. This attention to the situation means that it is not promising to follow a generic strategy that extends to all businesses. Rather, a personalized solution seems to be expected. Dora et al. (2014) discussed the obstacles facing small- and medium-sized companies to implement lean approaches and tools to achieve the original outcomes, considering the possible big advantages (Dora et al., 2014). The lean strategy aims to overwhelm the challenges of the food industry, fluctuating from a wide variety of goods that may die at dramatically diverse rates; a complex structure of the chain with slight coordination; multiple dynamic manufacturing methods, and, lastly, a product that could variate constantly, because of new patterns, However, a good level of service is also required by the retail and customer, amid issues such as these. Customers often expect fast distribution lead times, while manufacturing and shipping lead times are always far longer than those expected. The latter is also influenced by accelerated product degradation. This is compounded by predictions of consumer preference rather than special demands by the food and drink industry. Both those interactive dynamic aspects contribute to the organization, in demand to balance the danger of invasions, of large inventories with the resulting risk of significant food waste.

The global network of food supply chain: food leftover, damage, loss, and waste. The whole FSC, the first phase of agriculture to the last phase of consumption, is missed or wasted (Parfitt et al., 2010). In the literature, there are three key concepts of food waste. In the first place, the organization for food and agriculture (FAO) describes food waste as a salutary edible for human use, which occurs in all FSC areas that are disposed of, destroyed or deteriorated, or used by plagues at each stage. The second is that Stuart (2009) augments to FAO's concept that edible material which is deliberately food-fed to the animals or a by-product of food production which is aloof from the human food chain would also be comprised in the food waste (Stuart, 2009). Smil (2004) concluded that food waste contains the aforementioned concepts, but improves surplus feed, the difference between the energy worth per capita of food eaten, and the energy-benefit per capita of food required (Smil, 2004). The concept of Stuart gives a broader variety of possibilities for food surpluses and waste control, as it involves losses of food by feeding livestock and redirect by-products of food production. Therefore, the concept of Stuart is followed for this report. The decrease in comestible food mass in the human FSC is food waste or loss (Gustavsson et al., 2011).

Researches were carried out in developing and developed countries on the level of food loss and waste during the development and use of the FSC (Parfitt et al., 2010). These reports argue that there are significant disparities in information about worldwide food and waste losses. As far as half of all foods grown before and after they hit the market are missed or wasted, according to recent studies. "From field to the bucket," the losses in the field and the households for postharvest are measured at 2600 kcal/capita/day. Stuart (2009) reports that 30–50% of the food stocks are wasted by Europe and North America, sufficient to feed the starving world about three times (Gustavsson et al., 2011; Stop, 2020). It indicates that one-third of human food edible parts are missing or discarded by the worldwide FSC, which amounts to 1300 million tonnes a year. The supply of food losses and waste between developed countries, rich/poor producers, and consumers are distinct. Overall losses of food and waste in the developed world are greater than in the developing countries, deliberately an estimated annual food loss of 300 kg/capita in Europe and North America and an average annual food loss of 170 kg/capita in South and Southeast Asia. Many food losses happen in the early stage of the FSC in developed countries. This is due to insufficient technology for reaping, shipping constraints, and insufficient storage associated with harsh weather conditions. Food waste accounts for above 40% of overall food waste, and waste in the developing world at the consumption stage.

4.10 SUMMARY

Waste production rate is directly proportional to population growth rate. Increased growth in developing countries generates and absorbs an increased range of goods and services. The amount of waste produced is a duplicate of this vital expansion.

However, waste management is a big and ever-growing industry that demands analyzing and implication of advanced technologies at each step of waste management based on the new emergence of threats. The need to identify simple principles like recycling and disposal to properly coordinate waste management behavior has now become a matter of prime importance. Waste reduction is at the forefront of the hierarchy of waste; though, calculations are disreputably difficult. A dynamic and complex undertaking is the calculation, control, and appraisal of waste reduction. The dilemma is just how you calculate something that is not there. What can be assessed if it is not there is not always obvious. Monitoring and appraisal involve the compilation of data, and any approach has its problems (additions and drawbacks). It is too hard to assess which one is best. However, careful monitoring is an invaluable method for effective management and offers a valuable framework for assessment.

A necessity for a wide-ranging system of approaches to waste prevention has been underscored. The strategy recommends a pressure-oriented attitude instead of a conservative state- or an impact-oriented tactic to prevent waste. Strengthening steps to be taken to avoid and reduce the influences of waste generation, and waste disposal on the environment, is also important. As a principal phase toward a long-span waste prevention system, a prerequisite to broaden the boundary of systems for designing, producing, and consuming the waste has been emphasized. The challenges to implement the current projects for clearly defined objectives have been further shared global views on the dynamic links between the economic, social, technical, and ecological systems. Sustainable management of resources is a universal challenge that cannot be accomplished without society's active contribution together, including producers, consumers, and waste management establishments at acting at several levels within the range of local to global. Such a waste prevention paradigm might require fundamental changes in the current structures, behaviors, and governmental rules.

A certain target group is important and essential for good outcomes from any of the above strategies. Mass reduction and behavioral research can effectively quantify the effect of waste avoidance campaigns and can be acceptable tools for measuring and assessing waste prevention. Perhaps it is fairer to incorporate all of the methodologies with a particular focus audience at a given time. On the other hand, it is important to track and evaluate the avoidance of household waste in a manner that tackles all problems and the possible sensitivities. Finally, tracking and evaluating the prevention of waste aimed at allowing politicians, city governments, and professionals to define priorities, design their strategies, and measure changes in conduct. In this scenario, developed countries have designed affective policies and ensured their implementation, focused on SW administration, and trained human resource development and efficient technologies. Educating students and creating awareness in the society, allocation of sufficient budgets, and collaboration with respective developed countries in planning, research, and waste management processes can help the developing/less developed/under developed country to upgrade the existing waste management strategies. These actions will not only protect the environment, but also produce new jobs that will strengthen the economy.

KEYWORDS

- **behavioral change**
- **developing countries**
- **green life**
- **waste prevention**
- **waste production**

REFERENCES

Alam, P.; Ahmade, K. Impact of Solid Waste on Health and the Environment. *Int. J. Sustain. Dev. Green Econ. (IJSDGE)* **2013,** *2* (1), 165–168.

Al-Wabel, M.; Al Yehya, W.; Al-Farraj, A.;El-Maghraby, S. Characteristics of Landfill Leachates and Bio-Solids of Municipal Solid Waste (MSW) in Riyadh City, Saudi Arabia. *J. Saudi Soc. Agric. Sci.* **2011,** *10* (2), 65–70.

Avérous, L.; Pollet, E.; *Biodegradable Polymers*; Springer, 2012; pp 13–39.

Bortoleto, A. P. *Waste Prevention Policy and Behaviour: New Approaches to Reducing Waste Generation and Its Environmental Impacts*; Routledge, 2014.
Bortoleto, A. P.; Kurisu, K. H.; Hanaki, K. Model Development for Household Waste Prevention Behaviour. *Waste Manage.***2012,** *32* (12), 2195–2207.
Boschini, M.; Falasconi, L.; Cicatiello, C.; Franco, S. Why the Waste? A Large-Scale Study on the Causes of Food Waste at School Canteens. *J. Cleaner Prod.* **2020,** 246, 118994.
Butnariu, I.; Constantin, N.; Dobrescu, C.; Heput, T. Research on the Recycling of Pulverulent Waste from the Ferous and Non-Ferrous Industry in Order tu Reduced the Pollution. *Red* **2018,** *1* (5), 6–6.
Cecere, G.; Mancinelli, S.; Mazzanti, M. Waste Prevention and Social Preferences: The Role of Intrinsic and Extrinsic Motivations. *Ecol. Econ.* **2014,** *107*, 163–176.
Chalak, A.; Abou-Daher, C.; Chaaban, J.; Abiad, M. G. The Global Economic and Regulatory Determinants of Household Food Waste Generation: A Cross-Country Analysis. *Waste Manage.* **2016,** *48*, 418–422.
Christensen, T. *Sanitary Landfilling: Process, Technology and Environmental Impact*; Elsevier, 2012.
Cole, C.; Osmani, M.; Quddus, M.; Wheatley, A.; Kay, K. Towards a Zero Waste Strategy for an English Local Authority. *Resour. Conserv. Recycl.* **2014,** 89, 64–75.
Corbett, C. J.; Klassen, R. D. Extending the Horizons: Environmental Excellence as Key to Improving Operations. *Manuf. Serv. Operat. Manage.* **2006,** *8* (1), 5–22.
Cristóbal, J.; Castellani, V.; Manfredi, S.; Sala, S. Prioritizing and Optimizing Sustainable Measures for Food Waste Prevention and Management. *Waste Manage.* **2018,** *72*, 3–16.
De Menna, F.; Davis, J.; Östergren, K.; Unger, N.; Loubiere, M.; Vittuari, M. A Combined Framework for the Life Cycle Assessment and Costing of Food Waste Prevention and Valorization: An Application to School Canteens. *Agric. Food Econ.* **2020,** *8* (1), 1–11.
Dora, M.; Van Goubergen, D.; Kumar, M.; Molnar, A.; Gellynck, X. Application of Lean Practices in Small and Medium-Sized Food Enterprises. *Br. Food J.* **2014.**
Dubchak, S.; Bondar, O. *Bioremediation and Phytoremediation: Best Approach for Rehabilitation of Soils for Future Use*; Springer, 2019; pp 201–221.
Elbeshbishy, E.; Nakhla, G.; Hafez, H. Biochemical Methane Potential (BMP) of Food Waste and Primary Sludge: Influence of Inoculum Pre-incubation and Inoculum Source. *Bioresour. Technol.* **2012,** *110*, 18–25.
EPA, U. *Municipal Solid Waste in the United States: 2011 Facts and Figures*, 2013. https://archive.epa.gov/epawaste/nonhaz/municipal/web/html/.
EU. Guidelines on Waste Prevention Programmes, 2012.
Ferronato, N.; Torretta, V. Waste Mismanagement in Developing Countries: A Review of Global Issues. *Int. J. Environ. Res. Public Health* **2019,** *16* (6), 1060.
Folayan, M. O.; El Tantawi, M.; Ramos-Gomez, F.; Sabbah, W. Early Childhood Caries and Its Associations with Sugar Consumption, Overweight and Exclusive Breastfeeding in Low, Middle and High-Income Countries: An Ecological Study. *Peer J.* **2020,** *8*, e9413–e9413.
Folinas, D.; Aidonis, D.; Triantafillou, D. Malindretos, Exploring the Greening of the Food Supply Chain with Lean Thinking Techniques. *Procedia Technol.* **2013,** *8*, 416–424.
FSB. Federation of Small Businesses. Social and Environmental Responsibility and the Small Business Owner, 2006.
Gray, S. *Down the Drain: Quantification and Exploration of Food and Drink Waste Disposed of to the Sewer by Households in the UK*; Waste and Resources Action Program: Banbury, UK, 2009.

Gustavsson, J.; Cederberg, C.; Sonesson, U.; Van Otterdijk, R.; Meybeck, A. *Global Food Losses and Food Waste*; FAO: Rome, 2011.
Hung, Y.-T.; Wang, L. K.; Shammas, N. K. *Handbook of Environment and Waste Management: Air and Water Pollution Control*, Vol. 1; World Scientific, 2012.
IDB. *Government and Private Sector Discuss Ways to Ensure Belize's Sustainable Development*; IADB, 2017.
Iwata, T. Biodegradable and Bio-based Polymers: Future Prospects of Eco-Friendly Plastics. *Angew. Chem. Int. Ed.* **2015,** *54* (11), 3210–3215.
Jacobsen, H.; Kristoffersen, M.; Tsotsos, D. *Case Studies on Waste Minimisation Practices in Europe*, Vol. 2; European Environment Agency Copenhagen, 2002.
Jiang, J.-Q.; Zhou, Z.; Sharma, V. K. Occurrence, Transportation, Monitoring and Treatment of Emerging Micro-Pollutants in Waste Water—A Review from Global Views. *Microchem. J.* **2013,** *110*, 292–300.
Kanagaraj, J.; Senthilvelan, T.; Panda, R. C.; Kavitha, S. Eco-Friendly Waste Management Strategies for Greener Environment towards Sustainable Development in Leather Industry: A Comprehensive Review. *J. Cleaner Prod.* **2015,** *89*, 1–17.
Khan, M. A.; Ghouri, A. M. Environmental Pollution: Its Effects on Life and Its Remedies. *Res. World: J. Arts Sci. Com.* **2011,** *2* (2), 276–285.
Kovach, T.; Cho, R, Better Processes Make GOOD EATS: Food Industry Can Benefit from Lean Six Sigma Principles. *Ind. Eng.* **2011,** *43* (1), 36–41.
Kühn, S.; Rebolledo, E. L. B.; van Franeker, J. A. *Deleterious Effects of Litter on Marine Life*; Springer: Cham, 2015; pp 75–116.
Kumar, S. *Municipal Solid Waste Management in Developing Countries*; CRC Press, 2016.
Kumar, V.; Chandra, R. *Bioremediation of Melanoidins Containing Distillery Waste for Environmental Safety*; Springer, 2020; pp 495–529.
Lemly, A. D.; Skorupa, J. P. Wildlife and the Coal Waste Policy Debate: Proposed Rules for Coal Waste Disposal Ignore Lessons from 45 Years of Wildlife Poisoning. *Environ. Sci. Technol.* **2012,** *46* (16), 8595–8600.
Lestari, P.; Trihadiningrum, Y. The Impact of Improper Solid Waste Management to Plastic Pollution in Indonesian Coast and Marine Environment. *Marine Pollution Bull.* **2019,** *149*, 110505–110505.
Liu, X.; Liu, B.; Shishime, T.; Yu, Q.; Bi, J.; Fujitsuka, T. An Empirical Study on the Driving Mechanism of Proactive Corporate Environmental Management in China. *J. Environ. Manage.,***2010,** *91* (8), 1707–1717.
Lourenço, J.; Mendo, S.; Pereira, R. *Rehabilitation of Radioactively Contaminated Soil: Use of Bioremediation/Phytoremediation Techniques*; Springer, 2019; pp 163–200.
Melanen, M.; Kautto, P.; Saarikoski, H.; Ilomäki, M.; Yli-Kauppila, H. Finnish Waste Policy Effects and Effectiveness. *Resour. Conserv. Recycl.* **2002,** *35* (1–2), 1–15.
Mmereki, D.; Baldwin, A.; Li, B. A Comparative Analysis of Solid Waste Management in Developed, Developing and Lesser Developed Countries. *Environ. Technol. Rev.* **2016,** *5* (1), 120–141.
MOEW. *Ministry of Environment and Water*, 2020.
Okan, M.; Aydin, H. M.; Barsbay, M. Current Approaches to Waste Polymer Utilization and Minimization: A Review. *J. Chem. Technol. Biotechnol.* **2019,** *94* (1), 8–21.
Oke, A. Workplace Waste Recycling Behaviour: A Meta-Analytical Review. *Sustainability* **2015,** *7* (6), 7175–7194.

Palmiotto, M.; E. Fattore, V. Paiano, G. Celeste, A. Colombo, Davoli, Influence of a Municipal Solid Waste Landfill in the Surrounding Environment: Toxicological Risk and Odor Nuisance Effects. *Environ. Int.* **2014,** *68*, 16–24.

Papargyropoulou, E.; Lozano, R.; Steinberger, J. K.; Wright, B.; Ujang, B. The Food Waste Hierarchy as a Framework for the Management of Food Surplus and Food Waste. *J. Cleaner Prod.* **2014,** *76*, 106–115.

Parfitt, J.; Barthel, M.; Macnaughton, S. Food Waste Within Food Supply Chains: Quantification and Potential for Change to 2050. *Phil. Trans. R. Soc. B* **2010,** *365* (1554), 3065–3081.

Play, P. A. *End Plastic Waste*, 2020. https://www.plugandplaytechcenter.com/end-plastic-waste/.

Polanec, B.; Aberšek, B.; Glodež, S. Informal Education and Awareness of the Public in the Field of Waste Management. *Procedia-Soc. Behav. Sci.* **2013,** *83*, 107–111.

Rajmohan, K. V. S.; Ramya, C.; Viswanathan, M. R.; Varjani, S. Plastic Pollutants: Effective Waste Management for Pollution Control and Abatement. *Curr. Opin. Environ. Sci. Health* **2019,** *12*, 72–84.

RECAP. Waste Prevention Plan for Cambridge and Peterborough 2008–2022, 2008.

Rhodes, C. J. Applications of Bioremediation and Phytoremediation. *Sci. Progress* **2013,** *96* (4), 417–427.

Rim-Rukeh, A. An Assessment of the Contribution of Municipal Solid Waste Dump Sites Fire to Atmospheric Pollution. *Open J. Air Pollut.* **2014,** *3* (03), 53–53.

Salhofer, S.; Obersteiner, G.; Schneider, F. Lebersorger, Potentials for the Prevention of Municipal Solid Waste. *Waste Manage.* **2008,** *28* (2), 245–259.

Sandin, G.; Peters, G. M. Environmental Impact of Textile Reuse and Recycling–A Review. *J. Cleaner Prod.* **2018,** *184*, 353–365.

Schneider, F. Prevention of Food Waste in Residual Waste. In *Proceedings of ISWA Beacon Conference, Waste Prevention and Recycling, Vienna*, 2011.

Schneider, F.; Obersteiner, G. Food Waste in Residual Waste of Households–Regional and Socioeconomic Differences, 2007.

Schott, A. B. S.; Andersson, T. Food Waste Minimization from a Life-Cycle Perspective. *J. Environ. Manage.* **2015,** *147*, 219–226.

Smil, V. Improving Efficiency and Reducing Waste in Our Food System. *Environ. Sci.* **2004,** *1* (1), 17–26.

Šomplák, R.; Kůdela, J.; Smejkalová, V.; Nevrlý, M.; Pavlas, M.; Hrabec, D. Pricing and Advertising Strategies in Conceptual Waste Management Planning. *J. Cleaner Prod.* **2019,** *239*, 118068.

Stop. *Project STOP (Jembrana)*, 2020.

Stuart, T. *Waste: Uncovering the Global Food Scandal*; WW Norton & Company, 2009.

Tabasová, A.; Kropáč, J.; Kermes, V.; Nemet, A.; Stehlík, P. Waste-to-Energy Technologies: Impact on Environment. *Energy* **2012,** *44* (1), 146–155.

Tasaki, T.; Yamakawa, H. An Estimation of the Effectiveness of Waste Prevention by Using Point-of-Sales (POS) Data—The Case of Refills for Shampoo and Hair Conditioner in Japan. *Resour. Conserv. Recycl.* **2011,** *57*, 61–66.

The World Bank. *Solid Waste Management*, 2019 [cited Oct. 9, 2020]. https://www.worldbank.org/en/topic/urbandevelopment/brief/solid-waste-management#:~:text=In%20 2016%2C%20the%20worlds'%20cities,3.40%20billion%20tonnes%20in%202050.

Tian, H.; Gao, J.; Hao, J.; Lu, L.; Zhu, C.; Qiu, P. Atmospheric Pollution Problems and Control Proposals Associated with Solid Waste Management in China: A Review. *J. Hazard. Mater.* **2013,** *252*, 142–154.

Verma, R.; Vinoda, K. S.; Papireddy, M.; Gowda, A. N. S. Toxic Pollutants from Plastic Waste-a Review. *Procedia Environ. Sci.* **2016,** *35*, 701–708.
Walser, T.; Limbach, L. K.; Brogioli, R.; Erismann, E.; Flamigni, L.; Hattendorf, B.; Juchli, M.; Krumeich, F.; Ludwig, C.; Prikopsky, K. Persistence of Engineered Nanoparticles in a Municipal Solid-Waste Incineration Plant. *Nat. Nanotechnol.* **2012,** *7* (8), 520–524.
Wilcox, C.; Mallos, N. J.; Leonard, G. H.; Rodriguez, A.; Hardesty, B. D. Using Expert Elicitation to Estimate the Impacts of Plastic Pollution on Marine Wildlife. *Marine Policy* **2016,** *65*, 107–114.
Wilcox, C.; Van Sebille, E.; Hardesty, B. D. Threat of Plastic Pollution to Seabirds Is Global, Pervasive, and Increasing. *Proc. Natl. Acad. Sci.* **2015,** *112* (38), 11899–11904.
Wilson, D. C.; Parker, D.; Cox, J.; Strange, K.; Willis, P.; Blakey, N.; Raw, L. Business Waste Prevention: A Review of the Evidence. *Waste Manage. Res.* **2012,** *30* (9_suppl), 17–28.
Wilson, D. *The Evidence Base for Household Waste Prevention: How Best to Promote Voluntary Actions by Households*; Department for Environment, Food and Rural Affairs, 2005.
Wilson, R. A UK Framework for Waste Prevention, 2004.
Wilts, H. National Waste Prevention Programs: Indicators on Progress and Barriers. *Waste Manage. Res.* **2012,** *30* (9), 29–35.
World Meterological Organization. WMO Greenhouse Gas Bulletin, 2019.
WRAP. Household Food and Drink Waste in the UK. *Report Prepared by WRAP*; WRAP: Banbury, 2009.
Zaman, A. U.; Lehmann, S. The Zero Waste Index: A Performance Measurement Tool for Waste Management Systems in a 'Zero Waste City'. *J. Clean. Prod.* **2013,** *50*, 123–132.
Zokaei, K.; H. Lovins, A. Wood, Hines, *Creating a Lean and Green Business System: Techniques for Improving Profits and Sustainability*; CRC Press; 2013.
Zorpas, A. Environmental Management Systems as Sustainable Tools in the Way of Life for the SMEs and VSMEs. *Bioresour. Technol.* **2010,** *101* (6), 1544–1557.
Zorpas, A. A.; Lasaridi, K.; Voukkali, I.; Loizia, P.; Chroni, C. Household Waste Compositional Analysis Variation from Insular Communities in the Framework of Waste Prevention Strategy Plans. *Waste Manage.* **2015,** *38*, 3–11.

PART II

Waste Categories, Bases, Pollution Potential, and Management

CHAPTER 5

Environmental Sources and Threats of Plastic Pollution to Developing Worlds and Eco-Friendly Solutions

FATIMA AKMAL[1], MUHAMMAD IRFAN SOHAIL[1*], MUHAMMAD AZHAR[1], YASIR HAMEED[2], JIBBING XIONG[3], MUHAMMAD FARHAN[4], and AYESHA SIDDIQUI[5]

[1]*Institute of Soil and Environmental Sciences, University of Agriculture, Faisalabad 38040, Pakistan*

[2]*Ministry of Education, Key Laboratory of Environmental Remediation and Ecosystem Health, College of Environmental and Resources Science, Zhejiang University, Hangzhou 310058, People's Republic of China*

[3]*Jiangsu key Laboratory of Resources and Environment information Engineering, China University of Mining and Technology, Xuzhou 221116, China*

[4]*Department of Geodesy and Survey Engineering, College of Earth Science and Engineering, Jianging Campus Hohai University, Nanjing, P.R. China*

[5]*Department of Botany, University of Agriculture, Faisalabad 38040, Pakistan*

[*]*Corresponding author. E-mail: irfansohail9@gmail.com*

ABSTRACT

Plastic pollution is relatively contemporary, but an onerous challenge to both public and science for environmental management. Plastic debris is also referred as "techno-fossil" which is an amalgamation of long-persistence,

Waste Problems and Management in Developing Countries. Umair Riaz, Shazia Iqbal, & Moazzam Jamil (Eds.)

low density and tremendously varied size distribution. Plastic pollution has degraded each segment of the biosphere. The current global production of all kinds of plastics is above 381 million tons annual (Mt/yr). Moreover, the annual increased rate is 5%. This sprawl around each corner of the biosphere clearly indicates the aggressive growth of synthetic plastic production and mismanagement of its waste. The environmental release of plastic waste usually occurs from multiple sources, that is, wastewater treatment operations, incineration and degradation units, agriculture farms, etc. Although plastic does not behave like chemical pollutant, but plastic waste piling up in the environment is combination of varied sizes, that is, macro, micro, and nano scales which latter become inaccessible for fauna and flora. The contemporary research works have elaborated the hazardous physiochemical effects on biota especially in fresh water and terrestrial living organism. Although, the researchers have extensively elaborated the ingestion of plastics by aquatic fauna, further transferred into food webs. Yet, many questions remain open. New sources are being recognized constantly. As the plastic pollution is a global concern in context to biosphere degradation and economic threat, hence this challenge needs holistic policies, better assessments with doable technological strategies to tackle. Therefore, understanding the potential of pollution and risk to ecosystem, exploring the sources, nature of pollution and its dynamics in environment is a prerequisite. Moreover, most of the plastic wastes are generated in well-developed countries and pollution is generated as nerdy challenge in developing countries leaving extra burdens on the economy. This chapter elucidates the complete scenario of plastic pollution in the developing countries and possible management strategies.

5.1 INTRODUCTION

The use of plastic has increased dramatically with each passing year, but the swift production has occurred since 2004 (Geyer et al., 2017). Today, plastic seems inevitable in our lives, as its production surpasses most other synthetic materials production. Plastic, organic synthetic polymer commonly derived from natural gas and crude oil, has its prime market in packaging which accounts one-third of plastic production (Geyer et al., 2017) and remaining two-third is consumed by other sectors, that is, building and conduction, electronics, textile, transportation, and institutional products. Though for some, plastic boom and accelerated production might be an economic and scientific success, but for others, it is considered as a threat to natural environment. It is estimated that globally in 2015, 60–90 Mt of plastic waste was generated, and this can be tripled by year 2060 (155–265 Mt/yr) (Lebreton and Andrady, 2019). The leading contributors in plastic waste generation are mostly low- or middle-income countries. In 2015, Asia being the most populous continent which generated maximum plastic waste of 82 Mt, while Africa and Latin America generated 19 Mt of plastic waste (Waste Atlas, 2016).

The plastic products are not lasting long, shift in attitude from reusable to single use has become a global trend which increases incessantly. That is why plastic share in municipal waste increases from 1% to 10% in 1960 and 2005, respectively (Jambeck et al., 2015). In packaging, a large proportion is nonrecyclable and finally ends up in waste. As reported in different studies, 26–36% of plastic is in use for packaging (Ellen MacArthur Foundation, 2016). In 2015, out of total plastic waste, about 63% of plastic is poorly managed and disposed globally, Asia (65% of total waste is mismanaged) and Africa (88.5% of total waste is mismanaged) showed highest rate of ineffective waste disposal (Waste Atlas, 2016). According to Ellen MacArthur Foundation report, around 40% of plastic packaging is dumped to landfill and 32% is mismanaged and directly dumped into natural environment. And if this trend continues, by 2050, our seas would have more plastic than fish due to our poor management and negligence (Ellen MacArthur Foundation, 2016). In 2010 alone, 4–12 Mt of plastic debris ended up into aquatic ecosystem (Jambeck et al., 2015). And the five leading plastic waste contributors (China, Indonesia, Philippines, Thailand, and Vietnam) to ocean covered approximately 60% of the total waste (included the plastic scrap exported by developed countries) and out of the total oceanic plastic waste sourced from land: 75% is actually the result of "uncollected waste" and remaining 25% is due to the leakage from existing waste management systems (Ocean Conservancy, 2015).

Plastic is highly persistent, it decomposes slowly due to chemical resistance and stability, having undefined shelf life. Actually, all the plastic we have ever produced/synthesized is still present in our planet somewhere in open dumps, landfills, or in environment and we have no idea about its degradation timescale. Plastic is ubiquitous, we will find plastic where ever we look, found in water bodies (beaches, river beds to ocean floors), oceans creatures and even beneath the Arctic ice, loaded in rain, our tap water, beer, and even in the bottled water which is considered the safest water (Jambeck et al., 2015; Barnes et al., 2009). Likewise, plastic pollution though receives less attention in terrestrial environment, but it is found everywhere on the soil surface. That is why plastic is considered as an indicator of Anthropocene (period when humans started to dominate geological processes) due to its ubiquity (Zalasiewicz et al., 2016).

Conventional approaches available to treat waste management are ineffective in coping with plastic pollution. In general, there existed three fates of plastic debris. First, plastic can be dumped into managed landfills or directly left into open dumps, which is the most common practice in developing countries (DCs). Apart from land wastage (which could be productively used for agriculture) and financial encumbrance, landfill raises apprehensions related to hazardous chemicals (toluene, benzene, xylene, etc.) emission/generation (either leachate or gaseous) (Xu et al., 2011). Certainly, landfills are the storing problem for future which cannot be ignored. Second, thermally destroying the plastic, incineration with or without energy recovery. But again, the concern arises about the energy cost of procedure and emission of greenhouse gases. Therefore, incinerators operation, design, and emission control technology strongly decide the impact on health and environment (Geyer et al., 2017). Finally, plastic can be reprocessed into secondary material via recycling. Recycling delays the future plastic pollution if it replaces the primary material with secondary material. But recycling has its own limitations, that is, mixed and contaminated plastic would yield secondary plastic of low economic and quality value (Geyer et al., 2016; Zink et al., 2018). Globally, from the total collected plastic waste, substantial amount is dumped in the landfills (79%), and the remaining is incinerated (12%) and recycled (9%) (Geyer et al., 2017).

According to World Bank group, open dumping and burning are the most prevalent practice in low-income countries till date. As in North Africa and Middle East, open dumping covers 53% of total waste management (Kaza et al., 2018). It is pertinent to improve the waste collection and management/recycling systems in DCs, otherwise, the mismanaged plastic waste would become an unresolvable problem. There is a need to set regulations and establish plastic waste eradication/mitigations policies in DCs to improve and thereby limit the hazardous outcomes of plastic pollution to human health and control its abundance in natural environment.

5.2 PLASTIC WASTE: SOURCES, ACCUMULATION, AND POTENTIAL IMPACTS

Although plastic is a useful material and eases the daily life, but its waste is becoming a global issue increasing with greater pace causing problems especially in poor or DCs where plastic uses are manyfolds higher with poor waste management. Due to resistance against decomposition, it becomes a threat or challenge to the environment (UN environment, 2018). Seven different types of plastic materials are generally used, that is,

I. Low-density polyethylene (LDPE)
II. Polystyrene (PS)
III. Polyethylene terephthalate (PET)
IV. Polyvinyl chloride (PVC)
V. Polypropylene (PP)
VI. High-density polyethylene (HDPE)

There are six major sources of plastic pollution, such as (1) food wrappers and containers that are responsible for 31% of the environmental pollution for plastic waste, (2) bottle and container caps count 15.5%, (3) plastic bags can pollute the environment at about 11.2%, (4) Straws and stirrers account for 8.1%, (5) beverage bottles are responsible for 7.3% pollution, and (6) takeout containers contribute 6.3% of plastic pollution. Based on the size, plastic waste is divided in to three categories, that is, macro, meso, and microplastic materials. Depending on the size of plastic waste, microplastic (plastic parts $\leq$ 5 mm) yielded as a result of weathering of bigger plastic and/or used in abrasives, that is, cosmetics which is more influential and dangerous compared with macro or larger plastic waste.

5.2.1 PLASTIC POLLUTION

Plastic pollution is defined as the addition or accumulation of plastic waste materials, such as bottles, bags, and containers that is considered as one of the major threats of the present times because plastic material is not decomposed. Right now, oceans get more attention to deal with plastic waste. Presence of plastic waste was first reported 50 years back in 1972 (Carpenter and Smith, 1972). It is difficult to exactly measure the status of plastic waste. The estimates revealed that Norway and Switzerland produced about 24.9 megatons of plastic waste (Mudgal et al., 2011), but it is much difficult to distribute in sources. The estimation of plastic waste in oceans is much difficult. Plastic pollution has many bad impacts on humans and other living entities on land and in water (marine) as the waste releases different poisonous chemicals. In marine system, the most reported impact is entanglement of waste that is engulfed by wildlife. Alterations in the habitat of oceans and introduction of new species with waste are also the impacts of plastic pollution.

5.2.2 TOXIC CONSTITUENTS OF PLASTIC WASTES

Plastic waste is made of different constituents, but the prime concern is the biological toxicity of these materials during waste disposal. Several chemicals are used to give some properties to plastic, that is, phthalate, bisphenols, and retardants are used in flame. Such chemicals are categorized as toxic and have negative impacts on biological entities (human and animal) mainly disrupting the endocrine systems. Plastic is a polymer, that is, formed by the combination of different monomers. There are some toxic monomers that are carcinogenic (causing cancer) and affect the reproductive system. The mechanism behind how plastic chemicals affect organisms is not comprehensively explained in the literature. The prime hypothesized pathway is ingestion of micro/nanoplastic. Chemical entities in plastic could bioaccumulate in the body and can cause negative impacts on health. Plastic waste also attracts other contaminants especially in marine environment, for example, persistent organic pollutants (POPs). The mechanism of action of these pollutants is still ambiguous and needs more clarifications. The exposure of plastic chemicals is higher in the case of microplastic due to high surface area but the accumulation in circulatory system is lower as compared with large plastic particles, because microplastic passes more quickly through the digestive system (Science for Environment Policy, 2011).

5.2.3 PLASTIC WASTE ACCUMULATION AND STATE IN NATURAL ENVIRONMENT

Significant quantity of plastic is disposed in open ecosystem and/or designated landfills. Almost 10% of the municipal waste contains plastic materials (Barnes et al., 2009). The disposed plastic waste can contaminate terrestrial, freshwater, and marine habitats and affect their ecosystem. Although there is not enough data present on plastic pollution in terrestrial environment, but soil contamination with small plastic particles has been reported as a result of the spread of sewage sludge (Zubris and Richards, 2005), compost contaminated with plastic and glass (Brinton, 2005), and plastic material in streams, rivers, and oceans carried through rain water (Thompson et al., 2005). Additional efforts are needed to investigate the effects of plastic waste on natural terrestrial habitat and its impacts on soil physiochemical properties and on plant health.

Very first pollution case of plastic waste was reported in carcasses of seabirds sampled from shoreline in 1960 (Harper and Fowler, 1987). Mostly, plastic materials, such as bottles, containers, and toys are floating on the sea surface trapping air and the polymers are buoyant in water. Resultantly, seashore debris contains 50–80% of plastic waste (Barnes et al., 2009). Almost hundred thousand per square meter plastic items were found on some shoreline debris (Gregory, 1978), while 350,000 plastic items/km^2 were reported on ocean surface (Yamashita and Tanimura, 2007). It is reported that more than 10,000 per hectare plastic items were found in some seabed debris of Europe and debris including shopping bags were also found at 1000 m depth from the sea surface (Gregory, 2009).

Limited data are available for the presence of plastic items on ocean surface and in shoreline debris, but the decomposition of plastic waste is slower in water depths due to unavailability of light and low temperature (Barnes et al., 2009; Ryan et al., 2009). The more work is needed to explore different types and quantities of plastic items on the sea surface, in seashore debris, and on terrestrial environment. For this, United Nations Environment Program and OSPAR are working to introduce some SOPs to explore and quantify plastic waste in the environment (OSPAR Commission, 2007; Cheshire et al., 2009). It is reported that plastic waste is more abundant in seashores and beaches adjacent to populated areas and waste is buried in sediments (Barnes et al., 2009; Ryan et al., 2009). The plastic waste in environment will continuously increase unless until we will change our living style and minimize the use of plastic, but plastic waste that has already accumulated in the environment will persist for many years.

5.2.4 PLASTIC POLLUTION IMPACT AND SIDE EFFECTS

Plastic has become indispensable and global reliance on plastic is beyond imaginable. Persistence is something which makes this material appealing as well as detrimental in a parallel manner. Its unframed timeline of degradation made the usage of this material more worrisome. That is why, wherever we look or wherever we go, plastic can be seen everywhere in water, soils, forests, even in animals, and human bodies. Plastic pollution has become a nerdy environmental challenge today, as every species of living organism is facing the consequence. In addition, rapid boom in plastic production ignites the climate change problem, if this trend continues it is estimated that by 2050, plastic would pump gigantic amount of greenhouse gases into the atmosphere that could risk our limit (below 1.5 °C) of global warming.

5.2.5 IMPACT ON MARINE ENVIRONMENT

The impact of plastic pollution on marine environment is huge. It is estimated that almost 10 Mt waste is dumped into water bodies every year. Generally, it is assumed that there are five consolidated plastic waste areas including Pacific (Both South and North), Atlantic (Both South and North), and Indian ocean but plastic waste is distributed globally all over the marine ecosystem. The profusion of microplastic in aquatic environment is widely documented as reported by Obbard et al. (2014), the range of microplastic in Arctic ice cores is about 38–234 particles/m^3. Barnes et al. (2010) reported the availability of microplastic in Arctic and Antarctic surface waters. Lechner et al. (2014) documented that a single river contributed >1500 tons of microplastic into black sea every year and Sruthy and Ramasamy (2017) examined the Vembanad Lake of India, where the microplastic is found in the sediments ranging between 96 and 496 particles/m^2. In addition, in Wuhan, China the urban freshwater was studied and the studied samples of water showed the abundance of microplastic in the range of 1660.0–8925 numbers/m^3 (Wang et al., 2017). In short, plastic is found everywhere in the aquatic environment.

Now, as the abundance of microplastic is assured, it is of great concern that elevated microplastic abundance will lead to increased bioavailability and consequently increased the chances of animal encounter and ingestion. Plastic floating on water surfaces is a constant danger for fish and birds, as they ingest it mistakenly for food or get entangled in it. Among plastic waste, packaging proved most hazardous for marine lives. In an estimate, about 2249 aquatic lives got interacted with plastic waste which proved harmful for them. Marine animals do not ingest the plastic trash only but also consume toxic substances that are added in plastic as additives. According to IUCN endangered species Red List, out of 120 marine animals, 54 are acknowledged to be entrapped by plastic trash or have ingested plastic waste as food (Plastic Atlas, 2019).

5.2.6 IMPACT ON TERRESTRIAL ENVIRONMENT

The impact of plastic on terrestrial environment is relatively less studied and researched. But it is widely distributed in terrestrial environment as in the aquatic environment. Soil being the important component of terrestrial environment faced huge pressure due to plastic pollution. The abundance of microplastics in soil disturbs the soil–water relation, and as a result, affects the vital functions of soil, that is, water retention capacity, soil porosity, soil structure, and microbiota that play a key role in the maintenance of soil fertility (de Souza Machado et al., 2018). According to the scientists, the research conducted on the damages of plastic on terrestrial environment lagged about a decade behind than the research conducted on microplastic in oceans. But in point of fact, in soils, plastic pollution problem is about 4–23 times more than that in marine environment, because studies showed that out of 400 million tons/year of plastic waste, one-third of it became the part of terrestrial environment in one way or another (Plastic Atlas, 2019).

Globally, sewage sludge is widely used as fertilizer, which is one of the prime sources of microplastics deposition in soils especially agricultural fields, adding hundred thousand tons of microplastic. It is reported that in Portugal, 87% of the sewage sludge is applied to the agricultural lands either by composting or directly (Alvarenga et al., 2016). In Germany, application of sewage sludge in 3 years showed that 5 tons of microplastics/hectare were found in sampled soils which were picked by winds and detected in remote areas (Piehl et al., 2018). In Europe, about 0.43–0.63 Mt/yr a microplastic is piled through sewage sludge (Nizzetto et al., 2016).

5.2.7 IMPACT ON HUMAN HEALTH

Now the emerging concern is the entry of microplastic into the food chain of human via terrestrial and seafood ingestion. The plastic can be consumed by drinking and eating and may have detrimental effects on human health. In addition, microplastic pollution in tap water (83% of the samples showed microplastic contamination) (Kosuth et al., 2017) and bottled water (93% of studied samples showed contamination) has also been documented (Mason et al., 2018). According to University of Newcastle estimate, a person may consume about 5 g plastic/week equals to a credit card weight. According to a study in Canada, person who used bottled water may drink approximately 130,000 microplastic particles/year and person who used tap water may drink like 4000 microplastic down the throat. Though the numbers are worrisome, but still their possible impact on human health is little explored (Plastic Atlas, 2019).

The most used plastic polymer is PET (used in packaging, pipes, and plastic bottles etc.), which is considered as potential carcinogen (Li et al., 2016). In addition, another study revealed the reproductive abnormalities and cancer in rodents, invertebrates, and humans due toxic release of PS and PVC (another two commonly used plastic polymers) (Wang et al., 2016). PS also affected the cell morphology and cell viability of human gastric epithelial cells (Forte et al., 2016). Furthermore, numerous research findings explain the inhalation of micro/nanoplastics which cause adverse human health problems (Rist et al., 2018; Chan et al., 2017). There are different additives or array of chemicals added to the plastic for desirable properties (e.g., phthalates or bisphenol A), and these accompaniments in fact have potential to risk human health. Bisphenol A received considerable attention regarding human health as it interferes with the normal functioning of the liver, brain functioning, insulin resistance, reproductive system, and offspring development in the womb (Srivastava and Godara, 2017; Galloway, 2015). Likewise, phthalates caused birth defects and abnormal sexual development when exposed to humans (Cheng et al., 2013).

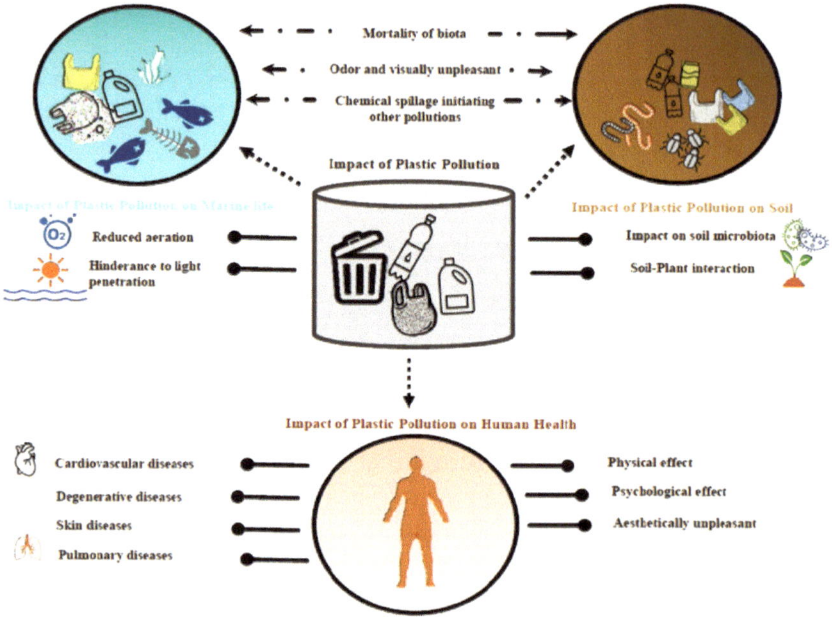

FIGURE 5.1 Impact of plastic pollution on ecosystem.

5.3 THE KEY DRIVERS AND MANAGEMENT TECHNIQUES ADOPTED FOR PLASTIC WASTE HANDLING IN DCs

5.3.1 KEY DRIVERS OF POOR WASTE MANAGEMENT IN DCs

Managing the wastes in any country is not an easy task. Managing the wastes is the issue which requires not only comprehensive scientific breakthrough and holistic approaches but also huge budgets and cooperation among numerous stakeholders (Bundhoo, 2018). The nerdiest challenge to globe is the management of plastic waste in special consideration to environmental pollution, social inclusion, and sustainable growth of economy. World produces over 400 million tons of plastics annually, much of which is "mismanaged" after use (Neufeld et al, 2016). Out of the 20 top-ranked plastic producer countries, 16 of them are upper middle or low-economic countries which lacked proper waste management infrastructure despite of their rapid economic growth (Jambeck et al., 2015).

Moreover, the recent pandemic of COVID-19 has had notable impacts on plastic use, waste management, and pricing. Global efforts to contain the pandemic have increased the circulation of single-use plastics for personal protection equipment (PPE). Restaurant takeaways and online grocery orders have promoted a rise in the use of disposable packaging (WEF, 2020). The key drive behind management strategies of plastic wastes is to educate and address the concerns accompanied with mismanaged plastic waste with respect to environment, landuse, economics, aesthetic, resource and health (Marshall and Farahbakhsh, 2013). The developing countries need rationalization and specific attention as unsustainable plastic waste management is the prime practice adopted by low-GDP countries. Low standards environmental legislations, poor economy, biased political, technical, and operational limitations are associated with DCs which augment the mismanagement of waste in these countries (Ferronato and Torretta, 2019).

The biggest challenge associated with DCs to effectively cope with the waste management issue is sustaining the economic growth. Attaining economic boom and sustainable growth intensively requires the reduction plans of ecological footprints, altering the "produce–consume–waste" systems of resources (SDGFG, 2016). Along the challenges of policy, there are severe challenges related to technical management of plastic waste (Carney Almroth and Eggert, 2019). Plastic is an intricate material, which makes its recycling even more difficult. Precisely, plastic is a complex material made up of polymers and supplemented with thousands of additives and chemical compounds to improve its characteristics, out of it many are known to be potentially toxic compounds. That is why these compounds could be very challenging in recycling as they alleviate the safety and quality of materials (Groh et al., 2018).

In low-income countries, governments and local authorities have failed to handle and manage the overwhelming waste produced by big cities, and therefore, limited resources resultantly perpetuate the inequality which is already being faced by the vulnerable population of the community (Marshall and

Farahbakhsh, 2013). Moreover, in less-developed countries, the prevailing unsustainable management practices of plastic waste worsened the situation by aggravating the environmental impacts and disease spread. These poorly managed waste practices and their implications on public health are rigorously ignored in most of the DCs. The combined effect of contagious and affluent diseases has put the poor countries in the "double burden." The behavior arrays and underlying attitudes of the people primarily determine the functioning and efficiency of waste management system, and these factors are founded by the indigenous social and cultural milieu (Schübeler, 1996). The impact of the public attitude and their responsiveness concerning waste management system is huge, starting from waste handling and grading/isolating, then intentions for volume reduction, recycling, development of collection services to inclination for investment on management programs (Yousif and Scott, 2007). Still in many parts of the world, the efficiency of waste management departments may be minimized or stopped as in many societies, the waste management is not perceived as a dignified job.

5.3.2 ONGOING PLASTIC WASTE MANAGEMENT PRACTICES IN DCs

Here in subsequent sections, we will discuss the existing practices adopted by low-income countries to handle the plastic waste.

a. Landfills/open dumping
b. Open burning
c. Incineration
d. Recycling

5.3.3 LANDFILLS/OPEN DUMPING/OPEN HEAPS

Landfilling or open dumping is the utmost common, old, easiest, unscientific, and nonengineered plastic waste management system. Management of used plastic materials in the DCs is still taking place through primary and traditional methods of municipal solid waste management system, that is, landfilling and dumping. Dumping of plastic waste has been always a priority technique for least developed nations over recycling due to low-budget practice. According to one estimate in 2010 alone, globally about 8 Mt plastic waste is being produced on land and via open landfills entered into the marine environment (Geyer et al., 2017). About 58–62% of plastic waste is dumped under landfills system globally (Geyer et al., 2017; OECD, 2018). Among all available management practices, open dumping covered 51% in Asia (Hoornweg and Bhada-Tata, 2012). In this system of disposal, the wastes are scantly disposed off in constructed dump sites, but the waste is disposed in an unacceptable way, usually the heaps of waste without any covering and safety guidelines.

Moreover, the environmental, health, and aesthetic consequences have never been the concern for municipalities and industries. In Asian DCs/cities, over and above 90% of the waste is straightly dumped in bare lands without any pretreatment (Srivastava et al., 2015; Narayana, 2009). Moreover, this open dumping does not mean that the dumps/waste is either compacted or covered by any material or mulches, such as soil, sheet, etc. It is left open for "waste scavengers" or scrappers belonging to poor societies. On the other hand, the DCs with comparatively stable GDP are shifting toward sanitary landfills, where waste is spread out in layers, then compressed and compacted to reduce the volume space and further different materials are added or soil is added to cover them up. They are simply engineered disposal sites to avoid and alleviate the pollution. Present-day sanitary landfill sites are equipped with leachate treatment and methane gas scavenging facilities (Garfi et al., 2009).

The major drawback associated with landfill disposal system is the wastage of abundant productive lands in peri-urban areas. Moreover, the aerosol (biochemicals) odors generated from these landfills are major concerns for public health. Additionally, the bad landscapes and aesthetic visuals of landfill in city environment do not represent a civilized society. Even if the landfills are constructed, yet they pose a threat of groundwater contaminations though leachates. The emission of greenhouse gases from bioplastic is also a prime concern for environmental sustainability beyond the borders. Overall, the losses to natural resources engendered through these open dumps and landfills are irrecoverable as the lands previously used as dumpsites are impossible to reclaim and remediate. Therefore, the landfills and open dumping disposal system of plastic management cannot be regarded as a sustainable and efficient system for any nation beyond the economy (Nkwachukwu et al., 2013).

One of the most hazardous problems regarding plastic waste landfills and dump sites is the spread of Dengue and other vector diseases originated from plastic wastes. Dengue is among the widespread diseases in tropical areas currently. Only in tropical regions, approximately 2.5 billion lives are at severe risk with millions of cases being reported each year. The outbreak of dengue flue in Asian and African countries was linked to plastic tires. It has been evaluated that the mosquito larvae which is dengue vector in many tropical and subtropical regions is sourced from used tires (Ferronato and Torretta, 2019; Yevtushenko and Tolstoj-Sienkiewicz, 2006). Although the governments in developing cities or countries are taking strict legislations to control the single-use plastic materials like polyethylene bags. Pakistan is at sixth position among world largest plastic waste producing countries, but it has banned the use of nonbiodegradable (polyethylene) bags for public usage at public stores in its swiftly growing metropolitan areas such as Islamabad, Lahore, and Faisalabad since August 14, 2020. A critical review presented by Xanthos and Walker (2017) has well summarized the policies and legislation on banning of single-use plastic globally.

5.3.4 SEMI-CONTROLLED OR OPERATED LANDFILLS

The constructed landfill sites or semi-landfill leachates disposal sites for waste management system usually operates like open landfill systems in DCs. To cut down the operational cost, the designed landfill systems are acting as a pollution source. In DCs, landfill system for plastic waste has never been designated. Usually these kinds of landfill systems receive wastes of multiple types without prior screening. The leachates collection systems for these landfill systems are inefficient, and further, the disposal of these leachates and recovery or treatment system for collected leachates hardly work in DCs. Anyhow, in some DCs, such waste disposal system contains a top covering which is usually a soil layer which somehow protects the environment from hazards.

5.3.5 OPEN BURNING

On the contrary to incineration technology, open burning of dumps and heaps of plastic is frequently practiced in DCs. The situation of plastic heaps or plastic waste as a constituent of municipal solid wastes dumping either temporary or making the permanent sites is more evident in poor section of municipalities and metropolitans, especially the slum areas possess these sites. The high-density population makes the waste management system more complicated. Therefore, the plastic heaps and dump sites in slum areas cause more pollution (Manaf et al., 2009). Open landfills and dumpsites usually turn into fire. Sometimes, these activities are led by the municipalities too, otherwise, there are several unknown reasons for igniting those dump sites. This open landfill system usually proceeds with open combustion. Combustion of plastic waste without any precaution and prior treatment or screening of waste generates abundance of xenobiotic causing significant health risk for humans (Tue et al., 2016).

The combustion of plastic wastes usually generated new toxic species as evident by the study of Minh et al. (2003) where the researchers found that open burning of plastic waste and municipal waste containing multiple type of plastic results in the generation of polychlorinated dibenzo-*P*-dioxin (PCDDs), polychlorinated dibenzofuran (PCDF), and polychlorinated biphenyls (PCBs) in dump facilities of Asia, that is, Cambodia, Philippines, India, and Vietnam. Especially, the smoke and odor generated from burning sites is the prime concern. In south Asian countries, such as India, Siri Lanka, Pakistan, Afghanistan, Nepal, the unrestricted open burning is adopted (Shakya et al., 2008). Moreover, this open burning of plastic is worsened in some politically unstable countries where waste tires are burnt during the political upsurges and rebellious outbreaks or protest.

5.3.6 INCINERATION

Incineration process refers to the thermal treatment of waste in a closed structure with emphasis on the reduction of bulk waste volume, yet the emission of pollutants remains a prime concern. The incineration implies the combustion process of unprocessed/and or screened waste under the controlled condition where combustion process operates at 850 °C or even higher temperature under open air access (DEFRA, 2007). Usually, the by-products produced during incineration depend on the type of waste being incinerated, but the common product of wastes are carbon mono and/or dioxides, sulfur dioxide, PM, PCDDS, PCDs, etc. and persistent materials. Heat is the ultimate output from incineration which now has attracted the DCs to focus on incineration process to develop as an alternate energy system. Though this system requires energy input, but it also yields significant amount of energy.

Incineration process is highly an exothermic process and heat losses are scanty. Incineration practice is though not environmentally sustainable, it reduces the mass transfer to aquatic environment. Many DCs are supporting this practice to turn the waste into energy. According to reports presented by Center for International Environmental Law (CIEL), Incineration of plastic waste has produced > 8.5 billion tons (Bt) of greenhouse gases during the year 2019 (Hamilton et al., 2019). About 8.3 Bt plastic and plastic containing light weight products have entered the anthroposphere. Moreover, half of the plastic produced so is the single-used plastic. Such kinds of products which are enjoyed for instant use, but these products last for thousands of years. Single-used plastic is abundant in DCs as compared with developed countries. About 4% of the total oil and gas production is consumed for incineration process worldwide. Till now, only 12% of the total plastic produced on earth has been incinerated which has significantly enhanced the carbon dioxide proportion into environment (Kistler and Muffett, 2019).

The incineration process in economically stable countries is much different from the incineration plants being operated in DCs. Incineration in DCs is much like open air combustion without compliance to safety guidelines and waste incineration directives set by government agencies. There are only few countries that are benefiting from this concept of "waste to energy." Therefore, it should be mandatory for incineration units to fully compliance with safety guidelines and work under the concept of "waste to energy" and do not produce secondary pollution into the atmosphere. Moreover, the constraint of disposal of slag or ash produced after incineration should be addressed through scientific and technical guidelines (Nkwachukwu et al., 2013).

5.3.7 RECYCLING

The plastic recycling is not only a wise management solution to pollution but also economically a marginal activity for DCs. Currently, the DCs with growing economic growth are investing on recycling option for wastes. The existing recycling rates for plastic waste are about 14–18% internationally. Even, the recycling rates of economically stable countries are also lower compared with other materials. The recycling rates are different across the geographical borders and depend on polymer types, infrastructure and technology. The nature of polymers used in the plastic material is the prime factor for controlling the recycling rates in any country.

Few polymers are more efficiently recycled, for example, the recycling rates for PET, HDPE, PP, PS, PVC are usually above 10%. On the contrary, the polymers, such as polystyrene (PS) and polypropylene (PP) are nonrecycled. The estimated recycling rates for different plastic materials in European Union is about 30% while recycling rates in rich countries are an average of about 10% while the data on recycling rates in DCs are usually not reported, yet the recycling management is growing in middle-income countries. The OECD claims that the current plastics recycling rates are about 20–40% in some cities in developing countries (OECD, 2018).

Contrary to high income countries where recycling is applicable in true sense, the recycling of plastic waste in developing and least developed nations refers to only mechanical recycling. In low-income countries, recycling is not being adopted in its true sense. In some lower middle-income countries, recycling is much more like a "mechanical recycling" or "reshaping of materials" and not the recycling where polymers are changed. Mechanical recycling refers to the use of plastic material to be converted into a new product, but the primary structure of polymers is still unchanged. Mechanical recycling usually involves the various processes, that is, grinding followed by washing, separation, drying, granulation, and finally the compounding. Sometimes, the virgin plastic is substituted with new materials. The molding and extrusion process are more widely adopted in DCs for recycling (Singh and Ruj, 2015).

Another significant technique of recycling in lower middle-income countries is carried out by the informal sector called "waste scavengers" or "waste pickers." This technique plays a prime role in plastic waste management system and provides livelihood to poor societies residing in the slum areas. Plastic scavenging is abundantly practiced in Bangladesh, Indonesia, India, Iran, Pakistan, and South African countries (Dhokhikah and Trihadiningrum, 2012). In some parts of the societies, this practice is being adopted as "art" where people reshape the materials into some art. The current estimates from the World Bank indicate that 2% of the whole population in DCs has adopted the profession of "plastic scavenging" (Garfi et al., 2009; SANDEC, 2008). Overall, recycling is the best way to combat the plastic pollution, but the major hurdle to recycling process is the presence of multiple polymers into a plastic material. Not all plastics are recyclable.

5.3.8 PLASTIC RECYCLING: TRENDS, CHALLENGES, ECONOMICAL CONCERNS, AND LIMITATIONS

The widespread use of plastic in accompaniment with poor management has created a massive plastic pollution problem. Till now, substantial amount of waste is being dumped in the landfills or natural environment and only 9% was being recycled (Rhodes, 2018). While in most studies, it is ascertained that mechanical recycling showed least environmental impacts compared with other plastic waste disposal techniques. A key strategy behind promoting recycling is to preserve natural resources and to keep the environment safe. It is an environment-friendly method, which not only reduces the waste generation which otherwise end up in landfill, but also decreases the quantity of raw material required for a new product.

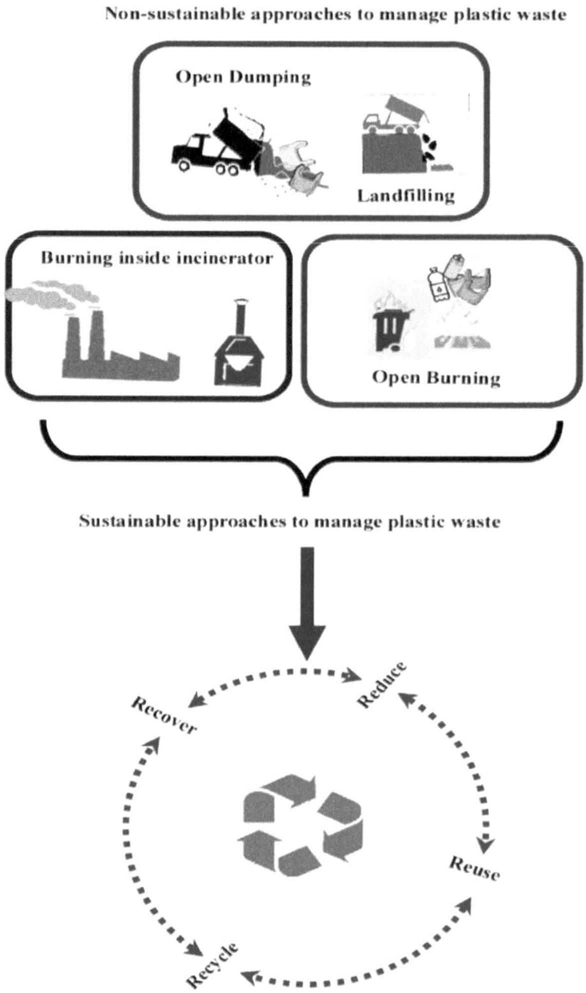

FIGURE 5.2 Sustainable and nonsustainable approaches to manage plastic waste.

5.4 CURRENT SITUATIONS IN PLASTIC RECYCLING

Plastic recycling is one of superior option for waste management as it costs limited threat to environment and encourages socioeconomic gains. Till today, only less than 10% of the plastic ever produced since 1950 is being recycled. Plastic recycling remained under-used and underdeveloped management technique, varying from country to country. Like in regulated developed countries, recycling rate is encouraged to be around 30%, and on the contrary, low-income countries with marginal industrial activities have 0% recycling rate (d'Ambrières, 2019). Unfortunately, in some developed countries as in the USA, despite the hype of recycling, only 1/10th of the plastic waste is recycled and the rest dumped into landfills (Plastic Atlas, 2019). In 2018, China being the biggest importer of plastic waste, China has set higher standards on imported plastics for recycling (0.5% contamination limits). This new limitation is impossible to achieve for most exporting countries, so, bulk of plastic waste (most of it is impossible to recycle) is anticipated to be illegally shifted to other Asian countries (Malaysia, India, Taiwan, Hong Kong, Vietnam, etc.) and African DCs since then (Plastic Atlas, 2019; Brooks et al., 2018).

At present, the European Union has set new targets for recycling industry, that is, packaging waste recycling target by 2025 is 65% and by 2030 is 70% (European Commission, 2016). And in DCs, the practice of recycling lags far behind compared with developed nations and portrayed grim reflection of plastic threat to the environment. In Indonesia, 70% of the plastic waste is poorly managed (via open dumping or open burning or discharge into ocean). Now, the government of Indonesia has launched the national action plan on alleviating plastic pollution which includes controlling plastic leakage in coastal areas by 70% and zero-plastic target by 2025 and by 2040, respectively (WEF-NPAPs, 2020). In India, plastic waste constitutes 10% of the waste, out of which 94% is recyclable having thermoplastic content (Central Pollution Control Board, 2015). In African countries, recycling is at the stage of infancy and has not been developed properly. In Nigeria, out of total uncontrolled plastic waste, only <12% is recycled (Babayemi et al., 2018). However, in some DCs, like in Bangladesh, plastic industry is booming with tax incentives to plastic recyclers and up to 69% of the plastic waste is recycled annually and exported as recycled plastic to other countries (Nasrin, 2019).

In DCs, primarily, the system of waste collection is in deplorable state and most of the waste materials continue to get dumped in unauthorized or unregulated sites. Informal recycling is likely to be organized and the informal sectors collect the waste according to local market demand for the materials. And the rest mismanaged plastic waste eventually enters into the marine environment as mentioned by Jambeck et al. (2015). The top 20 countries are classified on the basis of mismanaged plastic dumping into marine ecosystem.

5.4.1 ROLE OF PEOPLE AND THEIR SUPPORT

Role of people and their support could be very productive in avoiding plastic waste accumulation in the environment. Firstly, by ensuring the immediate reduction in plastic materials being used and then by reducing the possible waste that enter the environment by following adequate disposal method. It has been observed that plastic litter increased with increase in the use of disposable products and packaging. However, avoiding superfluous packaging and shifting to eco-friendly techniques is still possible. It is possible to retrieve this trend via improving recycling capacity, better education, awareness, support, and engagement (Thompson et al., 2009).

Education and awareness could prove a very effective tool in fight against plastic pollution. These campaigns are pressing needs to control plastic pollution especially in urban settings where consumption of disposable products increased immensely. There is proper evidence in Brazil where low literacy citizens responded well to these campaigns and accordingly rejected the use of microbeads products (Chang, 2015). The media and social networks have an important role in apprehending the sensitivity of issue by providing appropriate information. These educational activities could be proved as a long-term behavioral change strategy (Prata et al., 2019; Khan et al., 2019).

Both formal and informal education regarding environmental issues could be proved as a potential strategy to change consumers choices toward products. For instance, the demand of microbeads- free goods or low packaging plastic products would compel the companies to redesign their goods accordingly (Thompson et al., 2009; Ambrose et al., 2019). In DCs, participation of citizens in waste management is minimal. But it is possible to change the attitude of people via motivating and educating them as done by Jamaica government. The Jamaica government appointed environmental wardens from community members for the sole purpose of sensitizing their locality about community health and cleanliness (Monteiro and Kaza, 2016). There is need of educating people about waste disposal based on their types and to sort them accordingly. It is time to expand the domain and must include the education and awareness of people within the management program so that people have better understanding to manage waste and play their part to curb it.

5.4.2 ECONOMIC CONCERNS

The prime economic concern is the competition of recycled plastic with virgin plastic. With the high local market demand of virgin plastic and extra cost of recycling in developing nation, the recycled plastic could not get much attention. Lack of information/data regarding recycled/sustained plastic and social cost value of natural environment also act as an impediment in recycled plastic usage. In addition, finance is another issue regarding recycling. One way is to incorporate informal recyclers and scavengers in the management system by providing them decent income and secondly providing tax exemption or incentives to recycling plants in purchasing needed machinery to expand the recycling market (Babayemi et al., 2019).

In addition, plastic pollution along shorelines lessens the aesthetic value of tourist sites and consequently, the revenue of tourism got affected. As in Vietnam, the marine environment pollution due to excessive discharge of plastic waste has reached an alarming level (Danh and Hoi, 2019). Recycling of the single-use plastic waste (with energy recovery) would avoid the economic loss of about \$80–120 billion annually (Ellen MacArthur Foundation, 2016) and alleviate the environment cost by \$7.3 billion annually. Hence, to achieve the circular economy, plastic manufacturing companies can play a very important role in accompanying recyclers in optimizing the recovery processes of plastic, that is, improving packaging types and standardization for better sorting and separation of postconsumer materials. Moreover, the DCs direly need circular economy regarding the plastic waste management. Even the concept of "recycling" is not enough for sustainable and environment-friendly approach according to numerous economic and technology experts. The "recycling" also generates the waste somehow, the concept of "re-use and re-use" should be adopted to improve the economic sustainability which is termed as circular economy, which means no waste generation at all. Therefore, the countries with lower middle income must focus their policies and economy as well as technologies toward zero-waste generation by "re-use" of plastic-based good. There are some encouraging policies in few DCs where initiative and legislations has been in practice on ban of single use of polyethylene especially the shopping bags.

5.4.3 LIMITATIONS AND POSSIBLE ADVANCES

Recycling delays the disposal of plastic waste instead of avoiding it, unless cuts the rate of production of virgin plastic. In most of the developed countries (US and European countries), postuser plastic waste (46% in Europe) gathered for recycling is used to send to China and other Asian countries (mostly DCs) (Wilson et al., 2015). There, it competes with cheap and ample available virgin plastic. In addition, it is a well-known fact that in most of the DCs, there is no proper infrastructure for waste management and most of the plastic waste is mismanaged. Keeping the challenges in mind, there is no wonder why substantial amount of plastic waste is dumped up in natural environment (Geyer, 2020).

Moreover, recycling of plastic waste into secondary material has its own impact on natural environment. Though, these impacts are low in comparison to producing with primary resources, but still, they are not insignificant. Recycling is implausible to be a panacea, lest it alleviates the primary material production which so far proved to be an empty promise. As reported by Bishop et al. (2020), about 1–7% of the exported plastic waste is dumped into the oceans, indicating that the transferred plastic could provide potential route of plastic waste ending up into oceans, a hitherto unaccounted channel yet.

Mixed plastic waste recycling is another important challenge for recycling sector. Product design could play a very imperative role in assisting effective recycling of mixed products. As reported that mixed plastic cannot be recycled effectively due to impurities and cross-contamination present in it. Development of wider policies is needed for promoting environmental product designs to facilitate recycling activity. In addition, most collecting schemes available are for rigid packaging because flexible packagings are difficult to handle, collect, and sort. For effective mixed plastic waste recycling, there is a need of high efficiency sorting within recycle plants which make assure that the plastic types are organized and grated with high level of purity. Still, the requisite of each polymer recycle stream end market is required for further development (Hopewell et al., 2009).

It is time to reconsider and innovate the packaging designs of products especially soft drinks and food sector which covered about 32 Mt of plastic/ annually in packaging (e.g., prefer flexible packaging over rigid packaging) (Smithers PIRA, 2013). Countries should either reduce the amount of plastic being used or increase the recycling rate of postconsumer materials. There is a need of better legislatures and frameworks to set higher and better standards, especially in Asia, in improving waste collection system and then recycle the collected waste. New approaches have been introduced to replace plastic derived from fossil hydrocarbons to biodegradable plastic (plastic decomposed by microorganisms). Thus, the dumped plastic waste could decompose itself rather than accumulate in landfill or natural environment (Barnes et al., 2009). Currently, biodegradable plastic has only 4 million metric tons capacity for global plastic production according to European Bioplastics (2017). For future benefits, expansion in the use of biodegradable plastic will be suggested and emphasized against petroleum-derived counterparts. In addition, there is a need of an integrated, cost-effective, and holistic approach with the existing available resources. Intrusion of "3R" approach in waste management policy will help in sustaining environment and reserving natural resources for improving human well-being.

5.5 SUMMARY

In a nutshell, plastic played significant role in our lives and offer great well-being for future but our approaches toward its production, usage, and disposal methods are ineffectual. Unfortunately, with our ineffectiveness and mismanagement of plastic, the problem of plastic pollution is posing risk to organisms as well as to environment. Plastic pollution has many significant impacts on living entities both of land (including humans) and marine environment. As waste releases different poisonous chemicals. Definitely, there are solutions available which could only be achieved via combined actions. Generally, landfilling and open burning are the most used and conventional plastic management practices in DCs. Globally, from the total collected plastic waste, substantial amount is dumped in the landfills (79%), and the remaining quantity is incinerated (12%) and recycled (9%). It is high time that we should shift toward eco-friendly alternative approaches, which mitigate the environmental threats toward plastic pollution. The three Rs should be highly advocated, which could be used in combination with two other Rs that are energy recovery and redesign. These techniques should be used in combination and the waste produced later should only be disposed to landfill as landfill is proved to be a storing problem for future with its contaminant leaching problem. In carrying out these measures, everyone from individual (by choosing appropriate usage and disposal method) to industries (by redesigning and adopting end-of-life recyclability) and government (by setting policies and standards) should play a role in alleviating plastic pollution.

5.6 FUTURE PROSPECTS

Waste in any part of the globe is unstoppable, undesirable, and burden to economy unless the waste management strategies are not a part of robust and systematic circular economy. The pressing demand for current policies of circular economy is focused on the production of regenerative, reusable, and restorative nature. Moreover, there is an urgent need to focus on environmental cost on developed countries during trade, circulation, and reutilization in lower -middle-income countries. As the current investigations prove that the most alarming situation is the per capita generation which is many times higher in DCs as compared with rich countries. This situation is worse in middle-high-income countries of Asia. Therefore, the DCs need their policies and legislation to divert on reduction in per capita plastic waste generation to reduce the burden on management programs. The reinforcement of already existing legislations along with the holistic approaches of sustainable and integrated plastic waste management is necessary. Strategic cooperation and aids from developed world to DCs is also entailed for environmental and technological barriers to research and development on waste management programs in DCs. The existing innovations in technology, that is, the biodegradable plastic, etc. must be recognized as a practicable solution before these products gain complete access to market.

KEYWORDS

- **plastic pollution**
- **waste management**
- **environment**
- **landfill**
- **open burning**
- **recycling**

REFERENCES

Alvarenga, P.; Farto, M.; Mourinha, C.; Palma, P. Beneficial Use of Dewatered and Composted Sewage Sludge as Soil Amendments: Behaviour of Metals in Soils and Their Uptake by Plants. *Waste Biomass Valorization* **2016,** *7* (5), 1189–1201.

Ambrose, K. K.; Box, C.; Boxall, J.; Brooks, A.; Erikse, M.; Fabres, J.; Fylakis, G.; Walker, T. R. Spatial Trends and Drivers of Marine Debris Accumulation on Shorelines in South Eleuthera, The Bahamas Using Citizen Science. *Marine Pollution Bull.* **2019,** *142*, 145–154.

Babayemi, J. O.; Nnorom, I. C.; Osibanjo, O.; Weber, R. Ensuring Sustainability in Plastics Use in Africa: Consumption, Waste Generation, and Projections. *Environ. Sci. Eur.* **2019,** *31* (1), 60.

Babayemi, J. O.; Ogundiran, M. B.; Weber, R.; Osibanjo, O. Initial Inventory of Plastics Imports in Nigeria as a Basis for More Sustainable Management Policies. *J. Health Pollution* **2018,** *8* (18), 180601.

Barnes, D. K.; Galgani, F.; Thompson, R. C.; Barlaz, M. Accumulation and Fragmentation of Plastic Debris in Global Environments. *Phil. Trans. R. Soc. B* **2009,** *364* (1526), 1985–1998.

Barnes, D. K.; Walters, A.; Gonçalves, L. Macroplastics at Sea around Antarctica. *Marine Environ. Res.* **2010,** *70* (2), 250–252.

Bishop, G.; Styles, D.; Lens, P. N. Recycling of European Plastic Is a Pathway for Plastic Debris in the Ocean. *Environ. Int.* **2020,** *142*, 105893.

Brinton Jr, W. F. Characterization of Man-Made Foreign Matter and Its Presence in Multiple Size Fractions from Mixed Waste Composting. *Compost Sci. Utilization* **2005,** *13* (4), 274–280.

Brooks, A. L.; Wang, S.; Jambeck, J. R. The Chinese Import Ban and Its Impact on Global Plastic Waste Trade. *Sci. Adv.* **2018,** *4* (6), eaat0131.

Bundhoo, Z. M. Solid Waste Management in Least Developed Countries: Current Status and Challenges Faced. *J. Mater. Cycl. Waste Manage.* **2018,** *20* (3), 1867–1877.

Carney Almroth, B.; Eggert, H. Marine Plastic Pollution: Sources, Impacts, and Policy Issues. *Rev. Environ. Econ. Policy* **2019,** *13* (2), 317–326.

Carpenter, E. J.; Smith, K. L. Plastics on the Sargasso Sea Surface. *Science* **1972,** *175*, 1240–1241.

Central Pollution Control Board (CPCB), Assessment & Quantification of Plastics Waste Generation in Major Cities. Ministry of Environment Forest and Climate Change (MoEFCC), 2015.

Chan, F.; Rajaram, N.; House, R.; Kudla, I.; Lipszyc, J.; Tarlo, S. M. Potential Respiratory Effects from 3-D Printing. In *B58. Occupational Lung Disease: Case Studies, Epidemiology, and Mechanisms*; American Thoracic Society: New York, 2017; pp A3861–A3861.

Chang, M. Reducing Microplastics from Facial Exfoliating Cleansers in Wastewater through Treatment versus Consumer Product Decisions. *Marine Pollution Bull.* **2015,** *101* (1), 330–333.

Cheng, Z.; Nie, X. P.; Wang, H. S.; Wong, M. H. Risk Assessments of Human Exposure to Bioaccessible Phthalate Esters through Market Fish Consumption. *Environ. Int.* **2013,** *57*, 75–80.

Cheshire, A. C. et al. UNEP/IOC Guidelines on Survey and Monitoring of Marine Litter. UNEP Regional Seas Reports and Studies, No. 186; IOC Technical Series No. 83, 2009.

d'Ambrières, W. Plastics Recycling Worldwide: Current Overview and Desirable Changes. Field Actions Science Reports. *J. Field Actions* 2019 (Special Issue 19), 12–21.

Danh, N. T.; Hoi, H. T. Effects of Plastic Waste to Sea Environment in Vietnam. *IOP Conf. Ser.: Earth Environ. Sci.* **2019,** *351* (1), 012023).

de Souza Machado, A. A.; Lau, C. W.; Till, J.; Kloas, W.; Lehmann, A.; Becker, R.; Rillig, M. C. Impacts of Microplastics on the Soil Biophysical Environment. *Environ. Sci. Technol.* **2018,** *52* (17), 9656–9665.

DEFRA (Department for Environment, Food and Rural Affairs)-Incineration of Municipal Solid Waste, Waste Management Technology Brief, the New Technologies Work Steam of the Defra Waste Implementation Programme, 2007.

Dhokhikah, Y.; Trihadiningrum, Y. Solid Waste Management in Asian Developing Countries: Challenges and Opportunities. *J. Appl. Environ. Biol. Sci.* **2012,** *2* (7), 329–335.

Ellen MacArthur Foundation. The New Plastics Economy: Rethinking the Future of Plastics, 2016.

European Bioplastics, Bioplastics—Facts and Figures (European Bioplastics), 2017.

European Commission. Circular Economy Package, 2016; p 12.

Ferronato, N.; Torretta, V. Waste Mismanagement in Developing Countries: A Review of Global Issues. *Int. J. Environ. Res. Public Health* **2019,** *16* (6), 1060.

Forte, M.; Iachetta, G.; Tussellino, M.; Carotenuto, R.; Prisco, M.; De Falco, M.; Laforgia, V.; Valiante, S. Polystyrene Nanoparticles Internalization in Human Gastric Adenocarcinoma Cells. *Toxicol. Vitro* **2016,** *31*, 126–136.

Galloway, T. S. Micro-and Nano-Plastics and Human Health. In *Marine Anthropogenic Litter*; Springer: Berlin, 2015; pp 343–366.

Garfi, M.; Tondelli, S.; Bonoli, A. Multi-Criteria Decision Analysis for Waste Management in Saharawi Refugee Camps. *Waste Manage.*, **2009,** *29* (10), 2729–2739.

Geyer, R. Earth and Plastic. An Insider's Guide to a Rapidly Changing Planet, 2020; pp 214–220.

Geyer, R.; Jambeck, J. R.; Law, K. L. Production, Use, and Fate of All Plastics Ever Made. *Sci. Adv.* **2017,** *3* (7), e1700782.

Geyer, R.; Kuczenski, B.; Zink, T.; Henderson, A. Common Misconceptions about Recycling. *J. Ind. Ecol.* **2016,** *20* (5), 1010–1017.

Gregory, M. R. Accumulation and Distribution of Virgin Plastic Granules on New Zealand Beaches. *NZ J. Marine Freshwater Res.* **1978,** *12* (4), 399–414.

Gregory, M. R. Environmental Implications of Plastic Debris in Marine Settings—Entanglement, Ingestion, Smothering, Hangers-on, Hitch-Hiking and Alien Invasions. *Phil. Trans. R. Soc. B* **2009,** *364*, 2013–2025.

Groh, K. J.; Backhaus, T.; Carney Almroth, B.; Geueke, B.; Inostroza, P. A.; Lennquist, A.; Maffini, M.; Leslie, H. A.; Slunge, D.; Trasande, L.; Warhurst, M. Chemicals Associated with Plastic Packaging: Inventory and Hazards. *Peer J Preprints* **2018,** *6* (1), 7036.

Hamilton, L.A.; Feit, S.; Muffett, C.; Kelso, M.; Rubright, S. M.; Bernhardt, C.; Schaeffer, E.; Moon, D.; Morris, J.; Labbé-Bellas, R. Plastic & Climate: The Hidden Costs of a Plastic Planet. In *Center for International Environmental Law (CIEL)*; 2019.

Harper, P. C.; Fowler, J. A. Plastic Pellets in New Zealand Storm-Killed Prions (Pachyptila spp.). *Notornis*, **1987,** *34*, 65–70.

Hoornweg, D.; Bhada-Tata, P.; What a Waste: A Global Review of Solid Waste Management. Urban Development Series; Knowledge Papers No. 15; World Bank: Washington, DC, 2012.

Hopewell, J.; Dvorak, R.; Kosior, E. Plastics Recycling: Challenges and Opportunities. *Phil. Trans. R. Soc. B* **2009,** *364*, 2115–2126.

Jambeck, J. R.; Geyer, R.; Wilcox, C.; Siegler, T. R.; Perryman, M.; Andrady, A.; Narayan, R.; Law, K. L. Plastic Waste Inputs from Land Into the Ocean. *Science* **2015,** *347* (6223), 768–771.

Kaza, S.; Yao, L.; Bhada-Tata, P.; Van Woerden, F. What a Waste 2.0: A Global Snapshot of Solid Waste Management to 2050; The World Bank, 2018.

Khan, F.; Ahmed, W.; Najmi, A. Understanding Consumers' Behavior Intentions towards Dealing with the Plastic Waste: Perspective of a Developing Country. *Resour. Conserv. Recycl.* **2019,** *142*, 49–58.

Kistler, A.; Muffett, C. Plastic & Climate: The Hidden Costs of a Plastic Planet. Center for International Environmental Law, 2019. www.ciel.org/plasticandclimate.

Kosuth, M.; Wattenberg, E. V.; Mason, S. A.; Tyree, C.; Morrison, D. *Synthetic Polymer Contamination in Global Drinking Water*; Orb Media, 2017.

Lebreton, L.; Andrady, A. Future Scenarios of Global Plastic Waste Generation and Disposal. *Palgrave Commun.*, **2019,** *5* (1), 1–11.

Lechner, A.; Keckeis, H.; Lumesberger-Loisl, F.; Zens, B.; Krusch, R.; Tritthart, M.; Glas, M.; Schludermann, E.; The Danube So Colourful: A Potpourri of Plastic Litter Outnumbers Fish Larvae in Europe's Second Largest River. *Environ. Pollution* **2014,** *188*, 177–181.

Li, W. C.; Tse, H. F.; Fok, L. Plastic Waste in the Marine Environment: A Review of Sources, Occurrence and Effects. *Sci. Total Environ.* **2016,** *566*, 333–349.

Lord, R. Plastics and Sustainability: A Valuation of Environmental Benefits, Costs and Opportunities for Continuous Improvement, 2016.

Manaf, L.A.; Samah, M. A. A.; Zukki, N. I. M. Municipal Solid Waste Management in Malaysia: Practices and Challenges. *Waste Manage.* **2009,** *29* (11), 2902–2906.

Marshall, R. E.; Farahbakhsh, K. Systems Approaches to Integrated Solid Waste Management in Developing Countries. *Waste Manage.* **2013,** *33* (4), 988–1003.

Mason, S. A.; Welch, V. G.; Neratko, J. Synthetic Polymer Contamination in Bottled Water. *Front. Chem.*, **2018,** *6*, 407.

Minh, N. H.; Minh, T. B.; Watanabe, M.; Kunisue, T.; Monirith, I.; Tanabe, S.; Sakai, S.; Subramanian, A.; Sasikumar, K.; Viet, P. H.; Tuyen, B.C. Open Dumping Site in Asian Developing Countries: A Potential Source of Polychlorinated Dibenzo-p-dioxins and Polychlorinated Dibenzofurans. *Environ. Sci. Technol.* **2003,** *37* (8), 1493–1502.

Monteiro, E.; Kaza, S. Jamaica-Jamaica Integrated Community Development Project: P146460-Implementation Status Results Report: Sequence 03 (No. ISR20978, pp. 1–0). The World Bank, 2016.

Mudgal, S.; Lyons, L.; Bain, J.; Dias, D.; Faninger, T.; Johansson, L.; Dolley, P.; Shields, L.; Bowyer, C. Plastic Waste in the Environment–Revised Final Report for European Commission DG Environment. *Bio In-tell Serv*. Downloadable Http www Ec Eur Euenvironment waste studies pdf plastics Pdf, **2011,** *11*, 2047–2062.

Narayana, T. Municipal Solid Waste Management in India: From Waste Disposal to Recovery of Resources? *Waste Manage.*, **2009,** *29* (3), 1163–1166.

Nasrin, S. Plastic Recycling in Bangladesh. *Adv. Environ. Waste Manage. Recycl.* **2019,** *2* (1), 1–4.

Neufeld, L.; Stassen, F.; Sheppard, R.; Gilman, T. The New Plastics Economy: Rethinking the Future of Plastics. In *World Economic Forum*, Jan **2016,** *7*.

Nizzetto, L.; Futter, M.; Langaas, S. Are Agricultural Soils Dumps for Microplastics of Urban Origin? In *ACS Publications/ Environmental Science and Technology*, 2016.

Nkwachukwu, O. I.; Chima, C. H.; Ikenna, A. O.; Albert, L. Focus on Potential Environmental Issues on Plastic World Towards a Sustainable Plastic Recycling in Developing Countries. *Int. J. Ind. Chem.* **2013,** *4* (1), 34.

Obbard, R. W.; Sadri, S.; Wong, Y. Q.; Khitun, A. A.; Baker, I.; Thompson, R. C. Global Warming Releases Microplastic Legacy Frozen in Arctic Sea Ice. *Earth's Future* **2014,** *2* (6), 315–320.

Ocean Conservancy. Stemming the Tide: Land-Based Strategies for a Plastic- Free Ocean. Ocean Conservancy and McKinsey Center for Business and Environment, 2015. http://www.oceanconservancy.org/our-work/marine-debris/mckinsey-report-files/full-report-stemming-the.pdf

OECD. Improving Plastics Management: Trends, Policy Responses, and the Role of International Co-operation and Trade. OECD Environment Policy Papers, No. 12, OECD Publishing: Paris, 2018.

OSPAR Commission, OSPAR Pilot Project on Monitoring Marine Beach Litter. Monitoring of Marine Litter in the OSPAR Region; OSPAR Commission, OSPAR Publication: London, 2007; p 386.

Piehl, S.; Leibner, A.; Löder, M. G.; Dris, R.; Bogner, C.; Laforsch, C. Identification and Quantification of Macro-and Microplastics on an Agricultural Farmland. *Sci. Rep.* **2018,** *8* (1), 1–9.

Plastic Atlas. Facts and Figures about the World of Synthetic Polymers, 2019.

Prata, J. C.; Silva, A. L. P.; Da Costa, J. P.; Mouneyrac, C.; Walker, T. R.; Duarte, A. C.; Rocha-Santos, T. Solutions and Integrated Strategies for the Control and Mitigation of Plastic and Microplastic Pollution. *Int. J. Environ. Res. Public Health* **2019,** *16* (13), 2411.

Rhodes, C. J. Plastic Pollution and Potential Solutions. *Sci. Progress* **2018,** *101* (3), 207–260.

Rist, S.; Almroth, B. C.; Hartmann, N. B.; Karlsson, T. M. A Critical Perspective on Early Communications Concerning Human Health Aspects of Microplastics. *Sci. Total Environ.* **2018,** *626*, 720–726.

Ryan, P. G.; Moore, C. J.; van Franeker, J. A.; Moloney, C.L.. Monitoring the Abundance of Plastic Debris in the Marine Environment. *Phil. Trans. R. Soc. B* **2009,** *364*, 1999–2012.

SANDEC (Department of Water and Sanitation in Developing Countries). Global Waste Challenges. Situation in Developing Countries. Eawag: Swiss Federal Institute of Aquatic Science and Technology, Sweden, 2008.

Schübeler, P. Conceptual Framework for Municipal Solid Waste Management in Low-Income Countries. In *Collaborative Programme on Municipal Solid Waste Management in Low-Income Countries*; Wehrle, K., Christen, J., Eds., Vol. 9; St. Gallen Switzerland: UNDP/UNCHS/World Bank/SDC, 1996.

Science for Environment Policy. In Depth Reports. Plastic Waste; Ecological and Human Health Impacts, 2011.

SDGFG (Sustainable Development Goals Fund). *Goal 12: Responsible Consumption and Production*; United Nations Dev. Program; SDG Fund: New York, 2016.

Shakya, P. R.; Shrestha, P.; Tamrakar, C. S.; Bhattarai, P. K. Studies on Potential Emission of Hazardous Gases Due to Uncontrolled Open-Air Burning of Waste Vehicle Tyres and Their Possible Impacts on the Environment. *Atmosph. Environ.*, **2008,** *42* (26), 6555–6559.

Singh, R.K.; Ruj, B. Plastic Waste Management and Disposal Techniques-Indian Scenario. *Int. J. Plastics Technol.* **2015,** *19* (2), 211–226.

Smithers PIRA. Briefing: The Shift from Rigid to Flexible Packaging, 2013.

Srivastava, R. K.; Godara, S. Use of Polycarbonate Plastic Products and Human Health. *Int. J. Basic Clin. Pharmacol.* **2017,** *2* (1), 12–17.

Srivastava, V.; Ismail, S. A.; Singh, P.; Singh, R. P. Urban Solid Waste Management in the Developing World with Emphasis on India: Challenges and Opportunities. *Rev. Environ. Sci. Bio/Technol.* **2015,** *14* (2), 317–337.

Sruthy, S.; Ramasamy, E. V. Microplastic pollution in Vembanad Lake, Kerala, India: The First Report of Microplastics in lake and Estuarine Sediments in India. *Environ. Pollution* **2017,** *222*, 315–322.

Thompson, R. C.; Moore, C. J.; Vom Saal, F. S.; Swan, S.H. Plastics, the Environment and Human Health: Current Consensus and Future Trends. *Phil. Trans. R. Soc. B* **2009,** *364* (1526), 2153–2166.

Thompson, R.; Moore, C.; Andrady, A.; Gregory, M.; Takada, H.; Weisberg, S. New Directions in Plastic Debris. *Science* **2005,** *310* (5751), 1117–1118.

Tue, N. M.; Goto, A.; Takahashi, S.; Itai, T.; Asante, K. A.; Kunisue, T.; Tanabe, S. Release of Chlorinated, Brominated and Mixed Halogenated Dioxin-Related Compounds to Soils from Open Burning of E-waste in Agbogbloshie (Accra, Ghana). *J. Hazard. Mater.* **2016,** *302*, 151–157.

UN environment, World Environment Day Overview (WEF), 2018.

Wang, J.; Tan, Z.; Peng, J.; Qiu, Q.; Li, M. The Behaviors of Microplastics in the Marine Environment. *Marine Environ. Res.* **2016,** *113*, 7–17.

Wang, W.; Ndungu, A. W.; Li, Z.; Wang, J. Microplastics Pollution in Inland Freshwaters of China: A Case Study in Urban Surface Waters of Wuhan, China. *Sci. Total Environ.*, **2017,** *575*, 1369–1374.

Waste Atlas. D-waste. Waste Atlas, 2016. http://www.atlas.d-waste.com/.

WEF. White Paper; Plastics, the Circular Economy and Global Trade, 2020. https://www.weforum.org/whitepapers/plastics-the-circular-economy-and-global-trade

Wilson, D. C.; Rodic, L.; Modak, P.; Soos, R.; Carpintero, A.; Velis, K.; Iyer, M.; Simonett, O. Global Waste Management Outlook, UNEP, 2015.

World Economic Forum National Plastic Action Partnership, Radically Reducing Plastic Pollution in Indonesia: A Multistakeholder Action Plan, April 2020.

Xanthos, D.; Walker, T. R. International Policies to Reduce Plastic Marine Pollution from Single-Use Plastics (Plastic Bags and Microbeads): A Review. *Marine Pollution Bull.* **2017,** *118* (1–2), 17–26.

Xu, S. Y.; Zhang, H.; He, P. J.; Shao, L. M. Leaching Behaviour of Bisphenol A from Municipal Solid Waste Under Landfill Environment. *Environ. Technol.* **2011,** *32* (11), 1269–1277.

Yamashita, R.; Tanimura, A. Floating Plastic in the Kuroshio Current Area, Western North Pacific Ocean. *Marine Pollution Bull.* **2007,** *54* (4), 485–488.

Yevtushenko, A.; Tolstoj-Sienkiewicz, J. Distribution of Friction Heat between a Stationary Pin and Rotating Ring. *Arch. Appl. Mech.* **2006,** *76* (1–2), 33–47.

Yousif, D. F.; Scott, S. Governing solid waste management in Mazatenango, Guatemala: Problems and Prospects. *Int. Dev. Plan. Rev.* **2007,** *29* (4), 433–451.

Zalasiewicz, J.; Waters, C. N.; do Sul, J. A. I.; Corcoran, P. L.; Barnosky, A. D.; Cearreta, A.; Edgeworth, M.; Gałuszka, A.; Jeandel, C.; Leinfelder, R.; McNeill, J. R. The Geological Cycle of Plastics and Their Use as a Stratigraphic Indicator of the Anthropocene. *Anthropocene* **2016,** *13*, 4–17.

Zink, T.; Geyer, R.; Startz, R. Toward Estimating Displaced Primary Production from Recycling: A Case Study of US Aluminum. *J. Ind. Ecol.*, **2018,** *22* (2), 314–326.

Zubris, K. A. V.; Richards, B. K. Synthetic Fibers as an Indicator of Land Application of Sludge. *Environ. Pollut.* **2005,** *138*, 201–211.

CHAPTER 6

Industrial Waste, Types, Sources, Pollution Potential, and Country-Wise Comparisons

HAFIZ ABDUL KAREEM[1], SOBIA RIAZ[2], HALEEMA SADIA[3], and RIZWAN MEHMOOD[4]

[1]*College of Grassland Agricultural Sciences Oregon State University, OR 97331, USA*

[2]*Institute of Soil and Environmental Sciences, University of Agriculture Faisalabad 38040, Pakistan*

[3]*Department of Botany, University of Agriculture Faisalabad 38040, Pakistan*

[4]*Stored Grain Management Cell, Department of Entomology, University of Agriculture Faisalabad 38040, Pakistan*

**Corresponding author. E-mail: abdulkareem@nwafu.edu.cn*

ABSTRACT

Industrial waste can be defined as any leftover matter released from industrial activities either in gaseous, liquid, or solid form irrespective of a certain waste material which is classified as industrial or domestic waste without any developed and proper disposal systems. Industrial revolution is the renovation from agricultural economy to industrial economy after the last proto-industrialization stage from the mid-18th to early 19th century. Waste generation is increased at accelerated rates around the globe. In the year 2017, overall, 2.01 billion kg of waste material has been generated by world's cities with 0.74 kg per person per day. Waste generation rate is

Waste Problems and Management in Developing Countries. Umair Riaz, Shazia Iqbal, & Moazzam Jamil (Eds.)

anticipated to increase by 70%, that is, 3.40 billion tons of waste from 2016 to 2050 due to rapid growth of population and urbanization. Major pollution creating industries include tannery, pulp and paper, sugar, fertilizer, and textile. Because these industries produce heavy pollution which adversely affects the environment. Without proper treatment and unchecked release of industrial effluents into nearby water bodies, lakes, and natural streams, pollutes the said water sources. Significant quantity of industrial waste which includes metals, slag, ash, waste paint, and even radioactive material is produced by modern industrial actions. Different types of waste generated by industries include garbage from cafeteria, gravel and dirt, masonry and concrete, oil, chemicals, trash, scrap metals, solvents, trees and weed grass, lumber scrap, solid and liquid waste, chemical waste, and toxic industrial effluent. Waste material can cause serious health issues and may cause problems to our environment if it is not disposed of or not properly treated. Hence, we must follow standards and guidelines, it must be controlled only by specialized facilities and government authorized centers. Standards and guidelines are published by the Occupational Safety and Health Administration the Environmental Protection Agency for safe disposal and management of waste material is handled by specialists.

6.1 INTRODUCTION

Any unwanted materials that are attributed as unusable and are discarded after primary use without considering their secondary or tertiary use. Waste, as a term, is sometimes considered subjective as a particular material that is being discarded for one reason could not be waste for another use), whereas often considered objectively wrong, for instance, it is not suitable to classify the scrap as "waste" when we send leftover metallic waste for dumping in a landfill because they are still valuable and recyclable. Community waste, including domestic debris, chemical/harmful waste, and polluted water containing sewage (comprises feces and urine), surface runoff, and radioactive waste falls under the category of industrial waste. During industrial activities in mills and factories or mining operations, some of the materials are discarded after declaring as invaluable or waste, and this practice is being done from the time of industrialization. Industrial wastes are various types, such as metals, by-products, radioactive wastes, sandpaper, paper products, and other gaseous and liquid wastes like effluents. Rates of urbanization were intensified when the rural population started shifting from rural to urban areas, hence exerting pressure on the per capita available resources. This intense migration has caused mismanagement in waste management, and because of economic reasons, cities were not equipped to tackle waste at this level (Oteng-Ababio et al., 2013). Meantime, limited resource availability was another factor contributing to ineffective waste management policies and their implementation (Cointreau, 1982). In many African countries, the major contributors of wastes are households, industries, streets, parks, public areas, processing units, and manufacturing plants and institutions (Kaseva and Mbuligwe, 2005). The major problem occurs when the generated waste is directly disposed off without considering the human health and environmental impacts. One of the key contributors to improper waste management is the exponential growth in urban population due to intense migration from rural to urban areas, stressing resources (Yhdego, 1995). Accumulation of municipal waste is a major critical problem for the public health as well as for environmental issues in African and other developing countries (Achankeng, 2003; Adebilu and Okenkule, 1898). In many developing countries, one-third of the waste that is being produced is not collected (Zerbock and Candidate, 2003). This uncollected waste, when gets mixed with human feces and animal waste, is unsystematically dumped into open surface or in streets/drains and this dumping contributes to flooding, a site for insect and rodents breeding, and hence causes disease spread such as cholera. In many studies, it was concluded that the wastewater generation is directly related to the population size and several other activities of inhabitants. Small-scale industries, large processing units, personal business, import/export, domestic waste, agriculture, and other activities contribute differently to the wastewater generation. Hence, the exponential growth in population will also cause a direct increase in the wastewater generation. It is estimated that the rate of waste generation is 0.2 metric tons per capita once a year in Kampala only (Ngategize et al., 2001).

This chapter focuses on the industrial waste, specifically its sources, types, pollution potential, and the adapted waste management approaches globally. It also explains the figure of biological, physical, and chemical treatment methods depending on the source, nature, and composition of the waste.

6.2 TYPES OF INDUSTRIAL WASTE

Industrial waste can be divided into scarp lumber, gravel, scrap metal, solvents, masonry, oil, and concrete, chemical, plastic and might include leftover food materials of restaurants that can significantly impact the environment (Awuchi and Awuchi, 2019a, 2019b). Industrial waste might include gaseous, solid or liquid substances which could be hazardous, nonhazardous, or mirror entry. Hazardous waste covers the category of toxic, radioactive, reactive, ignitable, or corrosive elements while the elements classified under mirror entry category may vary according to multiple factors. These different kinds of industrial waste find their destination into sea by degrading the environment quality (Awuchi and Igwe, 2017). Industrial waste is usually mixed into the municipal waste, makes the precise assessment challenging. An estimation for the United States is as very high as 7.6 billion tons of industrial waste production annually (Oteng-Ababio et al., 2013).

6.3 SOURCES OF INDUSTRIAL WASTEWATER

6.3.1 FOOD INDUSTRY

Wastewater produced by agricultural and food processing plants has distinctive features. The public municipal wastewater handled by public is different from the wastewater generated from the treatment plants worldwide. Wastewater from food processing plants is nontoxic and decomposable, meanwhile having immense content of biological oxygen demand (BOD) and suspended solids (SS) (European Environment Agency, 2001). It is difficult to calculate the ingredients ratio in wastewater from both food and agriculture sources because of the alterations in pH and BOD in wastes from fruit, vegetable, and meat products as it depends upon on their time of harvesting, nature of postharvesting, storage methods, and food processing. Large volume of high-quality water is required during processing of food from raw materials. Cleaning up of vegetable alone creates a lot of water wastage with high loads of chemical residue and several allowed organic materials including surfactants. Conventional pollutants which can reduce oxygen value in water are mostly produced by dairy processing plants (SS, BOD) (U.S. Environmental Protection Agency, 2018). Slaughter house and meat processing units produce very strong organic wastes containing animal body fluids, such as blood, ocular fluid, and interstitial fluid. Pollutants generated from these sites include BOD, coliform bacteria, grease, SS, ammonia, oil, and organic nitrogen. Processing of foods from field to market, from market to kitchen, kitchen to consumer produces a lot of wastes which generally consists of plant organic materials and also contain other salt, coloring material, and flavorings ingredients (Awuchi and Igwe, 2017; Awuchi and Awuchi, 2019a). Sugar used in carbonated drinks and aspartame, sugar alcohols used in flavor beverages, etc. may wash off during manufacturing process and fetch up as part of the components of wastewater. Noteworthy amounts of oil and fats might be existing. Food raw material conveying, plant cleaning, bottling, product washing, and daily protocol activities in food processing industries cause a lot effluents discharge. Most of the food processing plants require on-site water treatment prior to effluent release on land or discharge to a sewer or a waterway system. Presence of high SS contents of organic materials in water wastage can increase the BOD and can cause plenty of troubles and can result in significant sewer surcharge fees. Rotating belt filtration (microscreening), sedimentation and wedgewire screening are generally the adapted strategies for the mitigation of suspended organic solids before discharge.

6.3.2 IRON AND STEEL INDUSTRIES

Iron is produced from its ore at a very high temperature involving strong reduction reactions in blast furnaces. The cooling waters used are certainly contaminated well with products specifically ammonia and cyanide. Coke is produced from coal at high temperature. Water is also used in the production of coke in the coking plants as a cooling agent as well as in the by-product's separation. The pollution in waste streams from industries includes the gasification products, such as cresols, benzene, ammonia, phenols, anthracene, naphthalene, cyanide, along with a variety of more complicated organic compounds which are collectively called polycyclic aromatic hydrocarbons (PAH) (Environmental Protection Agency, 2002). The alloys making, modification of steel or iron, and production of wire, sheet, or rods from it require hot and cold mechanical shifting stages in which portable water is consumed as lubricant and coolant. Water is polluted during these stages also and the contaminants include hydraulic oils, tallow, and the particulate solids particles. The final products of iron and steel before the onward sale must be free from rust. To remove the rust from its surface, *pickling* is carried out in strong mineral acid. This treatment prepares the iron for further chromium or tin plating or other surface treatment, such as painting or galvanization. Sulfuric acid (H_2SO_4) and hydrochloric acid (HCl) are the two acids commonly used for this treatment. Wastewaters include acidic rinse waters along with waste acids. Although several iron or steel production plants also have acid recovery plants (mostly those using HCl), and the mineral/inorganic acid is boiled away of iron salts. Hence, there remains large volume of highly acid ferrous chloride or ferrous sulfate to be disposed of. Many steel industry wastewaters are polluted with hydraulic oil, also known as soluble oil.

6.3.3 MINES AND QUARRIES

Slurries of rock fragments in water are the wastewater associated with mines and quarries. These wastewaters are generated when rainfall clean away the exposed surfaces and the haul roads as well as grading processes, and rock washing. Water volumes, particularly rainfall related to large sites, can be very high. Some advanced separation operations such as the washing of coal to isolate coal from the native rock using density gradients may produce wastewater polluted by fine particulate surfactants and hematite,. Some common pollutants are also oils and hydraulic oils. The minerals and rock particles in the indigenous rock formations eventually contaminate the water in the metal mines and ore recovery plants. Unwanted materials may reach the wastewater streams after desirable metals are being crushed and extracted. For metal mines, it often involves undesirable metallic elements, such as zinc and other materials like arsenic. High value metal extraction, for example, silver and gold can produce slimes containing very fine particles in which it is difficult to physically remove the contaminants. Moreover, the geologic formations that anchorage economically appreciated metals, such as copper, silver, and gold sometimes may comprise sulfide-type ores. The processing of these involves crushing rocks and forming very fine particles and extract the desired metals and hence the unwanted rock is stated as tailings. The tailings consist of a mixture of unwanted leftover metals and components of sulfide that subsequently forms sulfuric acid (H_2SO_4) when exposed to water and air, which naturally occurs when such waste is discarded.

6.3.4 BATTERY MANUFACTURING

Battery manufacturing includes a variety of assembling portable devices, such as power and electric tools or may also involve manufacturing devices for heavy duty machinery, such as cars, trucks, or other logistics. The contaminants produced at battery assembling units include lead, manganese, mercury, nickel, chromium, cobalt, copper, cadmium, cyanide, iron, oil and grease, silver, and zinc (U.S. Environmental Protection Agency, 2018), which must be treated (recycled).

6.3.5 ORGANIC CHEMICALS INDUSTRIES (ORGANIC COMPOUND REFINERIES)

The amount of polluants which is released through man-made organic substances extensively fluctuates from one plant to another. It depends upon the yield of manufactured products, such as insecticides, plastics, imitation fibers, gums, and bulk organic chemicals. These carbon-based substances might be C_6H_6, $C_{10}H_{8,}$ C_6H_5OH, $CHCl_{3,}$ C_2H_3Cl, and C_7H_8. Organic pollutants have been measured through biochemical oxygen demand as regulatory parameter understand the efficacy of a system to treat the leftover water in natural ways. The discharged metal discharged impurities could comprise Cu, Zn, Cr, Ni, and Pb.

6.3.6 ELECTRIC POWER PLANTS (POWER STATIONS)

The main source of generating surplus industrial wastewater has been coal-fired and fossil fuel power plants. Most of the generating stations produce surplus water with substantial quantities of metals, such as Cr, Pb, Cd, and Hg and also As, NO_3^-, NO_2^-, and Se. Leftover water streams contain the by-product of coal combustion and exhaust gas of sulfur and mercury. Wet scrubbers which control miasma usually transfer the grabbed pollutants to the unwanted watercourse (U.S. Environmental Protection Agency, 2018).

An extensively used management technique called coal ash basin has been used at coal thermal power stations. Ash ponds, also called coal ash basins or in other words, surface impoundment, are extensively used at the coal-fired processing units. This technology engages gravity property to settle massive suspended particulates from power stations wastewater. This treatment is not applicable to the dissolved chemicals. Hence different techniques have been used by power plants to control the pollutants according to the specific trash. These technologies consist of dry ash handling, chemical precipitation, biological treatment, evaporation crystallization system, and closed loop ash recycling (U.S. Environmental Protection Agency, 2019). Advanced techniques have assisted us with extreme efficacy in the treatment of exhausted gas of sulfur removed wastewater in electrodialysis and ion exchange membrane system to encounter the existing discharge levels which has been set by U.S. EPA (Power Mag. Electric Power, 2017). This methodology was parallel to other large-scale industrialized parameters.

6.3.7 TEXTILE INDUSTRIES (FABRIC INDUSTRIES)

Fabric industries as well as rug industrialists produce unwanted water during different processes which include yarn production, cleaning, and finishing. The contaminants produced by these industries include oil and grease, C_6H_5OH, Cr, S^{2-}, SS, and BOD (Environmental Protection Agency, 2017). Presence of residues of insecticides in swindles has precise problematic in handling surplus wastewater produced in fabric treating. If uncontaminated animal fats are found in the wastewater, they could be recovered for auxiliary execution for tallow production.

The fabric dyeing stations engender chemical water comprising color retardants and accelerators, natural and artificial dyestuff, wetting proxies, pH buffers, and guar. Handling is done through polymer-based flocking and settling agents and the monitoring parameters include heavy metals, grease, TSS, BOD, COD, and oil.

6.3.8 PETROLEUM REFINING AND PETROCHEMICALS INDUSTRIES (OIL REFINERIES)

The toxicants which are released from the petrochemical plants and oil refineries contain predictable amount of impurities, such as, NH_3, heavy metals, and other toxicants.

6.3.9 PAPER AND PULP INDUSTRIES (PULP AND PAPER CORPORATIONS

Emissions which arise from pulp and paper corporations contain a large amount of SS and BOD. While bleaching of wood pulp in paper production, these plants may generate chemical oxygen demand, C_6H_5OH, C_4H_4O, $C_4H_4O_2$, and $CHCl_3$ (Environmental Protection Agency, 2000). An imported method has been used as primary treatment in paper making for the removal of air bubbles and sedimentation by stand-alone paper mills, which may increase the BOD and COD levels. Biological handling, such as UASB and ASP could be obligatory. For mills which have extraordinary stacking of salt, tertiary treatment would be essential like common membrane treatment (i.e., reverse osmosis or ultrafiltration) to remove specific chemicals.

6.3.10 INDUSTRIAL OIL CONTAMINATIONS (ADULTERATION OF INDUSTRIAL OIL)

The main cause of oil contamination in leftover water brook would be automobile swab inlets, petroleum storage yards, workshops, transportation center, and power generation. This contaminated water is then released into resident cesspool or trade water scheme which must meet the indigenous ecological stipulations. The contaminants may include detergents, solvents, hydrocarbons, and grit.

6.3.11 WOOD PRESERVING (TIMBER CONSERVING)

Timber conserving plants produced conformist and lethal impurities, such as Cu, Cr, As, COD, C_6H_5OH, pH, SS, oil, and grease (U.S. Environmental Protection Agency, 2018).

6.4 LIFE CYCLE OF A PRODUCT (BIOLOGICAL CLOCK OF A PRODUCT)

The first step in the life cycle of a product would be design, manufacturing, circulation, and primary use. Then this is followed by the leftover pyramid stages, such as, reduce, recycle, and reuse. Every increased or decreased level in this life cycle of a product offers prospects for procedure intrusion to reform to curtail waste production to understand the requirements of product for its extensive consumption (United Nations Environmental Programme, 2013). The main aim to analyze the life cycle was to enhance the exploitation of inadequate resources of the world and to remove the useless material.

6.5 WASTE HIERARCHY (HIERARCHY OF WASTE OR CLASSIFICATION OF WASTE)

This ladder elucidates the three Rs rule which categorizes the tactics of waste management in a row with their effective terms of remaining abate. The main purpose of this waste pyramid was to extract extreme benefits through products and to produce minimum end residual (Albert, 2011). It can be epitomized as pyramid and the fundamental policies to be framed to prevent the lasting generation of end waste. The following step is to find the marginal consumptions alternative uses for the previously produced waste. After that recycling has been done. Material is recovered in this phase. The finishing achievement is the disposal of waste in landfills or ignition without any energy retrieval. The final step was last route to recover the prohibited waste (United Nations Environmental Programme, 2013). Although, the inhibition of all residual material was not about too possible, so, the highly effective method was aiming to prevent them through waste management. This hierarchy characterized the development of product through logical sequences. It also exemplified the concluding portions of life cycle flow chart of the product.

6.6 POLLUTER-PAYS PRINCIPLE

Polluter-pays guideline commands the contaminating party to pay for the effect on climate. with regards to the management of waste disposal, this as a rule suggested the necessity for squander generators to pay for the removal of unrecoverable substance.

6.7 RESOURCE EFFICIENCY

Resource efficiency indicates the fundamental conception that development and improvement of world economy cannot be supported at the present designs of productivity and utilization. All around, people extricate a larger number of resources to create products and ventures than the planet can restore it again (United Nations Environmental Programme, 2013). Resource efficiency is the decrease of ecological effect from production and utilization of these products, from the last crude material extraction to the last utilization and afterward removal.

6.8 WASTES MANAGEMENT IN MODERN ERA

Management of wastes are the activities and tasks needed to supervise or handle the waste disposal from its beginning to its last removal (United Nations Statistics Division - Environment Statistics, 2017), including its avoidance methodologies. This incorporates the assortment, transport, avoidance, management as well as removal of discarded materials, along with the guidelines of waste administration measures. Waste can be fluid, rigid, or gas and each kind of waste has various strategies for the treatment and removal. Management of wastes manages all kinds of waste, including mechanical, domestic, and natural. At times, waste may present danger to human well-being (U.S. Environmental Protection Agency, 2011). Waste management is proposed to decrease the severe impacts of wastes on human well-being, the climate, and to likewise improve cleanliness. The acts of waste administration are not uniform among nations both in evolved and evolving countries. Districts (countryside and cities) domestic, and modern areas may all adopt various strategies (Davidson, 2011). An enormous portion of waste administration rehearses to manage the city waste and wastewater which establish the main part of the waste made by mechanical, family unit, and business acts, even though vaporous waste is likewise vigorously created, particularly by factories and industries. Mechanical squanders are delivered because of the activities of industries which incorporate any material delivered futile during assembling cycle, for example, that of food and substance businesses, plants, mining activities, and industrial facilities. Most nations have sanctioned various laws to manage mechanical waste issues, yet consistence systems and severity differ. Implementation is regularly a challenge. Waste management is a significant part in a business' capacity to keep up ISO14001 accreditation. The ISO14001 standard urges organizations to improve their waste administration and natural efficiencies consistently by dispensing with squander through the acts of asset recuperation. This is done by the appropriation of asset recuperation rehearses like the reusing of materials, for example, food scraps, cardboard and paper, glass, metals, and plastic jugs. Reused materials can be offered to the development enterprises. Numerous inorganic waste streams might be utilized to create development materials. Cement and blocks might be reused as fake rock.

Waste management, including mitigation measures, is the actions and practices needed to handle or treat wastes starting from origin to final disposal (United Nations Statistics Division-Environment Statistics, 2017). This includes waste assortment, storage, prevention, processing and disposal along with wastes management process control and regulations. Waste may be liquid, solid or gas and various management and disposal methods exist for each form of waste. Waste management includes all forms of waste, whether agricultural, residential or organic. In certain situations, waste could pose a health risk (IFC, 2014). The goal of waste managing is to minimize the negative effect of waste on human health, the surroundings and aesthetics, as well as to improve hygiene. Among developed and developing countries, waste handling practices are not equal. Regions (rural and urban areas) and inhabited and manufacturing sector can take the diverse methods (Davidson, 2011).

Most of the waste handling practices deal through municipal solid waste (MSW) and wastewater, that result in a huge quantity of waste from households and profitable activities, while gaseous wastes are also heavily created particularly by factories. Industrial waste is generated by different industrial functions, including all of the materials that are discarded while processing, for instance, those of food and chemical industries, mills, and mining operations. In order to deal with industrial waste issues, most countries have adopted various laws, but enforcement regimes and strictness vary. Sometimes, compliance is a problem. Waste management is a significant component of the capacity of an organization or sector to retain ISO14001 accreditation. By reducing waste through resource recycling activities, the ISO14001 standard allows enterprises to enhance their waste management and environmental efficiencies annually. The implementation of resource recovery procedures such as the recycling of products, such as diet scrap, papers and cardboards, glass, metals, and artificial bottle is one way to do this. It is possible to supply recycled materials to the building industries. Many different sources of mineral wastes may be used to create co-waste.

6.9 WASTE MANAGEMENT CHALLENGES IN DEVELOPING NATIONS

Waste is the product of the public's collective failure, regulation, lack of funds, and professional support. For the effective implementation of any waste management activities, community awareness and appropriate information about waste handling and reuse of various forms of wastes, health hazards, eco-challenges and commercial concerns associated with waste handling is critical. To control it in a better way, everybody wants to grow better information of waste management problems and their actions. In this respect, basic training needs to be implemented by governments, which can be very successful. Bad awareness will make the best designed system doubtful.

The participations level of many people in waste handling is very less in developing countries, with inhabitants in city areas not actively participating in waste management process. It is very helpful for future opportunities and for the public's more active engagement in waste management activities. People should be trained to sort waste on the basis of its form, such as biodegradable wastes, harmful waste, green wastes, etc. Many peoples around the biosphere are not attentive of waste as recycled substantial, interestingly, many of them believe that biodegradable wastes are plastic. It is exceedingly difficult to involve individuals who are ignorant of the art of waste management.

In under developed nations, waste management is generally done by the poor laymen as a source of livelihood. In Asia, Latin America, and Africa, it has been estimated that 2% of the population relies on wastes to meet living necessities. A joint family organized, or single person manual searchers of waste are often involved with very inadequate helpful network and waste management practices and facilities with higher health hazards. Often, such activities prevent further education for their kids. It should be kept in mind, despite the bad results, this activity keeps them working and provides them with livelihoods in highly unemployed countries. But steps need to be taken to provide people engaged in waste management activities with healthier behaviors, social attitudes, to provide them with services to lessen their health-related risk, and improve their job performance.

People need to arise up powerfully against their local community offices or administration in evolving countries, where administration funding for waste management is scarce and if they see things are not improving and stacks of waste are piling up. To protect their climate, health, and keep life safe and stable for their children, they should protest.

6.10 RISK ASSESSMENT IN INDUSTRIAL WASTE MANAGEMENT

Risk evaluation is a structured mechanism by which possible threats that may be involved in a planned operation or undertaking (industrial waste management) are assessed. For effective collaborations between businesses, state regulators, the public, and other stakeholders, environmental risk communication skills are important. It is progressively more important for the entire stakeholders to recognize the knowledge about risk evaluation as more environmental protection decisions are taken on the basis of risk. The implementation of risk assessment on industrial waste includes three key interconnected steps: hazard detection, exposure assessments, and risk characterizations.

6.10.1 HAZARD IDENTIFICATIONS

The sources of the possible risks are identified and defined (e.g., chemical handled in a waste management unit).

The source of the potential risk has already been defined, that is, waste management units. However, for an exposure and risk, there need be a release of chemical from a waste handling facility. A variety of processes can release chemicals from waste treatment units, including vaporizations (where chemicals are released into the air in the vapor phase), leaching into groundwater (where substances leach down from above surface to groundwater sinkhole) and particulate emission (in which particulate matters adhere with chemicals and are discharged into the air when they turn to airborne).

Information characterizing the waste management unit is required to deal with these absolutions during the risk assessment. Critical parameters include the size of unit and position of the unit. Larger units, for instance, retain the ability to create better launches. Units to be found near the water table will generate more groundwater releases than those situated distantly from the water table. During a cold, rainy, non-windy environment, units present in a dry, warm, windy environment can generate larger unstable releases than other units.

6.10.2 EXPOSURE ASSESSMENT (PATHWAYS, ROUTES, AND ESTIMATION)

Determine the ways of experience and methods of exposure from the resource to a person.

Through a number of exposure mechanisms and processes, individuals and communities may come into contact with environmental pollutants. The mere presence in a waste management unit of a threat, such as toxic chemicals, does not signify the existence of a risk. The bridge between what is considered a danger and what actually poses a risk is exposure. Exposure evaluation requires the assessment of the possible or current mechanisms and degree of human interaction with toxic chemicals. When gathering all the data required for a full exposure assessment, the extent, frequency, length, and route of exposure to a chemical must be considered. An exposure estimation levels contain the identification of the probably exhibited inhabitants (receptors), ways of exposure, ecological media transporting the pollutant, amount of the contaminant at the receptor end, and exposure of the receptor.

All contact variables attained particular values in a deterministic exposure assessment. For instance, it may be presumed that the time span an individual stays in the same habitation adjacent to the capability is 30 years. Alternatively, single values can be replaced in a probabilistic analysis with prospect allocation functions to reflect the existent changeability spectrum, in addition to improbability. Testing the example of era in residence, it can be determined that 10% of the community people abutted to the competence remain in their habitat for fewer than 3 years, 50% fewer than six years, 90% less than 20 years, and 99% less than 27 years.

6.10.3 RISK CHARACTERIZATION

The standard of health details (i.e., cancer splash variables, mention quantities, and suggested applications) and the findings of the contact judgment (expected ingestion or dose by potentially exhibited peoples) are incorporated into the risk characterization process in order to arrive at quantitative cancer and noncancer risk estimates. There is a difference between estimated consumption of matters and mentioned dosage or suggested rates of concentration in order to describe the possible noncarcinogenic effects. The probabilities of a person developing cancer in their whole life are likely from the expected rate of ingestion and the distinct chemical value of the cancer gradient feature to characterize potential carcinogenic effects. The final calculation step is this procedure. This phase decides how and what the possible consequences are likely to be faced.

6.11 WASTE TRANSPORT AND HANDLING PRACTICES

Methods for collecting waste vary greatly between various regions and nations. Domestic waste disposal systems are offered on a daily basis by native administration bodies and by isolated commercial and industrials wastes companies. There is no formal system for waste collection in some areas and regions, particularly those in less-developed nations. In most areas of Canada, NZ, and the USA, European countries, and several other portions of the developing ecosphere, curbside collection is the most common form of disposal, where wastes are routinely picked at pauses by particular vehicles. That is generally done by curbside collection of wastes. Wastes may need to be collected or dumped at waste transfer stations in rural areas. The collected waste is then transported to suitable treatment facilities. In few countries, vacuity collections are cast-off where waste is elated through vacuum along some narrow bore tubes from home or commercial premises.

Such waste management approaches in North and Europe America are in use. Attached wastes are picked up from a variety of jurisdictions at curb side or wastes assignment station and processed into unusable waste and recyclables. These devices are accomplished of handling vast quantities of solids wastes, saving biodegradable, and transforming the remains into natural fuels and soil conditioners. In San Francisco, with its zero-waste target by 2020, the native administration has created by its own the Mandatory Recycling and Composting Ordinance (MRCO), forcing everybody in the urban to keep reusable and compostable out of the landfills. The three watercourses are picked up for the landfill-bound materials with curbside Fantastic three-bin scheme as black, blue for reusables, and green for compostable provided to businesses and residents and serviced by Recology (the sole refuse hauler in San Francisco).

The "Pay-As-You-Throw" scheme of the city charges clients by bulk of landfills-bound material that provide financing.

6.12 FINANCIAL MODELS FOR WASTE MANAGEMENT

The disposal of household waste in most developed countries is financed by local or national taxes, that may be related to revenue or goods prices. As a profitable facility, developed and commercial waste disposals are usually paid, usually as an integrated fee that covers the cost of disposal. Instead of the environmentally best approach, such as recycling and reuse, this method will allow the removal workers to go for the inexpensive removal choices, just like landfills. In many regions, such as Taipei, the city administration paid factories and homes for the amount of waste they generate. The wastes are picked up by the urban Councils if and only if it is stored in waste belongings provided by the municipality.

The Kampala Capital City Authority (KCCA) in Uganda has a waste management unit-in-charge of charging waste collection and disposal fees. The inhabitants of these areas are seeking as much as possible to reduce the production of waste. Such policies have successfully decreased the amount of waste generated by these cities and amplified the reprocessing rates. Morocco has also announced welfares from the $300 million hygienic landfills system being introduced. Although it might seem to be an expensive investment, the government estimates that it saved them additional losses of $440 million or the repercussions of deteriorating to dispose of wastes correctly (The Economist, 2018).

6.13 INTERNATIONAL WASTE MOVEMENT

Although the transport of waste in a given country is subject to national legislation, the transportation of waste across borders is also subject to international treaties. Hazardous waste has become a big problem for many nations. Ratified by 172 countries, the Basel Convention denounces the transport of risk wastes after industrialized to less industrialized states. The terms of the Basel Agreement have been incorporated into the leftover consignment legislation of the European Union. Though considered dangerous, radioactive wastes do not come under the authority of the Basel Agreement.

6.14 ENVIRONMENTAL IMPACT OF INDUSTRIAL WASTE

Due to the requirement of large quantities of water as an input to the production, development, or for machinery refrigeration, factories and energy plants are usually situated in the vicinity of the water bodies. Many sectors that are becoming industrialized do not yet have the tools or infrastructure to dispose of waste with smaller environmental impacts. Wastewater, both untreated and partly handled, is typically passed on into a nearby found water body. Marine environments and water dependable things, health, or intake water sources are directly impacted by metals, chemicals and wastes let out into bodies of water. Depending on the type of contaminant in marine environment, contaminants from the wastewater can cause the death of marine life or can give rise to many types of disease for those who eat these marine animals. Sea habitats are affected by chemicals and metals let out into water bodies. Efficient methods of eliminating waste properly are needed.

Nitrate and phosphate-containing wastewater also induces eutrophication, which can destroy existing water life. Research carried out in Thailand focusing on the sources of water pollution found that there was a strong relation between the highest concentrations of water contamination in the U-tapao River and industrial wastewater (Gyawali et al., 2012).

6.15 AIR POLLUTION

Air pollution arising from fossil fuel burning is another clear consequence of industrial waste. This affects the lives of many people because diseases are transmitted by it. As they have inadequate resources to address this unique problem, many environmental concerns have devastating effects on third world countries (Aivalioti, 2014). This also affects the consistency of the soil, as this huge issue must be solved by farmers. Additionally, one of the common air contaminants found in the air is nitrogen dioxide. Air contaminants have devastating effects on humans because they cause various diseases. Carbon monoxide can also cause the death if large amount reach the body beyond the amount that the body can handle. There have been a range of reports of sudden deaths of people and even an entire family in Nigeria and other parts of the world due to excessive inhalation of electric generator fumes. Ammonia also triggers plenty of complementary breathing.

The diseases that result from air pollution include eye, ear, mouth, or nose irritation. There is also a possibility that both bronchitis and pneumonia are particularly severe. Headaches, dizziness, and nausea from air pollution have been widely recorded by individuals (Society National Geographic, 2011). The WHO has claimed that the worst danger to human health is air pollution (Roser and Ritchie, 2017). There has been air pollution for a longer period. For humans, interior air contamination is also a possibility. The combustion of solid fuels, primarily from heating or cooking, causes this form of air pollution (Roser and Ritchie, 2017).

6.16 WATER POLLUTION

Water contamination is mainly a disturbing consequence of toxic waste. Large quantities of water that come into contact with toxic chemicals are used for most manufacturing processes. Typically, these chemicals are metals or toxic matters. Since bulk of the waste ends up in seas, lakes, or rivers, this adversely affects the climate. As a consequence, water becomes contaminated, posing a health risk to all. Farmers rely on this water, but if it is contaminated, it will pollute the crops that are grown. This affects the well-being of humanity because it continues to influence not only the lives of peoples but also animals if industrial corporations do not clean up their waste. The health of marine animals is compromised because this contaminated water endangers their lives. Water contamination can have detrimental effects on the human body with bacteria, viruses, and chemicals being the primary pathogens.

Cholera, typhoid, or Giardia are diseases that humans may be exposed to by drinking contaminated water (Denchak, 2018).

6.17 TREATMENT OF INDUSTRIAL WASTEWATER

Different types of wastewater pollution require a variety of pollution removal techniques (Recycling Guide, 2008).

6.17.1 LEVELS OF WASTEWATER TREATMENT

Wastewater treatment levels are generally classified as advanced major, secondary, and tertiary levels). The first stage of treatment, which includes removing a portion of SS from wastewater, is also the primary treatment. In the primary treatment, sedimentation and screening typically accomplish the separation process. Initial treatment effluents will usually contain significant carbon-based materials and will have relatively high biological oxygen demand levels.

The next stage of treatment requiring further treatment of the effluent is secondary treatment. In general, biological procedures accomplish the reduction of residual suspended solids and organic matter. Secondary treatment effluents often have a low biological oxygen demand (30 mg/L mean) and low SS (30 mg/L mean). During biological treatment, most readily biologically degradable organic materials are removed but 40–100 mg/L of dissolved and biologically resistant (called "refractory") organic material is removed.

These resulted materials may be the end products or artificial products of natural organic decay, such as oil, chlorophenol, PCBs, nitrocompounds, TCEs, synthetic detergents, pesticides, PCEs, etc.

Tertiary treatment is the third and the most advanced stage of treatment of wastewater. Precipitation, activated carbon adsorption, coagulation filtration, electrodialysis, reverse osmosis, advanced oxidation processes, ozonation, etc. are all unit operations and processes of tertiary treatment and are also used for the local treatment of industrial waste. These unit operations and processes are used to treat particular waste components, such as the removal of refractory organics, the removal of heavy metals, etc.

6.17.2 BRINE TREATMENT

Brine treatment includes removal of dissolved salt ions from the wastes stream. While there are parallels to the desalination of salt water or salty water, the treatment of industrialized brine can require unusual combinations of dissolved ions, such as hardness ions or other metals, requiring complex processes and equipment.

Usually, brine treatment systems are optimized to either reduce the final discharge volume for more economical disposal (as the expense of disposal is mostly volume-based) or to maximize the recovery of fresh water or salts.

To minimize energy consumptions, biochemical use, or bodily footprints, brine treatments systems can also be configured.

Brine treatment is typically used in the treatment of cooling tower blow-down, water produced from steam-assisted gravity drainage (SAGD), water produced from biofuels extraction, such as coal seam gas, fracture reverse flow of water, runoff of acid rock or acid extract, undo osmosis rejection, chlorinated alkaline wastewater, paper crush waste and flesh, drinks, and food production waste streams. Mechanizations for brine treatment can include membrane filtration processes like reverse osmosis, ion exchange processes, as electrodialysis or weak acid cation exchange, or processes of evaporation, such as brine concentrators and mechanical vapor recompression and steam crystallizers. Because of the increased salinity that is occurring by salt rigidity or natural pollutants, the accumulation of hydrocarbons or carbon-based material may block the osmosis membranes during the treatment of brine. Reverse osmosis possibly will not be feasible in this situation. For brine treatment, the processes of evaporation are the most widespread, because they allow the highest degree of concentration, as high as solid salt. They also generate the highest effluent purity including distillate quality. Evaporation processes are known to be more tolerant of organics, hydrocarbons, or hardening salts. Power utilization is however maximum and decay can be a concern, as concentrated salt water is the prime mover. Evaporation systems usually use binary substances or titanium.

6.17.3 BRINE MANAGEMENT

Brine supervision explores the wider framework of brine analysis which can include guidelines of management and legislation, commercial feasibility, ecological effects, reprocessing, conducting and transportation, limitation, consolidate as opposed to on-site-specific analysis, technology and economics, avoidance and cutback. Management of brines allocates various problems among drainage running in addition to further common management of waste.

6.17.4 OILS AND GREASE REMOVAL

Efficient removal of oil or greases depends on the characteristics of the oil with regard to the state of suspension and the size of the droplets, which in turns affects the option of strainer machinery. Free light oils, heavy oil that has a propensity to sinks, and combined oil, typically referred to as soluble oil, can be the oil found in industrial wastewater. In order to release the oils from their emulsion, soluble oils (or emulsified oils) will usually need cracking. This is done in many cases by lowering the water matrix's pH.

In water and oil, contamination typically occurs in four conditions, sometimes in grouping:

- Liberate oil-big surface-sitting oil drops.

Strong oil that lies at the bottom, which also adheres to solids such as dirt:

- Emulsified, in which oil drops are "chopped" a lot.
- Liquefied oil, in which the drops are completely scattered and not able to be seen. In industrial oily wastewater, emulsified oil droplets are the most common and are extremely hard to distinguish.

The technique for oil separation is based on the droplet size of the oil. Large drops of oil are quickly separated, including those that are free of oil contamination, but some separator systems work better than others as the droplets become smaller. Many separator systems would provide an optimal variety of sizes of oil droplets that can be handled effectively. This is referred to as the "micron ranking."

It is possible to study the water containing oil to find out the size of droplet by a video particle analyzer. On the other hand, in markets for oil droplet sizes, there are commonalities. In wastewater, larger droplets larger than 60 μ are frequently found in labs, re-fuel areas, and depots. In vehicle washing bays, meat processing and dairy production of effluent, and aluminum billet cooling towers, 20–50 μ oil molecules (droplets) are frequently found. In workshops and condensates, smaller droplets in the range of 10–20 μ are found.

Based on the oil droplet size, every technology of partition will have its specific operational curve detailing most ideal performance. Gravity pits or containers, American Petroleum Institute (API) oil-water platter sets or dividers, chemical analysis through dissolved air flotations (DAFs), centrifuges, media filters and hydrocyclones are the most common separators for oil and grease.

Many separator systems have optimal oil droplet size ranges that can be handled effectively. It is possible to test oily water to determine the size of droplet using a video particle analyzer.

6.17.5 SOLIDS REMOVAL

With solids retrieved as sludge or slurry, most solids are eliminated using simpler sedimentation techniques. Unique concerns arise from incredibly pure solids and densities containing solids similar to water density. Ultrafiltration or filtration can be required in such cases.

6.17.6 API SEPARATORS

Numerous oils can be recovered by a skimming system from open water surfaces. Oil skimming devices can often achieve the necessary level of water purity when a cheap and effective way is considered to extract grease, oils, and further hydrocarbons after water. Skimming is also a cost-effective way of extracting most oil before using chemical processes and membrane filters at sometimes. Since there is less oil to store, skimmers can prevent the filters from premature blindness and keep the chemical costs down. As higher viscosity hydrocarbons are involved in grease skimming, the skimmers must be well equipped with heaters that are sufficiently strong to retain grease fluid for proper discharge. An aerator or automatic apparatus can be cast-off to facilitate the removal of grease if floating grease forms clumps or mats. Hydraulic oils, however, and most oils that have deteriorated to a certain degree, may still have emulsified or soluble components that require further treatment to be removed. Instead of fixing it, emulsifying or dissolving oil using solvents or surfactants typically intensifies the problem, producing wastewater that is more difficult to handle. Wastewater typically contains gross volumes of oil and SS from important industries, such as chemical plant, natural gas production plant, oil refineries, and petrochemical plants. In order to distinguish the oil and SS from their wastewater effluents, these industries use devices such as the API oil–water separator. The name derives from the fact that the separators are constructed in compliance with the API (American Petroleum Institute, 1990) specifications.

The API separator is a separation system based on specific gravity. This system is designed to describe oil droplets' rising velocity based on their size and density using Stokes Law. The design is based on the contrast between the wastewater and the oil in specific gravity since the disparity between the water and suspended solids is greatly lesser than the exact gravity disparity. At the bottom of the API separator, the suspended solids settle as a sediment layer, the oil rise to the surface, and the washed wastewater is the middle layer among the solids and the oil layers. The oil layer is usually skimmed off, later disposed of or recycled, and a chain and flight scraper (or other related tool) and slush drain systems are used to extract the lower layer of the deposit. The water film leads to additional management for the further removals of any residual oils, and then to the removal of unwanted dissolved chemicals by some sort of biological treatment unit.

The API separator is alike parallel plate separator, but have sloped similar plates assembly (parallel packs). The parallel plates provide more space to coalesce into larger globules or balls for the droplets of suspended oil. The particular gravity between water and the suspended oil also depends on certain separators. The parallel plates, however, improve the degree of oil–water separations. The consequence is that parallel plate separators require considerably less space to achieve the same degree of separation compared to the traditional API separator.

6.17.7 BIODEGRADABLE ORGANICS REMOVAL

Decomposable carbon-based materials from animals or plants were typically likely to be handled by the complete traditional sewage handling method similar to trickling filter or triggered deposit (Recycling Guide, 2008). Problems may occur if wastewater is extremely concentrated (such as undiluted milk or blood) or is heavily contaminated with washing water. Treatment can be adversely affected by the use of disinfectants, pesticides, antibiotics, or cleaning agents.

6.18 REMOVAL OF OTHER ORGANICS FROM INDUSTRIAL WASTEWATER

It is very difficult to handle synthetic organic materials, including paints, pharmaceuticals, coke products, solvents, pesticides, and so on. Typically, the methods of treatment are unique to the substance under treatment. Distillation, ozonation, advanced processing of oxidation, vitrification, burning, adsorption, chemical immobilization, or removal of landfills are the treatment approaches. A variety of products like detergents might be subjected to biological degradation. In such case a changed method of wastewater treatment may be used.

6.18.1 REMOVAL OF ALKALIS AND ACIDS

Alkalis and acids may also be neutralized under regulated conditions. Neutralization causes precipitation that will need to be handled like solid residue that may also be poisonous. In most cases, gases can evolve, necessitating the gas streams to be treated. After neutralization, some other treatment types are sometimes needed. Waste stream rich in hardness ions can quickly and freely misplace the hardness ions in the accumulation of precipitates of magnesium and calcium salt from de-ionization procedures (Awuchi and Igwe, 2017). This precipitation process can result in serious pipe furring and can result in blockage of disposal pipes in extreme cases. Treatment is done by concentration of wastewater de-ionization and landfill disposal or by careful control of the pH of the wastewater released.

6.18.2 REMOVAL OF TOXIC COMPONENTS

Acids, alkalis, certain organic products, nonmetallic elements (like selenium or arsenic), poisonous components like metals (such as cd, th, pb, silver, etc.) are usually immune to biological processes except very dilute ones. By treatment with other chemicals or by modifying the pH, metals are often precipitated out. However, many are immune to mitigation or treatment and may require concentration followed by recycling or landfilling. Within the wastewater, dissolved organics can be incinerated using advanced oxidation processes.

6.19 COMMON WASTE DISPOSAL AND MANAGEMENT METHODS

6.19.1 INCINERATION

Burning is a form of waste removal in which dense carbon-based waste is exposed to burning to change it into both residues and gassy materials. For the disposal of both solid residues from the treatment of wastewater and industrial solid waste, incineration is useful. It reduces the amount of solids wastes by 80–95%. Burning and extra treatment systems for high-temperature wastes are often described as a current treatment. Wastes are converted by the incinerators into warmth, fume, ash, and steam. Burning is done by individuals on a small scale level as well as by business organization or governments on a large scale. It is used for the disposal of solids, gaseous, and liquid wastes. Incineration is known to be a practical way of removal of some risky waste materials (e.g., organic and medicinal waste). Incineration, owing to problems like as the release of gassy toxins, is a controversial waste disposal process.

In many other countries like Japan, where landuse is much more limited, the use of incineration is normal, because facilities typically do not require much area as landfill. Energy-from-wastes (EfW) or waste-to-energy (WtE) are wide terms for installations that burn waste to produce fire, electricity, or steam in a boiler or furnace. Incineration is not a perfect way at all times, then there are questions about the contaminants in the incinerator stacks' gaseous emissions. In particular, some highly determined biological compounds such as furans, PAHs, and dioxins that might be produced then that might have serious environment implications.

6.19.2 RECYCLING

It is a technique to reduce the waste by collection and re-using of the discarded materials attributed as waste that may contain a large variety of materials, such as plastic and empty containers of used beverages. Recycling is the reuse of parent material that can be again used for the production of new products. There are many ways of waste collection for recycling purpose. One of those methods is the Kerbside collection in which different collection waste bins and vans are used to collect waste separately. Waste is collected in two ways. Some communities use different bins to sort different types of wastes, such as metals, paper, and plastic waste before the collection is done. While in other communities, collection is done in such a way that all the waste materials are discarded in a single bin which is then separated later by the central facility and this method is called single-steam recycling. Aluminum is the major part of the materials that mare mainly recycled, such as steels products from aerosols and processed food cans, magazines and lighter paper, some expired steel furnishing materials or utensils, copper in the form of wires, rubber tires, jars, polyethylene and PET bottles, and paperboard cartons. Other recyclable materials are PVC, PS, and PP LDPE. All these products are feasible to recycle because their composition usually consists of single material. Electronic waste containing complex products is not easy to recycle because they need some deconstruction prior to recycling and it takes time and energy to disassemble such waste. The standards for waste accepted for recycling vary from city and country to country as every state has its own unique technique to tackle the recyclable waste.

The acceptance of waste for recycling is subjected to the product that is to be released as a by-product during processing. There was an international ban imposed by China in July 2017 on importing the 24 categories of recyclables as well as solid waste that may include textiles, mixed paper, and plastic that has affected directly and indirectly affected the exporters of China (Walker, 2018).

6.19.3 REUSE AND ENERGY RECOVERY (WASTE-TO-ENERGY)

It is a form of reuse in which a discarded material that cannot be recycled is converted into another form of energy by a range of processes, such as combustion, anaerobic digestion, landfill gas recovery (LFGR), and pyrolization (Zerbock and Candidate, 2003) and this process is referred to as water-to-energy. This method of waste handling is one of the nonhazardous waste management techniques. Reuse technology is eco-friendly in many ways such as generation of renewable energy by the conversion of nonrecyclable products into another form of energy such as heat or electricity, it may also work as an alternative of energy production using fossil fuels, hence reduces carbon emissions as well as it has minimal methane emissions as compared with landfills (Zerbock and Candidate, 2003). Almost 16% of the waste is handled through conversion into energy, worldwide (Czajczyńska et al., 2017). Waste can be used as input of energy production in two ways: the first method is the direct combustion of the materials to produce heat or electricity and the second is the conversion of materials into another type of fuel. This thermal management of waste materials has a wide range of usage as the heat produced can be used for heating or cooking purpose as well as it is also accountable for the electricity and steam production in the turbine.

Thermal treatment includes gasification and pyrolysis, that is, heating of waste at high temperature and oxygen-deprived environment over high pressure in a closed vessel. Pyrolysis is the process of converting solid waste into other forms of matter, such as liquid, solid, or gas products. Gas and liquid products are then subjected to combustion or chemical processing for the production of energy and other chemical materials. Char is the solid product from thermal treatment and it forms activated carbon upon further refinement. Synthetic gas (syngas) consisting of hydrogen and carbon monoxide (CO) is formed when organic materials are converted using gasification and advanced plasma arc gasification, and this gas is a major source of burning for electricity and steam generation. Hydrothermal monophasic oxidation is another method that can be used instead of pyrolysis, and it involves high temperature and pressure with intense water decomposition.

Biological Reprocessing

Biological reprocessing is basically the decomposition or recovery of organic materials from the organic waste products, for instance, paper products, food crumbs, and plant materials using digestion or decomposition method. The products of digestion or decomposition are further decomposed and recycled that are aimed to be used in agriculture or landscaping sector. Moreover, these processes also form waste gases specifically methane, a possible principal fuel source for the production of heat and electricity. Major goal of biological processing is to enhance or speed up the natural decomposition of organic matter.

Pyrolysis

It is the decomposition of organic materials by using both heat and chemicals in stoichiometric oxygen void conditions and the resulting gaseous products are several hydrocarbon fumes.

Pyrolysis is the method of forming recovered fuels by converting a large variety of industrial and household wastes. Pyrolysis uses a large variety of waste input such as plant waste, food waste, and tires that are possible substitute of fossil fuels. Pyrolysis includes the vibration of feed stock molecules at such a high frequency that results in molecule disintegration, and this rate increases with a parallel rise in temperature. Hence the process of pyrolysis speeds up. Temperature is maintained at 430°C if the end product is to be used for industrial purpose, whereas solid charcoal and gases are products of slow pyrolysis. Pyrolysis is the method of converting waste biomass into useful liquid fuel, such as plastic and wood waste which can possibly generate fuel. The only residue of pyrolysis are the materials that cannot be converted into gases, such as glass, sand, pyrolysis coke, and metals. Pyrolysis is a safer approach as its end products, such as chlorine, alkali metals, and sulfur are less toxic as compared with incineration. Though, some by-products of pyrolysis are toxic gases, such as SO_2 and HCl depending upon the nature of waste to be treated (Chen et al., 2014).

Resource Recovery

Resource recovery is the process of altering the discarded material into another useful form. It is the process of recovering or separating the constituents or resources from waste that can be further used as energy source and this whole process is done specifically at resource recovery facility (Government of Montana, 2012). Resource recovery is eco-friendly and an economic technique (Walker, 2018). Resource recovery is the economical and cost-effective technique, reduces the quantity of waste to be disposed, eco-friendly as well as spares area for landfills (Yhdego, 1995). Resource recovery could be an alternative of waste management as it works on the principle of Life Cycle Analysis (LCA). It is concluded that many extensive studies have favored the following channel for the handling of mixed solid waste, that is, the supervision, source segregation, and assemblage that is tailed by anaerobic digestion like incineration/composting of the biodegradable components as well as reuse/recycling of nonorganic portion. There are many materials/products that are considered to be waste and thrown away can be recycled or reused or converted into another useful form, this is how resource recycling is advantageous. Waste material from wood pebbles and similar packing items can possibly be recycled into valuable products to be used in horticulture. Walkways, corridors or arena surfaces can be covered by such recycled chunks.

Proper and wise waste management results in a wide range of advantages such as:

1. Economical: Better waste management results in better economy by different ways, such as recycling, reusing, and treatment of waste can establish markets, also supports the efficient manufacture and consumption of items as well as resource recovery and reuse techniques account for valuable materials to be recovered. This whole process has intense potential for jobs and business opportunities.
2. Environmental: Solid waste management has positive impact on environment as it decreases the adversarial effects of waste by the means of reuse, recycling, composting, and resource extraction results in sustainable resource use. It is also a key toward better air and water quality as well as significantly decreases greenhouse gas emissions.
3. Intergenerational Equity: Successive generation could have healthy environment, fair, and extensive society as well as strong economy if we follow effectual waste management approaches.
4. Social: Proper waste management can reduce the adverse health impacts, and the results are better for local community with fair socioeconomic conditions by creating new employment sources,

eventually, it supports and uplifts the local community by eradicating poverty in third world countries.

Liquid Waste Management

Municipal wastewater has been increasing with a rate of 0.1–30.8 kg per capita equivalent per year (kb/p.e/year) with parallel increase in industrialization (Syed Shatir et al., 2017). The end product of wastewater treatment is the sewage sludge. Other possible liquid waste management approaches are composting, incineration, and landfill.

The Avoidance and Reduction Method

The major approach toward waste management is the minimization of waste production called waste reduction. Following are the practices that can be done to reduce the waste generation. Prefer to repair the damaged or broken parts of machinery over buying a new one, promote re-using second-hand items, constructing the new items in such a way that they can be reused, refilled or converted into another form after one use, make sure to remove the food or liquid residue from cans or wrappers, and to discourage the public use of disposable items (Recycling Guide, 2008) as well as to craft items with less weight but achieving same purpose.

Physical Treatment

Physical treatment comprises altering the physical properties of waste, such as its state (i.e., gas, liquid, and solid), size, density, and shape. In physical treatment, chemical composition of the waste does not change. One type of actual treatment, immobilization, includes exemplifying waste in different materials, for example, cement, plastic or paste, to keep fragments from leakage or volatilizing. Hardly any instances of physical action incorporate filtration, immobilization, carbon absorption, volatilization/evaporation, sanitization, crushing, chopping, and so on.

Chemical Treatment

Chemical treatment includes changing the chemical composition, structure, and properties of waste through chemical reactions. Chemical treatment can comprise either other materials (reagents) mixing, elevating warmth heating, or a mixture of these two.

Recycling of misused elements can be done by chemical treatment. Some of the chemical treatments include precipitation, oxidation, neutralization, reduction, ion transfer, acid discharge, thermal desorption, incineration, extraction, etc.

Biological Treatment

Biological treatment might be categorized into aerobic and anaerobic forms. It is imperative to provide oxygenated environment for better functioning of microorganisms and fine decomposition in aerobic biological treatment. As a result of decomposition, the organic matter and other constituents are altered into various available forms, such as CO_2, water, sulfates, nitrates, elementary organic produced, and cellular based biomass (i.e., development and reproduction of cells). In contrast, in anaerobic condition, the transformation of organic constituents and nitrogenous compound into oxygen, methane gas, and other plant available forms can be done in the absence of oxygen. An enclosed air tight digester unit is required to perform anaerobic biological activity. Following are few examples which comprises:

- Aerobic:
  Activated sludge
  Aerated lagoon
  Trickling filter
  Rotating biological contactor (RBC)
- Anaerobic digestion

Here Below Are the Discussions of Above-Mentioned Techniques; Ion Exchange Chromatography

Anionic and cationic undesirable substances present in the wastewater can be removed easily by using ion exchange chromatography. Hydrogen or sodium and hydroxyl ions are exchanged by cations and anions, respectively. Ion exchange resins are found in natural or synthetic group constitution combined with functional groups. In the wastewater treatment the operating ion exchange resins are mostly synthetic gums which are formed by the organic complex polymerization to a permeable three-dimensional form.

If ion exchange adhesives exchange positive ions, they are called *cationic* and if they exchange negative ions, they are called as *anionic.* Some cation exchange resins contain *acidic* functional groups, for example, sulfonic compounds, while various resins that exchange anion contain *basic* functional groups, just like amine. The degree of ionization of the functional groups defines the potency of basic or acidic nature like the circumstance with dissolvable acids or bases. Thus, a resin is a strong acid cation exchange resin if it contains sulfonic acid groups. In other types of ion exchange resins, the most common functional groups for weak acid are carboxyl (-COOH), quaternary ammonium (R_3N+OH^-) for strong base, and for weak base, amine (-NH_2 or -RNH) groups are present.

Oil and Grease Separation (Flotation)

In the tank, the unbound oil stays on its surface and later removed out during oil separation. The basic principle behind gravity separator is to eliminate the unbound (free) oil molecules comprising the size of more than 0.15 mm. Stokes' law will be applicable if the Reynolds number is lower than 0.5. Usually, wasted amount of oil that is 50 mg/L are attained. Temperature fluctuations and the specific gravity of the oil can affect the hydraulic weight of a cross-flow corrugated plate separator. Nominal flow rates are fixed for a temperature of 20°C and a specific gravity of 0.9 for the oil. A hydraulic loading of 0.5 $m^3/m^2/h$ of actual plate area will usually result in the separation of 0.06-mm droplets. The process of acidification can be used to break the oil emulsions before separation, which includes the polymers used to break emulsion, or salts addition of iron or alum, or the use of emulsion-breaking polymers.

Coagulation

For clarification of waste from industries having suspended and colloidal solids, a technique of coagulation may be used. Waste material of paperboard can effectively be coagulated by using alum in low dosages. Rapid settling of flock is carried out by using polyelectrolyte and silica. Coagulation may also be used to clarify the emulsified oil containing waste sludge.

Oil droplets are about 10–5 cm in water and these are stabilized by ion adsorption.

Anion active species and soaps are emulsifying agents. By the use of salts, for example, calcium chloride ($CaCl_2$), emulsion can be broken down. Emulsion can also be broken down by lowering the pH.

Precipitation-Heavy Metals Removal

Heavy metals from industrial waste are removed through precipitation. Through the addition of sodium hydroxide or lime, heavy metals are precipitated as hydroxides to pH of lowest solubility. The pH of lowest solubility may vary among metals in consideration. For example, zinc and chromium are less soluble at 10.3 and 7.5 pH, respectively. It is crucial to primarily process the waste matters to get rid of the compounds or substances that will affect the metal precipitation, when we treat industrial wastewater containing metals. The complexes are formed by ammonia and cyanide that is why their pretreatment is necessary to restrict the removal of metals.

Filtration is necessary to remove the flock carried by the process of precipitation, to meet the low requirement of effluents. Concentration of effluents is reduced to the level of 0.5 mg/L or even less due to filtration. To precipitate the chromium waste with lime, it must be reduced from hexavalent to trivalent C^{+3} first. Ferrous sulfate, Sulfur dioxide and sodium metabisulfite are common reducing agents used for the reduction of chromium waste.

Solid Removal

Simple techniques of sedimentation can remove most of the solids as sludge or slurry. But there is a problem with extremely pure solids and of solids having density very close to water density. For the removal of such solids filtration and ultrafiltration techniques are used. Although, the polyelectrolyte or alum salt flocculation technique also can be used.

Activated Sludge Process

Industrial sewage and wastewater are passed through a biochemical process which is called as activated sludge process (ASP), to generate a biologically oxidized floc material by using microorganisms and oxygen (air). Generally, the process of activated sludge includes an aeration tank to inject air and to mix it in wastewater. A settling tank also known as settler or clarifier which allocates the waste material or slurry to resolve properly. In aeration tank, a portion of the waste material or sludge is recycled and the remaining sludge materials are recovered for additional treatment and absolute dumping.

6.20 SUMMARY

After rapid industrialization in mid-18th to early 19th centuries, the first transformation from an agricultural to economical industrial economy is known as industrial revolution. Exponential growth in world's population exerted pressure on natural resources as well as on the economy to produce more livelihood resources. Since then, a large number of industries were established to meet the dire need of capita which indirectly affected the environment in many ways. Industrial waste and effluents are being discarded openly or in freshwater bodies, or smoke is released into the air without any further treatment. Hence, industrialization has adverse effects on the environment causing air, water, soil, and noise pollution. However, in many countries, industrial waste is treated prior to dumping using different physical, chemical, or biological methods that convert the waste into less toxic form.

KEYWORDS

- **landfill gas recovery**
- **energy-from-waste**
- **polycyclic aromatic hydrocarbons**
- **steam-assisted gravity drainage**
- **suspended solids**
- **carbon monoxide**
- **waste-to-energy**

REFERENCES

Achankeng, E. J. Globalization, Urbanization and Municipal Solid Waste Management in Africa. In Proceedings of African Studies Association of Australasia and the Pacific: African on a global stage, 2003.

Adebilu, A. A.; Okenkule, A. A. Issues on Environmental Sanitation of Lagos Mainland Nigeria. *Environmentalist* 1998, *9* (2), 91–100.

Aivalioti, M. New Opportunities in Industrial Waste Management. *Waste Manage.* **2014,** *34* (10), 1737–1738. DOI: 10.1016/j.wasman.2014.07.006

Albert, R. The Proper Care and Use of a Garbage Disposal. Disposal Mag, 2011.

Awuchi, C. G. Industrial and Community Waste Management: Global Perspective 2020. *Am. J. Phys. Sci.* (Online) **2020,** *1* (1), 1–16. ISSN 2519-5549.

Awuchi, C. G.; Awuchi, C. G. Impacts of Plastic Pollution on the Sustainability of Seafood Value Chain and Human Health. *Int. J. Adv. Acad. Res.* **2019a,** *5* (11), 46–138. ISSN: 2488-9849.

Awuchi, C. G.; Awuchi, C. G. Physiological Effects of Plastic Wastes on the Endocrine System (Bisphenol A, Phthalates, Bisphenol S, PBDEs, TBBPA). *Int. J. Bioinf. Comput. Biol.* **2019a,** *4* (2), 11–29. http://www.aascit.org/journal/archive?journalId=809

Awuchi, C. G.; Igwe, S. V. Industrial Waste Management: Brief Survey & Advice to Cottage, Small and Medium Scale Industries in Uganda. *Int. J. Adv. Acad. Res.***2017,** *3* (1), 26–43. ISSN: 2488-9849.

Chen, D.; Yin, L.; Wang, H.; He, P. Pyrolysis Technologies for Municipal Solid Waste: A Review. *Waste Manage.* **2014,** *34* (12), 2466–2486. DOI: 10.1016/j.wasman.2014.08.004.

Cointreau, S. Environmental Management of Urban Solid Wastes in Developing Countries: A Project Guide. Urban Development Technical Paper No. 5; International Bank of Reconstruction and Development/The World Bank: Washington, DC, 1982.

Czajczyńska, D.; Anguilano, L.; Ghazal, H.; Krzyżyńska, R.; Reynolds, A. J.; Spencer, N.; Jouhara, H. Potential of Pyrolysis Processes in the Waste Management Sector. *Therm. Sci. Eng. Progress* **2017,** *3*, 171–197. DOI: 10.1016/j.tsep.2017.06.003.

Davidson, G.. Waste Management Practices: Literature Review; Dalhousie University—Office of Sustainability, 2011.

Denchak, M. *Water Pollution: Everything You Need to Know. Our Stories*; Natural Resources Defense Council Environmental Protection Agency: New York, 2018.

Government of Montana. Resource Recovery. Government of Montana, 2012; Grand Traverse County. What Is Resource Recovery? Grand Traverse County, 2006.

Gyawali et al. Effects of Industrial Waste Disposal on the Surface Water Quality of Utapao River, Thailand. *2012 Int. Conf. Environ. Sci. Eng.*. **2012,** *32*, 5.

Kaseva, M. E.; Mbuligwe, S. E. Appraisal of Solid Waste Collection Following Private Sector Involvement in Dar es Salaam. *Habitat Int.* **2005,** *29*, 353–366.

Ngategize, P.; Moyini, Y.; Aryagaruka, M.; Mwigaga, M.; Kigonya, A. Solid Waste Management Strategic Plan for Mpigi District Local Government. Final Report. Mpigi District Local Government, 2001.

Oteng-Ababio, M.; Melara Arguello, J. E.; Gabbay, O. Solid Waste Management in African Cities: Sorting the Facts from the Fads in Accra, Ghana. *Habitat Int.* **2013,** *39*, 96–104. doi:10.1016/j.habitatint.2012.10.010

Recycling Guide. Removing Food Remains to Reduce Waste. Recycling Guide, Feb 14, 2008.

Roser, M.; Ritchie, H. Air Pollution. Our World in Data. Society, National Geographic Society, 2017.

Syed Shatir, A. S-H.; Wang, Y.; Hu, S.; Su, S.; Xiang, J. Thermochemical Processing of Sewage Sludge to Energy and Fuel: Fundamentals, Challenges and Considerations. *Renew. Sustain. Energy Rev.* **2017,** *80*, 888–913. doi:10.1016/j.rser.2017.05.262

U.S. Environmental Protection Agency. *Introduction to the National Pretreatment Program*; U.S. Environmental Protection Agency: Washington, DC, 2018; pp 1–1, 1–2. EPA 833-B-11-001

U.S. Environmental Protection Agency. Resource Conservation and Recovery Act Overview. EPA. Feb 6, 2019.

United Nations Environmental Programme. Guidelines for National Waste Management Strategies Moving from Challenges to Opportunities; United Nations Environmental Programme, 2013. ISBN 978-92-807-3333-4.

United Nations Statistics Division—Environment Statistics, 2017. unstats.un.org.

US EPA, OA. Clean Air Act Requirements and History. US EPA. US EPA, OAR, (2015-05-29). Evolution of the Clean Air Act. US EPA; USEPA (2014). Energy Recovery from Waste. USEPA, 2014.

Walker, T. R. China's Ban on Imported Plastic Waste Could Be a Game Changer. *Nature* **2018,** *553* (7689), 405–405.

Yhdego, M. Urban Waste Management in Tunzania Issues, Concepts and Challenges; towards a Community Based Approach. *Resour. Conserv. Recycl.* **1995,** *21*, 213–226.

Zerbock, O.; Candidate, M. S. Urban Solid Waste Management: Waste Reduction in Developing Nations; School of Forest Resources and Environment Science, Master's International Program; Michigan Technological University, Working Paper, 2003.

CHAPTER 7

Livestock Waste, Types, Sources, Pollution Potential, and Country-Wise Comparisons

SOBIA RIAZ[1*], RIZWAN MEHMOOD[2], HAFIZ ABDUL KAREEM[3], and HALEEMA SADIA[4]

[1]*Institute of Soil and Environmental Sciences, University of Agriculture Faisalabad 38040, Pakistan*

[2]*Stored Grain Management Cell, Department of Entomology, University of Agriculture Faisalabad 38040, Pakistan*

[3]*College of Grassland Agriculture, Northwest A&F University, Yangling, Shaanxi 712100, China*

[4]*Department of Botany, University of Agriculture Faisalabad 38040, Pakistan*

**Corresponding author. E-mail: sobiariazuaf@outlook.com*

ABSTRACT

Livestock industry has benefitted human being in many ways, providing food, livelihood and contributing to economy. Meanwhile, it is the main cause of persistent organic compounds, odor and harmful gases. Subsequently, the sustainable management of livestock waste is very important to reduce the production of noxious gases hence preventing the environmental issues. Farm-based livestock waste is generally divided into three groups: waste water, soil, and slurry. Waste in a desiccated form is directly used as a fertilizer or supplement in fields or might be used as a fuel source of burning, whereas solid waste is managed by treating with different approaches such

Waste Problems and Management in Developing Countries. Umair Riaz, Shazia Iqbal, & Moazzam Jamil (Eds.)

as composting of incineration. Major sources of livestock wastes include run-off from dairy (comprising animal excreta and dung, wash water, and waste feed), dairy compost, and wastes from poultry operations (fertilizers, leftover feed, wastewater, and bedding material) and various other types of waste from livestock shed run-off. Livestock and farm activities affect the environment, soil, water, and air directly or indirectly. Meanwhile, the major concerning impacts are aquifers contamination, organic and inorganic chemicals released through industrial and agricultural operations, effluents, microbes, noxious gases and odor, microorganisms contaminating the air, soil, and water, particulate matter, and bio-aerosols. From the last couple of decades, there is an increasing trend witnessed in gradient of greenhouse gases emission mainly from agriculture and livestock sector. Along with this inclination of production and consumption rates, people are more concerned and aware of the climate changes led by anthropogenic activities hence following a number of biological, physical, and chemical approaches to treat the waste or convert it into a less toxic form.

7.1 INTRODUCTION

Dairy industry or livestock industry has benefitted human beings in many ways such as production of egg, milk, and meat; meanwhile, it has adversely impacted the environment by generating bulk quantity of wastes due to poor handling of waste. Increased urbanization, betterment of living standards or lifestyles, rapid increase in population and per capita income are the major driving forces behind the demand of animal products and this demand may continue to increase in near future. To benefit themselves from the enhancing demand of animal products, livestock owners have increased the rate or production and hence the huge quantity of waste too. Conversely, a proper management of the waste produced has a negative impact on the surrounding from the environmental and human health perspectives. During past few decades, developing countries have shown a significant increase in the production of livestock goods and major trend was observed in industrial farms surrounding urban centers. These livestock farms around the urban areas with high population density may cause adverse environmental and health hazards. Increasing demand of livestock products depicts the dire need of efficient laws for the management of livestock activities and productions and hence to make these practices more economically and environmentally sustainable for waste dealing (FAO.org).

Specifically in developing countries, there was an exponential increase observed in the livestock and dairy products during 1980–2004. In developing countries the production rate was grown to threefold with 5% annual rate of increased production because people in these regions consume about half of the world's meat portion. Even though, meat consumption rate is two to four times greater per person in developed countries as compared with developing countries. Due to intense change in location and kind of production of dairy of livestock waste has even worsened the circumstance in developing countries. Conventional farming system consisting of a small herd of livestock in conjunction with crops have put together enormous manufacturing processes alongside farm animals. There is an obvious shift in livestock-keeping practices from conventional grass-dependent buffalo and other animals to mass feed-dependent pigs and poultry (FAO.org).

Now, farmers are being benefitted from the nearest main markets for eggs and meat because livestock and other poultry activities have been clustered in or along the mega cities. While in 1990 there was an exponential increase in widespread production and consumption activity of livestock up to 80% with direct increase in growth, specifically in the Asia region. Such extensive, wide-ranging production activities sometimes result in clustered mass environmental issues. Large-scale industries are responsible for huge accumulation of nutrients in the environment as concentrated feed as well as their waste production rate is higher than the recycling rate since adjacent land area is badly contaminated by fertilizers absorption. A hub of joint livestock activities may act as a source of air, soil, and water pollution as well as threatening public health. This pollution threat is more evident in areas where livestock and the production activities are clustered adjacent to cities with higher growth or water reservoirs. Sewage is sometimes directly disposed off into nearby water resources or kept in huge pools leading to leakage of effluents and contamination of aquifers or other groundwater resources. Harmful gases from industrial livestock activities contribute to acid rain and formation of greenhouse gases as well as exposing inhabitants to disgusting odors (FAO.org).

7.2 SOURCES OF LIVESTOCK WASTES

Generally, three main classes of chemical contaminants are discussed such as organic, inorganic, and radioactive, and each category further contains several more classes of contaminants. Anthropogenic activities release numerous toxic chemicals that are persistent in nature and are not readily degraded or destroyed and hence pose serious health and ecological issues as they tend to accumulate in the environment as well as in the human body. In the agriculture sector, there is discriminate use of chemicals fertilizers and pesticides over a huge area to enhance the crop yield. This unfair use of chemicals often results in their entrance in soil, water, and environment and may pose significant threats. For example, the contamination of groundwater with nitrates is led by the over-use of nitrogen fertilizer hence percolating the root zone. Such agricultural activities may give rise to serious environmental, soil, air, and water issues.

Source: A livestock Farm at Nowshehro Feroz Sindh (All rights reserved by Sobia Riaz).

In general, the inorganic chemicals-based fertilizes are applied regularly at least once in a year and contain in genus of reducing quantities N, P, K, and metals. About 50–200 kg/ha amount of these chemicals is applied in a year, as N, P, or K. Micronutrient fertilizer (e.g., B, Fe, Zn, Cu, Mo) amendments required in lesser amount are also applied on a regular basis in the fields of agriculture. The average rates of these chemicals are 0.5–2 kg/ha that is applied in each 2–5 years to agriculture land, in their particular elemental form. The other application of soil amendments to agriculture fields that are also based on inorganic chemicals is categorized in the third group. Application of these amendments to agriculture area in regular intervals is done for two causes: (1) enhance soil texture, reduce salinity and to lessen or enhance soil pH and (Ali, 2004) to refill macronutrients such as K^+, Mg^{++}, Ca^{++}, and

SO4¼. The deficiencies macronutrients are controlled by applying these chemicals between the range of 50–500 kg/ha. To manage salinity and soil pH, usually the range from 2000 to 10,000 kg/ha is applicable. In two main forms of N the soil and plants minerals can act as sinks. On the other hand, many animals, plants, humans (fertilizer application), and atmosphere can add more amount of N (NO^{3-}) that directly pollutes the groundwater. On a larger scale, animal feeding setup comprises feedlots specifically for swine, beef, and production of poultry, fish farms, and dairies. This kind of setup acts as a point source for some chemicals. In waste disposal practices, unfettered waste of animal is led to contamination of some common chemicals like nitrate, phosphate, and ammonium. In an animal waste, from 1000 to 50,000 mg/kg concentrations of these three chemicals are found. Phosphate (>1 mg/L) present in minute amount can be enormously venomous to standing water bodies because it led to microbial growth that causes eutrophication. Different pharmaceutical chemicals are used for the animal treatment and they pose serious health hazards for humans. These health concerns are discussed with depth in the next section. Some pharmaceutical products that are normally used for animal treatment not only are the source of potential contamination but also exert sever impact on human health.

7.3 HEALTH HAZARDS

Methemoglobinemia is also known as blue baby syndrome which occurred in infants due to groundwater pollution with high level of nitrate. In the digestive system, the conversion of nitrate into nitrite occurs which led to this Methemoglobinemia syndrome. There is protein in blood which carries oxygen named as oxyhemoglobin which reacts with nitrite and forms methemoglobin. After the reaction the forming product is not able to carry oxygen and ultimately rebates the ability of oxygenated blood. Subsequently, oxygen dispossessions of oxygen occur in tissues of body. Digestive and respiratory difficulties along with blue coloration of mucous membrane can happen to Methemoglobinemia-suffering infants. In an agricultural system pesticides that are organic in nature are most commonly applied once a year although in a lesser amount than fertilizers. Conversely, synthetic pesticides contain very harmful materials that are poisonous to plants even when applied at a low concentration. For solid form of pesticides they are usually dissolved into water or oil for better management at the time of application. To control insects, gaseous forms of pesticides are used and known as fumigants. Arsenic boron and sulfur are some toxic forms that are used as primary acting agent in those chemicals that are applied in close vicinity to humans. Water solubility, mobility, toxicity, and environmental persistence of organic pesticides are controlled by their chemical structure.

A concession between toxicity and resistance is the main requirement of pesticide development that moves it to a further step. It also required a chemical structure that reduced the toxic effects and was fairly soluble in water. The upcoming pesticides are required to be more soluble in water and sustained attention to their toxic properties so that resistance of these chemical in the surroundings could be decreased. This type of pesticides has minimum life spans in the atmosphere and rarely bio-accumulates in animals or humans. Conversely, the water bodies contain these chemicals, when misused. For example, nowadays in ground and surface water resources the mostly found pesticides members are related to triazine family. On the other hand, pesticides that are chlorinated are found rarely in water and also present in sediments and soils. Many residues in significant amount produced by animals, are gentle for environment during unbolt situations with small concentrations of animals. But large-scale production systems of animals have produced intense sources of animal-derived pollutants in the last 100 years.

7.4 AIR EMISSIONS

Livestock farming is not only the producer of animal manure but also a significant emitter of pollutants present in air. These pollutants contain a specific amount of non-CO_2 greenhouse gases (CH_4 and N_2O), ammonia, and odor. The soils and waters acidification and more quantity of nitrates present in drinking water are main hazards of ammonia emission. One of the main reasons of global warming is the production of greenhouse gases due to which global warmth increases by increasing the amount of various gases in the atmosphere. The soil application of manure and releasing the odor from animal shelters are increasingly considered an irritation in highly populated countries as the degree of livestock processes (Melse and Timmerman, 2009). Particularly, a large amount of various compounds emit from huge livestock systems, including volatile organic compounds, ammonia, methane, particulate matter, nitrous oxide, hydrogen sulfide, and endotoxins. The coverage to these multifaceted is linked with many airborne disorders, for example asthma-like syndrome, bronchitis, mucous membrane irritation, asthma and chronic disruptive pulmonary disease, as confirmed in studies of farm workers (May et al., 2012). The endotoxins and irritant agents along with organic aerosols cause respiratory illness among swine employees and the people surrounding the farms (Leytem et al., 2011). Several emissions of manure such as ammonia can pass through from the direct farm site, thus producing harmful effect at the regional level. The major amount of gases releasing from animal husbandry is formed from accumulated or freshly dumped faeces and urine, during the activity of microbes (Martinez et al., 2009). Gases emissions from livestock sector are mentioned in Figure 7.1. In livestock agriculture the stored manure of the entire livestock species in anaerobic environments or the enteric fermentation in ruminants' digestive tracks is associated with methane emissions. In both cases, methanogenic archaea (require anaerobic microorganisms) are primarily responsible for the production of methane. In a day the total production of methane per animal will rely mostly on the intake feed level, the forage-to-concentrate ratio of the diet, carbohydrates digestibility in diet, addition of fat amount in diet, and the occurrence of whichever feed supplements that can change the microbial community of rumen. (Dammgen et al., 2012). According to Intergovernmental Panel on Climate Change (IPCC) report,1 kg methane quantity has 63 times heating 1 of CO_2 effect (over 100 years of calculation, methane-warming effect is 21–23 times more than the effect of CO_2). The soil fertilized by solid manure or dung storage system is usually linked with emission of N_2O in livestock agriculture (Junior et al., 2013). N_2O and CH_4 are major gases of greenhouse that are concerned with global warming fact.

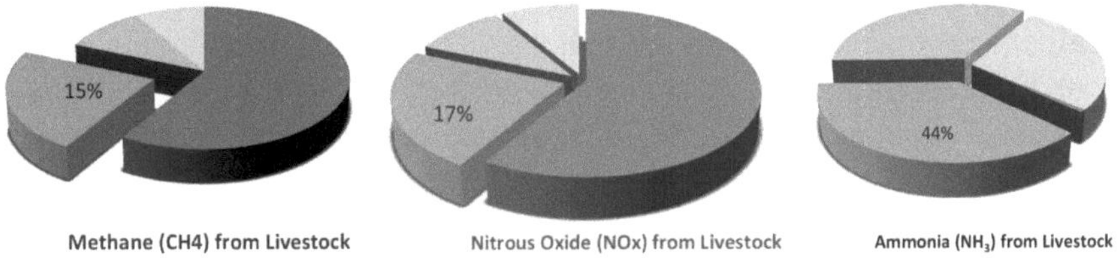

FIGURE 7.1 Contribution of livestock industry in Nh_3, NO_x, and CH_4emissions.

Moreover, they are involved in chemical reactions of plants in troposphere that verify ozone concentrations. Particularly, livestock is considered a huge cause of ammonia emission. In the faeces of livestock, urease enzyme hydrolyzed the urea in urine which eventually led to the formation of ammonia.

There are many factors on which volatilization of ammonium in the liquid phase to ammonia gas depends such as pH, temperature, and wind speed; therefore, ammonia emissions are highly variable from an animal feeding facility. The production of nitrogen by growing-finishing pigs as urine and faeces substituent, for example, the excretion, is divided into the part that ends up in the liquid manure form which is removed from the storage at farm and the part as ammonia that is emitted to the atmosphere, either from manure storage tanks or by means of the circulating air of the pig house. A significant amount of all nitrogen that is excreted (one quarter) is emitted in the form of ammonia with the exhaust air and three quarters end up in the liquid manure storage. Because one-fourth of all nitrogen is released into the atmosphere, it can be established that abatement technologies for ammonia release from these animal houses, for example, by end-of-pipe treatment of exhaust air, are of major importance when we are trying to close nitrogen nutrient cycles in sustainable livestock farming (Melse and Timmerman, 2009).

Poison emissions can occur from livestock through the locomotion of animals on dry lots, through the exhaustion of animal-home ventilation systems, or through the on-site or indirect movement of farm machinery through the formation, in part, of secondary aerosol particles from the ammonia emissions from animal manure (Cambra-Lo et al., 2010). The production of photochemical ozone (O_3) is partly motivated by VOC, as VOC can lead to NO_2 oxidation (collectively called NO_x), and O_3 production can lead to the presence of sunlight (Ling and Guo, 2014). The odors are considered to be responses to certain organic volatile compounds, but these odorants are not generally connected to pathogens or faecal indicator microorganisms. The microbial metabolism of various organic materials, such as feed, bedding, and excreta, generates odorous volatile compounds in all animal operations. Various VOCs and fixed gases from swine installations have been reported by Schiffman et al. (2000) (Millner, 2009). Some odorous compounds (NH_3 and H_2S) can accumulate quickly and become respiratory risk for workers and animals in confined indoors with intensive animal stocking densities. Over and above the perimeter of animal facilities, complaints regarding odor often lead to illness (health symptoms).

7.5 IMPACTS ON ENVIRONMENT

As the world populations have been increasing at an exponential rate so the domestic animals have been increased with the same pace with 50% increase in ruminants, 280% in dairy and poultry, and 200% in pork animals. Pork is responsible for huge production rate of 45 million tons/year, while poultry and cattle accounts for proportion of 29 and 42 million tons/year, respectively. Moreover, it is obvious that if the demand of livestock and dairy products keeps growing with the current pace, producers will have to increase the production rate of milk by 13 million tons and of meat by 7 million tons per annum. Hence in developing countries, this huge upsurge will demand upgradation of already practiced adaptive strategies for the management and treatment approaches of livestock waste. It is estimated that world population might reach to 9 billion sooner than 2050 as per capita income increases it leads to an increased rate of animal products such as meat and global dairy consumptions by 58% and 74%, respectively (FAO.org).

The countries with higher demand of livestock and dairy products have a greater number of livestock. Livestock sector has a wider range of environmental impacts (Ilea, 2009). Most concerned environmental impact that livestock industry posed is climate change (McMichael et al., 2007). Livestock industry contributes to 18% of greenhouse gases (GHGs) release as it is stated in the FAO report (third chapter) by Steinfeld et al. (2006). Around 7516 million tons/year of carbon dioxide (CO_2) is discharged into the environment by livestock industry (Ilea, 2009). Goodland and Anhang (2009) (Goodland) disagreed from this calculation and stated that the livestock sector contributes up to 51% of GHGs emission and the estimate for CO_2 is 32,564 million tons globally. This gap between the data of FAO and other literature is because FAO used outmoded data sources ranging around years 1964–2001. However, if we compare the pollution potential of livestock and electrical industry, even at 18% of GHGs emission by livestock industry the former has less pollution potential. On the other hand, transportation industry causes only 13% of emissions and in this case too, the industrial sector is more polluting (Russels6). Carbon dioxide (CO_2), methane (CH_4), nitrous oxide ($N_{-2}O$), and ammonia (NH_3) are the major forms of emissions from the livestock industry (Leytem et al., 2011). Carbon dioxide (CO_2) is one of the major forces behind global warming which is being released by domesticated animals (Goodland). The continuous increase in livestock-keeping practices has raised the chances of CO_2 emission to exceed 565 gigaton by 2030, as warned by researchers. Another greenhouse gas that is being emitted by the livestock industry at 68% rate is nitrous oxide (N_2O); it is persistent in nature and can stay for about 150 years in the environment as well as having 296 times more high global warming potential and ozone deterioration rate as compared with carbon dioxide. Another environmental concern related to the livestock industry is acidification of atmosphere and acid rain that results from ammonia emissions (64%). Thirty-five to forty percent of emission of methane gas with global warming potential 23 times greater than carbon dioxide is being contributed by the livestock industry globally.

It was stated by the United States Environmental Protection Agency that contribution to methane release in the environment has increased by 50% and 37% from both pigs and cattle, respectively (Russles S.).

TABLE 7.1 List of Pollutants Released from Livestock Waste.

Pollutants	Sources
Total dissolved solids	Total dissolved solids (TDS) may include Sodium (Na), Calcium (Ca), Magnesium (Mg), Potassium (K), and Phosphorus (P) forms: animals waste containing >10,000 mg/L
Organic Carbon	Excessive amounts of soluble carbon together with soluble P can quickly reduce O_2 availability in water by raising the biochemical oxygen demand. Residual pesticides: Used to control pests in animal facilities
Residual Metals	Copper (Cu) and Arsenic (As) from animal diets and pesticides
Pharmaceuticals	Antibiotics, growth regulators.
Gases	From waste storage facilities and waste disposal activities, Greenhouse including (CO_2, N_2O), toxic (NH_3, H_2S), Odors—H_2S, mercaptans, indoles, org-sulfides

Source: FAO.

All the activities in the livestock industry do not have same pollution potential as some production practices may need more widespread land such as about one-third of world's total area that is being occupied for livestock raising (Ilea, 2009). Higher consumption and requirements of dairy products along with less area availability are the major causes of deforestation. Forests are being cleared to meet the land demand for the agricultural and livestock practices. During 1990–2000, around two-folds of the area of Portugal was cleared for the livestock and agricultural activities (Ilea, 2009). Fodder production is another cause of cutting forests. Almost 40% crops is being harvested globally for fodder consumption. Half of the quantity of crops consumed by animals as feed can be enough to overcome the food insecurity globally by feeding all of the population under starvation (Ilea, 2009). More than 137 plants, unique insects and animal species are under threat of extinction due to deforestation (World Bank: World Bank Working Paper No: 22). It is record mass extinction in a duration of 65 million years as stated by Ceballos et al., livestock industry and livestock production is the root of many environmental issues such as groundwater and surface water contamination, air and soil pollution. Meanwhile, livestock production in some of the arid and semiarid areas is a highly well-managed system. Livestock production is efficient and a more managed system as compared to cultivation if animals can move; the latter has less recorded data in terms of feeding communities and sustainability particularly under threat of varying climatic conditions. Another environmental concern from agricultural activities is the farming of staple crops, for example rice (Thornton et al., 2007).

The livestock production also results in another environmental concern that is resource wastage, being more precise, the water wastage (FAO, 2006). In United States alone, total water consumption of animals or agriculture sector is about 55% while private residents consume only 5% of total consumption (Worm et al., 2006). A study was conducted to measure the total consumption rate of resources like water, soil, and fertilizers and it was concluded that vegetarian/omnivorous food has more adverse environmental impacts as compared with vegan food (Jacobson, 2006). A vegan user conserves 4164 L water, 2.8 m^2, 20 kg crops, 10 kg carbon oxide (CO_2), and an animal (Ruini et al., 2015).

However, a huge amount of waste is generated due to livestock production activities as 116,000 pounds of waste is generated per second in United States only. Haines and Staley stated that waste generated by 2500 milking cows is equal to the waste from 411,000 residents. Hence, general awareness for livestock industry and production regarding these environmental concerns should be risen in public.

7.6 SOIL POLLUTION

Since decades, livestock waste had been frequently dumped and disposed off on agronomic land without any appropriate nutrient-managing plan. This improper disposal leads to leaching of hazardous elements, over-fertilization of agricultural soil and adjacent area, hazardous run-off hence posing extra risks to surrounding environment. Frequent nonjudicial application of fertilizers/manure beyond the limit of crop demand leads to results in the accumulation of macro-nutrients including nitrogen (N), potassium (K), and phosphorus (P), as well as of heavy metals, specifically copper (Cu) and zinc (Zn) that affect animal health directly by grazing or indirectly by crop feeding (13–15). Endocrine disrupting compounds (EDCs) are another group of prime importance present in livestock waste. These compounds can be found in animal feaces or may symbolize a group of drugs applied in the feed of specific animals. They pose serious health threats to both animals and humans in exposed individuals or their offsprings by hindering the functioning of their hormonal signaling (Edstrom et al., 2011). Endocrine disruptors can impersonate the role of hormones like androgenic and estrogenic as well as obstruct the hormone receptors to change the product of internal signaling processes. The key concern related to EDCs is their performance and adverse effects they start exhibiting at very low concentration of parts per trillions or nano-gram/L. Major sources of EDCs in animal wastes are steroids as well as arsenate contributing for arsenic (Edstrom et al., 2011). The chief outcome of soils overloaded with EDCs or nutrients is the disturbance in soil-water and air elements.

7.7 WATER QUALITY IMPACTS

Sediments, nitrogen, microorganisms, and phosphorus are the major water quality parameters of concern. All of these pollutants are available in livestock waste and can pose significant pollution to surface and groundwater. Water pollution occurs through a variety and complex series of mechanisms. Other farming practices such as nature and management of manure including rate and method of application directly influence the nature and availability of pollutants. Livestock industry contributes to water contamination in two ways: (1) Leaching or runoff of different minerals or chemicals from animal products consumption or production and (2) when these products are directly dumped into nearby water reservoirs. When green manure or animal manure is applied unjustifiably it results in excess nutrients accumulation in plant and soil and eventually these nutrients leech down to groundwater of lost over surface runoff. Mainly, nitrogen (N) and phosphorus (P) are the most concerned nutrients from agriculture with highest potential of water pollution (EEC, 1991). Soil organic matter plays a vital role in immobilization of nitrogen and phosphorus, but a large portion of these nutrients cause surface water pollution by runoff. Free ammonia (NH_3) is more toxic in aquatic ecosystem as compared with ammonium salt (NH^{4+}) because it poses serious threats to some of the species of fishes. For example, a concentration of 5 mg/L or above might have adverse effects on Salmon fish (Cliver, 2009). It was stated by many authors especially by European Union (2014) that livestock waste disjoins many nutrient cycles, promoting indirect short-term nutrient loss or a direct loss that might be long term, relevant to bioaccumulation of nutrients. Serious health hazards including water-borne diseases have been reported due to the consumption of contaminated portable water from livestock and agriculture sources (FAOSTAT, 2013;Gilchrist et al., 2007).

This contamination proves that manure containing huge bacterial mass enters the soil–water system through addition in surface water. There are a number of methods suggested to manage and handle the manure application and their subsequent adverse impacts to safeguard streams and rivers. These methods are not practicable in some of the high-risk areas; hence, farmers are bound with a few options.

7.7.1 WATER POLLUTION

Potential pollutants from manure and other livestock sources use the following routes to be transported from the treated areas' surface:

1. In solution of suspensions
   This category includes some of the organic nitrogen species; for example, uric acid, nitrates, soluble phosphorus, and carbon can be transported in the form of suspension solution.
2. Sorption or sorbed on soil particles
   Ammonium (NH^{4+}) and the phosphorus (P) elements might get sorbed with soil particles and transferred to another area by the process of erosion.
3. In particulate form
   Organic carbon (C), phosphorus (P), and nitrogen (N) can be transported in the form of particulates.

Microorganism can use both ways of transportation, that is, in solution form or by sorption (Edwards and Daniel, 1992). All the potential pollutants that appear in the soil-waste system are controlled by the surface run-off and erosions. The degree of surface and groundwater quality pollution potential from different manures application depends upon a variety of factors affecting erosion and surface runoff such as: rainfall intensity and duration, soil, soil texture, roughness properties of the surface, topography of the land, and so on. In case of water pollution, the risk is directly associated with management practices including time of application and rate of loading. Flow of livestock waste containing a large variety of pollutants may affect the groundwater quality directly or indirectly. As a whole, groundwater pollution is the combination of subprocesses including saturated and unsaturated water movement, diffusion, sorption/desorption, and many more. Management and transformation practices of livestock activities directly impact the nature and amount of pollutants that in return have significant effects on hydraulic properties of soil-waste system and hence on subsurface transportation.

Among others, the most concerned pollutants for researchers are soluble nitrogen and its other forms (specifically NO^{3-}), microorganisms, soluble phosphorus forms, and salts.

7.8 PUBLIC HEALTH FEARS AND FOOD SAFETY

Cattle waste may contain different pathogenic micro-organisms (bacteria, viruses, or parasites), which may pose a health risk in agricultural land after spreading. While certain pathogens are obligatory parasites, others can survive for a long time in the environment. Although there is always a great deal of concern about the potential spread of zoonotic diseases to people, it is livestock disease that accounts for by far the largest number of confirmed illnesses (Burton, 2009). In the case of a disease outbreak in farm animals, manure and related effluents are typically an important vector of transportation and there is a very higher rate of dissemination in the absence of good bio-security and good waste management, with the real risk of transmitting disease from the farms. Evidence of food plants that are contaminated by pathogenic agents has certainly been published and, in some cases this relates to animal manures as fertilizers (Ackers. et al., 1998;Dufour, 2012). Applying dung in any raw food crop can only increase a range of other common pathogens such as assalmonella or campylobacter to increase their risk of contamination. There are still relatively small cases of food crop contamination, but the perceived risk from the retailers and the general public increases the use of this manure. The health risk for people in the near neighborhood of manure (farm staff), however, is high, as can the internal atmosphere (e.g., inhalation of the dust, high levels of ammonia, etc.) as well as the direct contamination of the waste materials present (Burton, 2009). In addition to the plant, land expansion practice and the resulting aerosols would have the main effect upon local communities. This operation creates a nuisance of the odor, but not confused with an infectious dose of a zoonotic disease. There have been several cases of water pollution from zoonosis, but each of these is also a very significant human death case (Guan and Holley, 2003; Cooley et al., 2007). Food pollution is most frequently recorded by zoonosis when manure contaminates water later used for irrigation or contact with food. In the production of leafy vegetables eaten fresh, the irrigation route is especially significant. There can also be a risk of water being used in washing such as vegetables and other foods (Cliver, 2009).

7.9 LIVESTOCK WASTE TREATMENT TECHNOLOGIES

For the management of livestock waste there is the need of development of a closed-loop system that not only increased the livestock waste value but must also be acceptable socially, economically and to the environment with its no negative effects. Carbon sequestration, energy retrieval, nutrient recycling, and water reuse, especially in the regions facing the problem of drought stress, is the principal goal (Harrington and McInnes 2009). Applicable protocols, and needs and demands of particular farm lead to the treatment objectives of manure or livestock waste. But nutrient management, manure stabilization, pathogen reduction, utilization and proper storage, odor reduction, and reduction in the release of hydrogen sulfide, ammonia, and greenhouse gases are the common objective of manure treatment.

Manure treatment results:

1. The reduced manure mass makes it easy to transport and apply to fields or crop lands.
2. Less land is required to apply manure due to reduced nutrient's amount. To use livestock waste as fertilizer, "modified" manure is produced which has right agronomic proportions of nutrients through treatment technology.
3. Controlled antibiotics and pathogens. Because of the reduction in antibiotics and pathogen concentration in livestock waste, it diminishes biosecurity and risks to human health.
4. Release of useful by-products. There is the production of energy (heat, electricity), organic materials, and fertilizer by treatment process which can be used on farm or sold to gain income for farm management (Ogejo, 2009).

Manure treatment principles are grouped into physical, chemical, and biological processes. Livestock waste treatment technologies are typically amalgamation of these rudimentary processes. Generally, the common methods of treatment fall under the biotechnological treatments category. On farm level, only anaerobic treatment (biogas) or composting can be practical due to cost reasons. Anaerobic treatment method is becoming popular and important particularly; that is why while doing this treatment hygienic measures must be considered. In case of bursts of notifiable contagious diseases the treatment of livestock waste by the physical and chemical processes is useful. There will be unswerving results of both these processes. To avoid the disease spread via the environment, disinfection of livestock waste by the chemical process is a preferred way of treatment and it also depends on type of infective agents (Martinez et al., 2009).

7.9.1 MICROBIOLOGICAL PROCESSES

Anaerobic bio digester for animal manure treatment is one of the most broadly used and popular technologies. Biogas produced from anaerobic bio digester can also be utilized for energy production in the form of electricity or heat (Sommer, 2013; Cantrell et al., 2008). Swine producers can get supplementary benefits by using processed sludge and effluents as fertilizers in soil, and gain carbon credit as well, which can be accomplished through financial institutions and broker companies. However, certain restrictions are present in anaerobic and bio digester system. Disposing of manure through bio digester is a partial solution of the problem, because it is considered that bio digester do not eliminate phosphorus and nitrogen from effluents, nor they condense the crop area required to fascinate these nutrients. In anaerobic digestion, carried out by microorganisms biogas, generally carbon dioxide and methane is produced due to break down of complex livestock waste. Hydrolysis, fermentation, and methanogenesis are three main processes that occur in anaerobic digestion. Hydrolysis breaks down the complex compounds into soluble compounds. These soluble compounds are then available to fermentative microorganisms (acetogenic bacteria and acidogenic bacteria) which convert them into acetic acid, alcohol, volatile fatty acids, and off gases containing CO_2 and H_2. Methane (60–70%) and carbon dioxide (30–40%) are then produced by the metabolism of these intermediate products. Liquid portion separated from digestate by a solid liquid separator can be utilized directly as fertilizer or may be transformed into secondary fertilizer by struvite crystallization and ammonia stripping (Hidalgo et al., 2014). Biogas production is increased by combined digestion of livestock waste and other biomass substrates in mixed reactors, besides this it offers many other advantages like management of organic waste and animal manure. Codigestion of digestible waste of organic nature generated after food industry and animal manure is very imperative due to socioeconomic aims and for biogas plant's economy (Nieto et al., 2012).

It is stated that anaerobic digestion is increased by the pretreatment of manure that subsequently enhances the biogas production (Thornton et al., 2007). Anaerobic digestion system is a two-phase process and each digester has different bacteria depending upon selection and enhancement. In the first phase, acideogenic bacteria break down the complex pollutants in the form of volatile fatty acids, whereas in the second phase, acetogenic and methogenic bacteria transform these fatty acids into methane (CH_4) and carbon dioxide (CO_2). To avoid the over-burdening and pile-up of toxic by-products, the acidification phase is controlled by this two-phase arrangement that subsequently increases the steadiness of the treatment. Furthermore, the first phase might serve as metabolic buffer favoring the factors that inhibit the formation of methanogenesis but aiding the acideogenic phase such as: lowers the pH, increases organic rate, reduces hydraulic retention time (HRT), and inhibits the pH shock methanogenic bacteria (Hidalgo et al., 2012). Composting is also a substitutive method used to handle and manage the livestock waste (Kunz et al., 2009). In this process, liquid manure is converted into a solid form by evaporation of water. This treatment decreases the volume of manure meanwhile enhancing the nutrient contents, but it does not decrease the manure application of the total crop area. Hence, composting is one of the finest waste management technologies with minimum manure transportation costs. Such practices may be economically beneficial for the farm owners because it increases the income by selling an organic compound having high agronomic cost.

Some physical and chemical properties of animal manures hinder the process of composting hence decreasing the efficiency of treatment such as surplus of moisture, minimum porosity, high pH index, higher Nitrogen (N) concentration for organic carbon (C) that is subsequently responsible for lowering C/N ratio. Consequently, accurate composting management is necessary to attain best quality compost. For this purpose, different aeration approaches such as bulking agents, process controlling options, and substrate conditioning feedstock are being observed to increase the compost quality as well as to decrease the composting time (Liu et al., 2011). There are different methods of composting such as in windrows or reactors, in open air or under any shade/roof, based on the process parameters needed for efficient hygenization.

All the waste management technologies following the aerobic processes have surplus supply of oxygen to manure to support the microbes to:

1. Support biological oxygen demand (BOD) or chemical oxygen demand (COD)
2. Decompose organic content of the manure
3. Degrade nitrogen and phosphorus content by biological uptake

4. Handle nitrogen by the process of denitrification (conversion of manure into gaseous form of nitrogen)

The biological aerobic treatment of manure was established with the objective of excessive nitrogen removal, but later a series of supplementary steps were appended to manage the excess phosphorus content as well. The most extensive treatment procedure consists of: separation of liquid and solid phase by a filtering system followed by a mechanical separation in which solid phase is subjected to composting whereas liquid phase undergoes biological treatment. A single reactor is used for biological treatment that is repeatedly supplied with oxygen and other favorable conditions to carry out the processes of denitrification and nitrification (Ali, 2004). The economic cost to run this treatment is because of constant supply of oxygen required in filtration and nitrification of raw material (manure). Hence, anammox (anaerobic ammonium oxidation) treatment with lower oxygen requirements can be a possible alternate of aerobic treatment with high cost (Magrı et al., 2013).

Separated liquid fractions are treated through sequencing batch reactors (SBRs), most efficient treatment plants for the activated sludge. A single reactor is used for the series of deferred cycles that are done in different reactors when carried out in continuous flow treatment and it permits the assimilation of oxygen-deprived anaerobic anoxic-oxic environment. SBRs are more flexible because it has capacity to increase or decrease the time required in different stages rather than a fixed reaction time as compared with continuous flow configuration. Better efficiency results in increased removal of nitrogen (N) depending on desired nutrient demand for discharge or it may enhance the nitrogen removal efficiency (Scaglione et al., 2013). Meanwhile, some scientists are not in favor of accepting aerobic manure treatment (involving nitrification and denitrification) as a sustainable approach because this treatment involves continuous energy consumption, releases nitrogen into environment (NO_x), and the end process of denitrification is di-nitrogen gas (N_2) that cannot be consumed by crops or plants; hence, it destroys the nitrogen in the form of ammonia, nitrates, and nitrites (Melse and Timmerman, 2009). On the other hand, waste management solutions with low technology such as soil filter system or constructed wetlands are preferable over such aerobic treatments because while treating livestock waste they exploit natural sources including sun, land, seeds, wind, etc. (Martinez et al., 2009; Harrington and McInnes, 2009). Such low-technology solutions are economical with lower treatment costs, but efficiency is also compromised.

7.10 PHYSICAL PROCESS

7.10.1 LIQUID–SOLID SEPARATION

Manure can be split up into two separate categories to reduce the handling or transportation cost and enhance its fertilizer value:

1. Liquid part that is to be used on farm.
2. Dry part of manure that is nutrient rich and can be used as fertilizer or may be transported to other areas for further treatment.

This separation plays a vital role in lessening the odor emission as well as helps to yield energy-rich material to be used in biogas production or for incineration (Hjorth et al., 2010). Liquid–solid separation also encourages the removal of nutrients from farm manure as well as the addition of secondary step of sedimentation or other settling techniques such as screening, centrifuging or belt presses play significant role in preventing accumulation of manure particles in subsequent chemical or biological processes involved in the multi-stage advanced treatment system (Christensen et al., 2013; Riaño and Garcıá-González,2014). Sedimentation using a thickener (consisting of a container that is cylindrical at the top and conical shape at the bottom) is the cheapest and most economical technology for the separation. Use of a rake of vibration can be done for the enhancement of settling as well as transferring of solid particles settled on the upper portion on the container (Christensen et al., 2013).

A particular separation efficiency can be attained by enhancing the gravitational force that in return decreases the time duration needed for settling and this is achieved by using centrifuge force in a vertical or horizontal centrifuge. Solids can be separated from liquids by screens and filters separators with the application of gravity. Mostly, filter separators are actually screens containing rotating perforated cylinder with an area to load waste material and a removal system of scrapers. Liquid is passed out of screen and then poured off.

Another approach working under the principle of phase separation is pressurized filtration and the tools for applied pressure used in this process are screw press or press auger. Generally, the hierarchy of mechanical separators for the separation efficiency of dry material and phosphorus is as: centrifuge>sedimentation>non-pressurized filtration>pressurized filtration. Separation efficiency can be increased by treatment with flocculants prior to separation (Bolan et al., 2009) and the most commonly practiced mechanical separation techniques are filter belts or screens usually used for flocculated slurry (Guerdat et al., 2013). Meanwhile, we cannot ignore the pros of separation techniques as animal slurry separation may give rise to another issue such as an alteration in ratio of plant nutrients to heavy metals in vegetative parts. It was revealed in recent studies that some of heavy metals such as zinc (Zn), copper (Cu), and cadmium (Cd) were converted into solid fraction when treated under solid–liquid separation combined with flocculation (Møller et al., 2007; Kwon et al., 2013). Moreover, the extracts in separators such as polymers and aluminum may cause significant environmental issues (Nahm, 2005). Once the initial liquid–solid separation is done, all the dissolved nutrients like potassium (K), phosphorus (P), and nitrogen (N) are separated and concentrated through membrane filtration of liquid phase. This separation process forms a liquid phase with excessive nutrient and hence pure water that can be reused for safe spill in the environment (Hjorth et al., 2010). Areas having higher density of livestock are under potential threat of excessive nutrients accumulation. Membrane separation technology could be a possible alternate for the collection of manure nutrients in small amount that is further transported to different regions depending upon nature of use. One of the most effective solid–liquid separators are microfiltration and ultrafiltration because they have maximum efficiency to segregate the nutrients attached with particles as phosphorus. Also, nanofiltration or reverse osmosis can be used for the isolation of nutrients from ammonia and potassium (Masse et al., 2013).

7.11 THERMAL TREATMENTS

It is a kind of physical treatment involving high temperature to disrupt the organic matter bonding and to convert the resulting products immediately into hydrocarbon fuels, synthetic gas, or charcoal residuals. Prolonged reaction time is needed in biological treatment whereas multiple and complex end-products can be formed within minimum time during thermal conversion (Wu et al., 2012). Combustion, pyrolysis, gasification, and liquefaction are different stages of thermal conversion. Such treatment is considered to be environment friendly in many ways such as: it reduces pollution potential of waste, decreases volume of waste, has less odor problems and energy production from manure. As thermal processes need intense heat supply heat recovery is an important part to make such processes practicable (Cantrell et al., 2008).

Combustion directly converts the energy of manure into the heat though it is not a suitable approach to store energy for a longer period of time.

Furthermore, the particulate matters in the form of ash have not any feasible reuse or recycling (Edstrom et al., 2011). Pyrolysis involves intense heat and anaerobic conditions for the conversions of organic ratio of manure into volatile gases including condensable tars (oxygenated hydrocarbons) and noncondensable vapors that in return make pyrolytic oil or bio-oil (Ro et al., 2010). Char is produced by slow pyrolysis of animal waste that is economically beneficial for farmers because of energy production and generation of carbon credit from carbon sequestration. Green coal can be a potential feedstock source for different existing plants such as gasification and combustion plants; also it enhances soil fertility when used as an amendment (Liu et al., 2013).

Lignocellulosic portion of the biomass is hydrolyzed by the liquefaction and transformed into lighter organic oils (bio-oils). As compared with pyrolysis, liquefaction needs pressurized environment (5–20 MPa) and usually requires low temperature (250°C–350°C). In a batch reactor with carbon monoxide-rich environment carrying temperature range from 285°Cto 350°C, volatile solids are converted into oil at a boiling point, but if there is slurry in reactor, temperature may rise over 100°C on atmospheric pressure. All the volatile organic compounds (VOCs) as ammonia and free fatty acids and water molecules tend to evaporate at this temperature. Regaining of energy utilized for the evaporation of the water is done through the condensation of the vapors phase. This process has higher energy consumption as evaporation of per ton of water denotes 670 kWh of energy needed (Hjorth et al., 2010).

7.12 CHEMICAL PROCESS

Major chemical processes for the treatment of farm manure may include chemical precipitation, chemical coagulation, chemical oxidation, chemical stabilization, chemical disinfection, and ion exchange. Some heavy metals, nitrogen (N), and phosphorus (P) can be removed or recovered by combining chemical method with physical processes. Mechanical liquid–solid separation can be done using chemical pretreatments such as flocculation, coagulation, and precipitation. Recently, studies have shown removal efficiency of physiochemical processes as: suspended solids removal (32%), total organic carbon (78%), and nutrients (82% N and 50% P) when farm manure was treated with an oxidation step following the coagulation/flocculation (Chelme-Ayala et al., 2011). A significant decrease in coliform was also observed. Manure can be treated through chemical disinfection using different substances such as caustic soda, per-acetic acid, lime wash, calcium cyanamide, and so on. Usually, such treatments are to be used during an epidemic and are not routinely practiced. To make chemical disinfection more effective it is necessary to cautiously mix the slurry and stir well before addition of chemicals.

7.13 GLOBAL PRODUCTION, GENERAL TRENDS, AND PERSPECTIVES

The current world population is unevenly distributed economically and socially. Five billion people live in the "developing world" whereas the "developed world," sometimes called "The West," represents less than one billion people.

As the economy of these countries improves, their demand for animal products is likely to increase, along with a greater disparity between rural and urban economies resulting in a greater urban population growth. The disparity between rural and urban economies results from the fact that the price of farm products has not increased since the early 1970s, while all other costs have. As a result of a greater income, a diet richer in meat, milk, and eggs can improve human nutrition. In contrast to the developed world where many people eat too much animal products, most people in developing countries eat too little. Since 1960, the world population has doubled while animal numbers have increased by 50% for ruminants, 200% for pork and 280% for poultry. From the global distribution of major animal types (Figure 7.1), cattle numbers are more evenly distributed than those for swine. Asia still contributes about one-third of the cattle production and more than half of the swine production. For swine production, Europe's share is about 20%. The regional concentration of pigs and pig meat production shows that the 10 largest producers account for 74% to the global pig stock (Windhorst, 2006).

In terms of meat, pork accounts for the largest proportion at 80 milliontons per year compared to 50 milliontons each for cattle and poultry. The total annual global meat production is estimated between 200 and 230 milliontons. Almost half of the world's chicken population is concentrated in Asia. The regional concentration in chicken meat and egg production is particularly high: more than 64% of the global chicken meat production is concentrated in 10 leading countries, almost half coming from the USA, China, and Brazil. The trends in world population (Delgado et al., 1999; Ehui et al., 1998) and the concurrent growth in demand for livestock products suggest that by 2020, annual production will have to grow by another 200 billion

L of milk and 100 milliontons of meat (Figure 7.2). Such a large increase will require more than the simple adaptation of current livestock farming practices as they exist in Europe and North America.

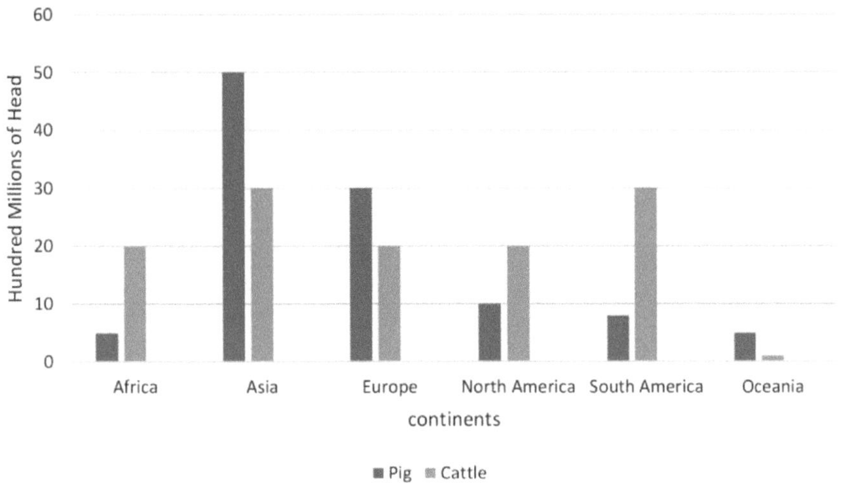

FIGURE 7.2 Estimates of pig and cattle numbers by continents in 2004.

Source: FAO—Database, from Windhorst (2006).

7.14 SUMMARY

Since the industrial revolution, the farming and livestock activities have been upgraded to the industrial level. The production as well as consumption rate has increased and there is more waste produced than to be recycled. Livestock waste may consist of poultry shed run-off, wastewater, fertilizers, leftover food, animal faeces or excreta, solid, slurry or the chemicals from pesticides and so on. This bulk quantity of waste has affected the soil, water, air, and environment as well as human health in many ways. To minimize these adverse effects and due to awareness regarding consequences of man-made climate change and environmental, waste is now being treated prior to disposing off. There are a number of approaches used for the waste recycling or treatment depending upon the nature of waste. Chemical processes such as chemical oxidation, chemical stabilization, chemical disinfection, physical processes including liquid–solid separation, and thermal processes consisting of combustion at high temperature to break the bonds are the major waste management techniques to tackle the livestock waste.

KEYWORDS

- **industrialization**
- **farming**
- **industrial operations**
- **wastewater**
- **slurry**
- **anthropogenic sources**
- **thermal process**

REFERENCES

Ackers, M.; Mahon, B.; Leahy, E. An Outbreak of *E. coli* O157: H7 Infections Associated with Leaf Lettuce Consumption. *J. Infect. Dis.* **1998,** *177* (6), 1588–1593.

Ali, S. A Socio-Ecological Autopsy of the *E. coli* O157:H7 Outbreak in Walkerton, Ontario, Canada. *Soc. Sci. Med.* **2004,** *58* (12), 2601–2612.

Bernard, E.; Larkin, R.; Tavantzis, S.; Erich, M.; Alyokhin, A.; Sewell, G.; Gross, S. Compost, Rapeseed Rotation, and Biocontrol Agents, Significantly Impact Soil Microbial Communities Inorganic and Conventional Potato Production Systems. *Appl. Soil. Ecol.* **2012,** *52*, 29–41.

Bezanson, G.; Ells, T.; Prange, R. Effect of Composting on Microbial Contamination and Quality of Fresh Fruits and Vegetables-a Mini-Review. In: International Symposium on Organic Matter Management and Compost Use in Horticulture, ISHS. *Acta Hort.* **2014,** *1018*, 631–638.

Bolan, N.; Laurenson, S.; Luo, J.; Sukias, J. Integrated Treatment of Farm Effluents in New Zealand's Dairy Operations.*Bioresour. Technol.* **2009,** *100*(22), 5490–5497.

Burton, C. Reconciling the New Demands for Food Protection with Environmental Needs in the Management of Livestock Wastes. *Bioresour. Technol.* **2009,** *100* (22), 5399–5405.

Burton, C.; Turner, C., Eds. Manure Management-Treatment Strategies for Sustainable Agriculture, 3rd ed.; Silsoe Research Institute: Bedford, 2003.

Cambra-Lo, pez, M.; Aarnink, A.; Zhao, Y.; Calvet, S.; Torres, A. Airborne Particulate Matter from Livestock Production Systems: A Review of an Air Pollution Problem. *Environ. Pollut.* **2010,***158*, 1–17.

Cantrell, K.; Ducey, T.; Ro, K.; Hunt, P. Livestock Waste-to-Bioenergy Generation Opportunities.*Bioresour. Technol.* **2008,** *99*, 7941–7953.

Christensen, M.; Christensen, K.; Sommer, S. Solid-Liquid Separation of Animal Slurry. In *Animal Manure Recycling: Treatment and Management*; Sommer, S., Ed.; Wiley: New York, 2013; pp 105–130.

Cliver, D. Disinfection of Animal Manures, Food Safety and Policy. *Bioresour. Technol.* **2009,** *100* (22), 5392–5394.

Combalbert, S.; Bellet, V.;Dabert, P.; Bernet, N.; Balaguer, P.; Hernandez-Raquet, G. Fate of Steroid Hormones and Endocrine Activities in Swine Manure Disposal and Treatment Facilities. *Wat. Res.* **2012,** *46* (3), 895–906.

Cooley, M.; Carychao, D.; Crawford-Miksza, L.; Jay, M.; Myers, C.; Rose, C.; Keys, C.; Farrar, J.; Mandrell, R. Incidence and Tracking of *Escherichia coli* O157:H7 in a Major Produce Production Region in California. *PLoSOne* **2007,** *2*, e1159.

Dammgen, U.; Amon, B.; Hutchings, N.; Haenel, H.; Rosemann, C. Data Sets to Assess Methane Emissions from Untreated Cattle and Pig Slurry and Solid Manure Storage Systems in the German and Austrian Emission Inventories. *Arch. Natur. Schutz. Land schafts for schung* **2012,** *62* (1–2), 1–20.

Delgado, C.; Rosegrant, M.; Steinfeld, H.; Ehui, S.; Courbois, C. *Livestock to The Next Food Revolution*; International Food Policy Research Institute, Food and Agriculture Organisation of the United Nations, International Livestock Research Institute: Washington, DC, 1999.

Dufour, A., Ed. *Animal Waste, Water Quality and Human Health*; IWA Publishing: London, 2012.

Edstrom, M.; Schußler, I.; Luostarinen, S. Combustion of Manure: Manure as Fuel in a Heating Plant. In*Baltic Forum for Innovative Technologies for Sustainable Manure Management*; Technical Report, 2011.

Edwards, D.; Daniel, T. Environmental Impacts of On-Farm Poultry Waste Disposal—A Review. *Bioresour. Technol.***1992,** *41* (1), 9–33.

EEC/91/676, O.J. NL 375.Protection of Waters against Pollution Caused by Nitrates from Agricultural Sources, Dec 31, 1991; p1.

Ehui, S.; Li-Pun, H.; Mares, V.; Shapiro, B. The Role of Livestock in Food Security and Environmental Protection. *Outlook Agric.***1998,** *27* (2), 81–87.

European Union. Endocrine Disruptors Website: How the European Union Uses the Precautionary Principle to Tackle Endocrine Disruptors, 2014.http://ec.europa.eu /environment/endocrine/definitions/endodis_en.htm(accessed Feb 7, 2014)

FAOSTAT,2013.http://faostat3.fao.org. http://www.fao.org/3/a-a0261e.pdf

Garg, M.; Makkar, H. Balanced Feeding for Improving Livestock Productivity: Increase in Milk Production and Nutrient Use Efficiency and Decrease in Methane Emission. *Anim. Prod. Health.* **2012,** 173.

Gilchrist, M.; Greko, C.; Wallinga, D. The Potential Role of Concentrated Animal Feeding Operations in Infectious Disease Epidemics and Antibiotic Resistance. *Environ. Health. Perspect.* **2007,** *115*, 313–316.

Goodland, R.; Anhang, J. *Livestock and Climate Change: What If the Key Actors in Climate Change Are…Cows, Pigs, and Chickens?* World Watch: Washington, DC, 2009; pp 10–19.

Guan, T.; Holley, R. Pathogen Survival in Swine Manure Environments and Transmission of Human Enteric Illness—A Review. *J. Environ. Qual.* **2003,** *32*, 383–392.

Halden, U.; Schwab, J. Environmental Impact of Industrial Farm Animal Production. A Report of the Pew Commission on Industrial Farm Animal Production, 2008.

Harrington, R.; McInnes, R. Integrated Constructed Wetlands (ICW) for Livestock Wastewater Management. *Bioresour. Technol.* **2009,** *100* (22), 5498–5505.

Hidalgo, D.; Corona, F. A.;lamo, J.;Aguado, A. Resource Recovery from Anaerobic Digestate: Struvite Crystallisation versus Ammonia Stripping. In *Abstracts of the International Congress on Water, Waste and Energy Management*, Oporto. July 16–18, 2014.

Hidalgo, D.; Sastre, E.; Gomez, M.; Nieto, P. Evaluation of Pre-treatment Processes for Increasing Biodegradability of Agro-Food Wastes. *Environ. Technol.* **2012,** *33* (13), 1497–1503.

Hjorth, M.; Christensen, K.; Christensen, M.; Sommer, S. Solid-Liquid Separation of Animal Slurry in Theory and Practice. A Review. *Agron. Sustain. Dev.* **2010,** *30* (1), 153–180.

Hooda, P.; Edwards, A.; Anderson, H.; Miller, A. A Review of Water Quality Concerns in Livestock Farming Areas. *Sci. Total. Environ.* **2000,** *250*, 143–167.

Ilea, R. C. Intensive Livestock Farming: Global Trends, Increased Environmental Concerns, and Ethical Solutions. *J. Agric. Environ. Ethics*. **2009,** *22*, 153–167.

Jacobson, M. F. *More and Cleaner Water. Six Arguments for a Greener Diet: How a More Plant-based Diet Could Save Your Health and the Environment*. Center for Science in the Public Interest: Washington, DC, 2006.

Junior, C.; Cerri, C.; Dorich, C.; Maia, S.; Bernoux, M.; Cerri, C. Towards a Representative Assessment of Methane and Nitrous Oxide Emissions and Mitigation Options from Manure Management of Beef Cattle Feedlots in Brazil. *Mitigat. Adapt. Strate. Global Change* **2013,** *20*(3).

Kunz, A.; Miele, M.; Steinmetz, R. Advanced Swine Manure Treatment and Utilization in Brazil. *Bioresour. Technol.* **2009,** *100* (22), 5485–5489.

Kwon, S.; Jang, Y.; Owens, G.; Kim, M.; Jung, G.; Hong, S.; Kim, K. Long-Term Assessment of the Environmental Fate of Heavy Metals in Agricultural Soil after Cessation of Organic Waste Treatments. *Environ. Geochem. Health* **2013,** 1–11.

Leytem, A.; Dungan, R.; Bjorneberg, D.; Koehn, A. Emissions of Ammonia, Methane, Carbondioxide, and Nitrous Oxide from Dairy Cattle Housing and Manure Management Systems. *J. Environ. Qual.* **2011,** *40* (5), 1383–1394.

Ling, Z.; Guo, H. Contribution of VOC Sources to Photochemical Ozone Formation and Its Control Policy Implication in Hong Kong. *Environl. Sci. Policy* **2014,** *38*, 180–191.

Liu, J.; Xu, X.; Li, H.; Xu, Y. Effect of Microbiological Inoculaon Chemical and Physical Properties and Microbial Community of Cow Manure Compost.*Biomass.Bioenerg.* **2011,** *35* (8), 3433–3439.

Liu, X.; Li, Z.; Zhang, Y. Energy Balance Analysis on the Slow Pyrolysis Process of Cattle Manure. *Appl. Mech. Mater.* **2013,** *392*, 531–534.

Magrı, A.; Beline, F.; Dabert, P. Feasibility and Interest of the Anammox Process as Treatment Alternative for Anaerobic Digester Supernatants in Manure Processing—An Overview. *J. Environ. Manage.* **2013,** *131*, 170–184.

Martinez, J.; Dabert, P.; Barrington, S.; Burton, C. Livestock Waste Treatment Systems for Environmental Quality, Food Safety, and Sustainability. *Bioresour. Technol.* **2009,** *100* (22), 5527–5536.

Masse, D.; Gilbert, Y.;Saady, N.; Liu, C. Low-Temperature Anaerobic Digestion of Swine Manure in a Plug-Flow Reactor. *Environ, Technol.***2013,** *34* (18), 2617–2624.

May, S.; Romberger, D.; Poole, J. Respiratory Health Effects of Large Animal Farming Environments.*J. Toxicol. Environ. Health B.* **2012,** *15* (8), 524–541.

McMichael, A. J.; Powles, J. W.; Butler, C. D.; Uauy, R. Food, Livestock Production, Energy, Climate Change, and Health. *Lancet* **2007,** *370*, 1253–1263.

Melse, R.; Timmerman, M. Sustainable Intensive Livestock Production Demands Manure and Exhaust Air Treatment Technologies. *Bioresour. Technol.* **2009,** *100* (22), 5506–5511–10.

Millner, P. Bioaerosols Associated with Animal Production Operations. *Bioresour. Technol.* **2009,** *100* (22), 5379–5385.

Møller, H.; Jensen, H.; Tobiasen, L.; Hansen, M. Heavy Metal and Phosphorus Content of Fractions from Manure Treatment and Incineration. *Environ. Technol.* **2007,** *28*, 1403–1418.

Nahm, K. Environmental Effects of Chemical Additives Used in Poultry Litter and Swine Manure. *Crit. Rev. Environ. Sci. Technol.* **2005,** *35*, 487–513.

Nieto, P.; Hidalgo, D.; Irusta, R.; Kraut, D. Biochemical Methane Potential (BMP) of Agrofood Wastes from the Cider Region (Spain). *Water. Sci. Technol.* **2012,** *66* (9), 1842–1849.

Ogejo, J. Selecting a Treatment Technology for Manure Management. Virginia Cooperative Extension, 2009; pp 442–306.http://pubs.ext.vt.edu/442/442-306/442-306.html(accessed Feb 11, 2014).

Riaño, B.; Garcıá-González, M. On-Farm Treatment of Swine Manure Based on Solid-Liquid Separation and Biological Nitrification- Denitrification of the Liquid Fraction. *J. Environ. Manage.* **2014,** *132*, 87–93.

Ro, K.; Cantrell, K.; Hunt, P. High-Temperature Pyrolysis of Blended Animal Manures for Producing Renewable Energy and Value-Added Biochar. *Ind. Eng. Chem. Res.* **2010,** *49*(20), 10125–10131.

Ruini, L. F.; Ciati, R.; Pratesi, C. A.; Marino, M.; Principato, L.; Vannuzzi, E. Working toward Healthy and Sustainable Diets: The "Double Pyramid Model" Developed by the Barilla Center for Food and Nutrition to Raise Awareness about the Environmental and Nutritional Impact of Foods. *Front. Nutr.* **2015,** *2*, 1–6.

Russell, J. B.; Rychlik, J. L. Factors That Alter Rumen Microbial Ecology.*Science* **2001,** *292*, 1119–1122.

Scaglione, D.; Tornotti, G.; Teli, A.; Lorenzoni, L.; Ficara, E.; Canziani, R.; Malpei, F. Nitrification Denitrification via Nitrite in a Pilot-Scale SBR Treating the Liquid Fraction of Co-digested Piggery/Poultry Manure and Agro-Wastes. *Chem. Eng. J.* **2013,** *228*, 935–943.

Schiffman, S.; Walker, J.; Dalton, P.; Lorig, T.; Raymer, J.; Shusterman, D.; Williams, C. Potential Health Effects Odor from Animal Operations, Wastewater Treatment, and Recycling of Byproducts. *J. Agromed.* **2000,** *7*, 7–81.

Sommer, S. Animal Manure from Waste to Raw Materials and Goods. In *Animal Manure Recycling: Treatment and Management*; Sommer, S., Ed.; Wiley: New York, 2013; pp 1–4.

Steinfeld, H.; Gerber, P.; Wassenaar, T.; Castel, V.; Rosales, M.; de, Haan, C. *Livestock's Long Shadow*; FAO: Rome, 2006.

Thorne, P. Environmental Health Impacts of Concentrated Animal Feeding Operations: Anticipating Hazards Searching for Solutions. *Environ. HealthPerspect.* **2007,** *115*, 296–307.

Thornton, P.; Herrero, M.; Freeman, A.; Mwai, O.; Rege, E.; Jones, P. and McDermott, J. Vulnerability, Climate Change and Livestock-Research Opportunities and Challenges for Poverty Alleviation. *SAT e. J.* **2007,** *4*(1), 1–23.

Windhorst, H. W. Regional Patterns of Livestock and Poultry Production in Europe. In *Livestock Production and Society*; Geers, R., Madec, F., Eds.; Wageningen Academic Publishers: The Netherlands, 2006; p 307.

Worm, B.; Barbier, E. B.; Beaumont, N.; Duffy, J. E.; Folke, C.; Halpern, B. S.; Sala, E. Impacts of Biodiversity Loss on Ocean Ecosystem Services. *Science* **2006,** *314*, 787–790.

Wu, H.; Hanna, M.; Jones. D. Thermogravimetric Characterization of Dairy Manure as Pyrolysis and Combustion Feedstocks. *Waste Manage. Res.* **2012,** *30*(10), 1066–1071.

CHAPTER 8

Radioactive Wastes: Management by Potential Treatments

ARYADEEP ROYCHOUDHURY* and SWARNAVO CHAKRABORTY

Department of Biotechnology, St. Xavier's College (Autonomous), 30, Mother Teresa Sarani, Kolkata 700016, West Bengal, India

Corresponding author. E-mail: aryadeep.rc@gmail.com

ABSTRACT

The by-products left out after the exploitation of radioactive materials for the production of nuclear power in nuclear reactors, generation of nuclear weapons, and the utilities of nuclear fission reactions and other forms of nuclear applications in the medical and research fields are usually termed as radioactive wastes. Radioactive wastes pose havoc threats to the environment and also to different forms of life existing on this planet. The use of radioactive materials has increased every year since last few decades, leading to the massive accumulation of radioactive wastes in nature. Due to improper disposal of these wastes, the exposure of human and other living beings to the harmful irradiation from these radioactive wastes has led to the progressive rise in health issues and other potential dangers in the society. Thus, it has become necessary to ponder upon the management and proper handling of these wastes to control the rise in the number of casualties each year due to exposure to these radioactive wastes irradiation. Therefore, this paper tries to focus on the necessary treatments that are involved in the proper processing, storage, and discarding of harmful radioactive wastes to render these wastes relatively harmless to the environment.

Waste Problems and Management in Developing Countries. Umair Riaz, Shazia Iqbal, & Moazzam Jamil (Eds.)

8.1 INTRODUCTION

Any form of waste containing radionuclides at higher concentrations or activities more than the threshold of clearance from the environment is termed as radioactive waste. These wastes constitute the leftover debris from radioactive materials used in nuclear power plants, research laboratories, medical sectors, and in the development of nuclear weapons (Khelurkar et al., 2015). Radioactive wastes are generated from a range of sources, which include naturally occurring radioactive materials (NORM) as the minor source, along with industries associated with the generation of nuclear fuel and nuclear weapons, and other research and hospital sectors as some of the major sources (Khelurkar et al., 2015). The use of radioactive materials has progressively increased during the past few decades and thereby increasing the rate of production of radioactive wastes and their associated hazardous effects on human health and environment. Currently, it has become mandatory to curb the long-term ill effects of radioactive wastes, by the proper storage, processing, treatment, and disposal of these wastes (Figure 8.2). Thus, it is evident that extensive research pertaining to radioactive waste management is a must (Weber et al., 2009).

8.2 TYPES OF RADIOACTIVE WASTES

According to the Environmental Protection Agency (EPA), the radioactive wastes can be classified into six major categories (Figure 8.1), which are as follows:

1. Low-level wastes (LLW): these wastes usually constitute the contaminated industrial wastes.
2. High-level wastes (HLW): these wastes are associated with Spent Reactor fuel and other forms of potent radioactive wastes generated from reprocessing industrial sectors.
3. Transuranic wastes (TRUW): generated as a by-product in the generation of nuclear weapons.
4. Spent nuclear fuel (SNF): associated with many reactor units.
5. Uranium mill tailings: due to uranium ore milling and mining.
6. Naturally occurring radioactive materials (NORM): usually constitute all forms of radioactive elements occurring in the natural environment. However, human intervention has led to the increment in the levels of these materials in the nature, as compared with the level

that is expected to occur under unaltered environmental conditions. A range of industrial activities tops the list in their contribution to the levels of NORM, including oil industries, coal smelting, gas industries, smelting, and mining of metals.

However, this classification can vary from one country to another, where according to the form, wastes of radioactive origin can be classified as gaseous or liquid or solid radioactive wastes or can simply be classified on the basis of the elements constituting the waste, such as uranium (U), neptunium (Np), actinium (Ac), plutonium (Pu), and thorium (Th) (Wang et al., 2010).

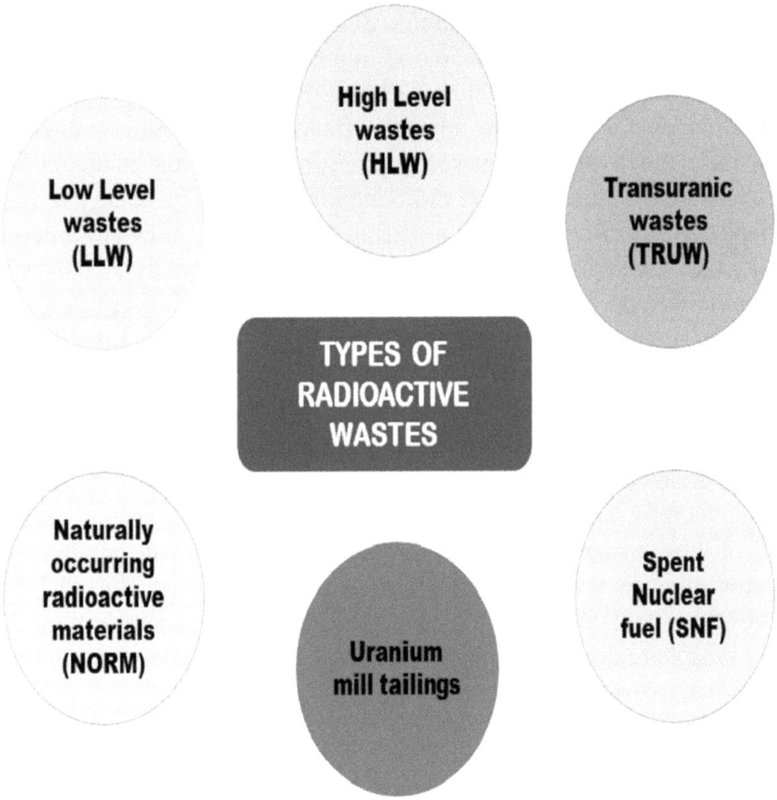

FIGURE 8.1 Classification of radioactive wastes.

8.3 POTENTIAL HAZARDS DUE TO RADIOACTIVE WASTES

Radioactive wastes pose serious threats to life, including plants, animals, and human extensively. The presence of radionuclides dispersed in the rhizospheric soil and surface water can lead to exposure of animals and plants to lethal doses of radiation, leading to havoc damages. The massive ill effects of radiation are observed in humans. The physical effects of radiation exposure may be acute or chronic or even long-term effects. Acute effects of radiation are associated with exposure to massive doses of radiation in a very small period of time, while those associated with chronic effects are due to long exposure with very low doses of radiation. On the other hand, long-term effects of radiation are rather manifested late, and usually occur due to radiation exposure for a period of more than 6 months. The common ailments associated with the long-term effects include aplastic anemia, blood cancer (leukemia), cataracts, malignant tumors, and birth defects in fetuses (Hox and Roos, 2008).

Radioactive wastes pose adverse effects on the environment as well. Most radioactive wastes that seep underground have the tendency to emit heat, by virtue of the decaying radionuclides that radiate energy in the form of heat energy. This heat built up beneath the soil mimics an underground medium field attached heat source that causes unnecessary rise in temperature underground. This elevated temperature modulates the viscosity and density of the underground fluid and adversely altering the flow of the fluid underground. This alteration severely affects the core composition of many chemical substances and elements present in the underground medium field. Migration of underground fluid causes migration of elements like mercury and elevation of temperature might generate new cracks or seal preexisting cracks, thereby altering the stress field existing underground. Although, most of these migrations occurs in the local environment, however, due to extreme migration, the contamination can reach the groundwater table. By virtue of this contaminated groundwater due to leaching of radionuclides, the entire biosphere gets adversely affected and displays havoc consequences on human lives as well (Liu and Wentai, 2019).

8.4 TREATMENT OF RADIOACTIVE WASTES

8.4.1 PHYSICAL AND CHEMICAL TREATMENTS

According to the forms of the radioactive waste, that is, solid, liquid, or gas and on the levels of the radioactive nuclide present in the waste, radioactive waste treatment methods can be classified as radioactive waste liquid treatment, solidification treatment, and radioactive waste gas treatment.

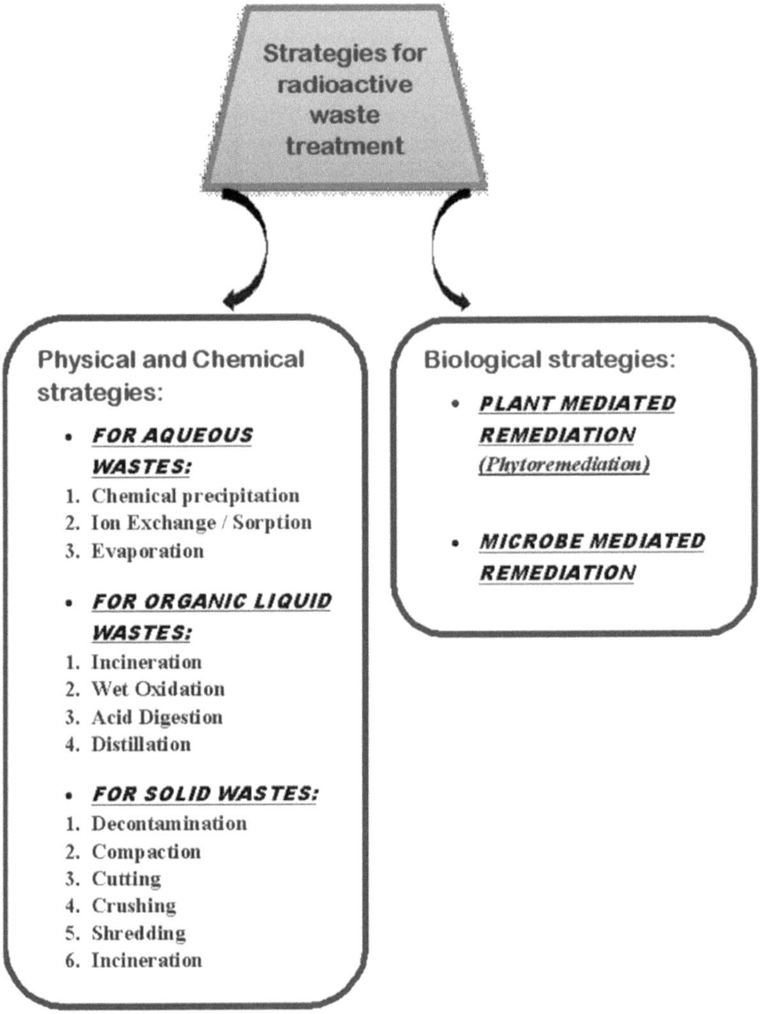

FIGURE 8.2 Strategies for treatment of radioactive wastes.

8.4.1.1 TREATMENT OF RADIOACTIVE WASTE GAS

The treatment of gaseous radioactive wastes chiefly employs an air filter device, where the efficiency of the device is proportionally modulated depending on the degree of radionuclide emission, to bring down the radiation emission within the limit of the radiation safety regulations. The device works by absorbing most of the radionuclides in the air and after the filtration process, the rest of the almost inert gas is released into the atmosphere. Industries dealing with activated metals and other contaminated materials often generate radioactive aerosols. One of the most widely used air filtering devices includes the high efficiency particulate air (HEPA) filter, which prevents the leakage of these radioactive gas contamination into the environment. However, aerosols of size below 300 nm are not efficiently removed by these filters because HEPA filters have the threshold filtration efficiency of around 300 nm and they cannot effectively function below this range. Moreover, the efficiency of these filters falls with time and there comes a need for its replacement with a new filter. Therefore, it is mandatory to check the efficiency of the filters from time to time to determine the need of its replacement (Lee et al., 2020).

8.4.1.2 TREATMENT OF RADIOACTIVE AQUEOUS WASTE

1. Chemical precipitation: This method is widely employed in the removal of intermediate to low level radioactive aqueous wastes generated by nuclear power stations, fuel processing units, and research facilities. One of the advantages of this precipitation process is its cost-effectiveness, which makes this process quite widely accepted. Moreover, this method can efficiently deal with large volumes of liquid wastes containing lower levels of radioactive elements to those containing relatively higher concentrations of radioactive particulates. The main chemical treatments involve a series of pretreatment methods including oxidation and decomposition of the complex organic molecules, alterations in the pH and alterations in the valency state of the complex species. Sometimes, certain carriers can be used for the active precipitation of the contaminating particulates present in the liquid effluents (IAEA, 2001).
2. Ion exchange/sorption: In order to remove soluble radionuclides from radioactive liquid effluents from research laboratories, generation of radioactive isotopes and nuclear fuel cycles, this method plays an appreciable role. This method carries out the transfer of the radioactivity of a larger volume of liquid effluent into a much smaller volume of solid matrix. Basically, the radioactive cations

or anions from the liquid are transferred to a rather insoluble matrix that has ionizable polar groups attached to it. If the ionic groups attached are negatively charged then the cations from the solution gets exchanged, but if those are positive, the exchange mainly deals with anions. The greatest suitability of this process is its reversibility, which allows the regeneration of the ion exchangers after the recovery of the radioactive wastes. The method can be operated either in continuous mode or in batches and can utilize a wide range of ion exchangers including natural ion exchangers, such as clay, cellulose, charcoals, collagen, or synthetic ion exchangers like zeolites and hydrous oxide gels of metals or different forms of organic resins (IAEA, 2001; IAEA, 2002).

3. Evaporation: It is one of the most primitive methods, yet quite effective in addressing low, intermediate, and high-level radioactive wastes. This method employs an evaporation equipment to remove or concentrate radioactive salts and heavy metal contamination, but can efficiently work on lower volumes of effluents. The greatest drawback of this equipment is its unsuitability to deal with effluents having huge levels of salts, as demands lots of energy expenditure making the process rather expensive. Under these circumstances, this equipment generates a lot of heat, thus increasing the chances of fire and explosions (IAEA, 2001).

8.4.1.3 TREATMENT OF RADIOACTIVE ORGANIC LIQUID WASTE

1. Incineration: The main application of the process of incineration is involved in the process of downscaling or reducing the volumes of solid and liquid radioactive effluents to ease in the disposal of these effluents posttreatment. The process is administered at a high temperature that helps in the combustion of the effluents and generates a series of gases as end products, such as carbon dioxide, sulfur dioxide, hydrogen chloride gas, oxides of nitrogen, and lots of water. Due to the generation of these polluting gases as end products, other filtering units for the efficient treatment of these gases called off-gas filtering system. Therefore, for the setting up of this extra filtering apparatus, the process is rather expensive (IAEA, 2006; Chang, 2001).
2. Wet Oxidation: This process is used for the transformation, immobilization or degradation of the of any organic component of the radioactive effluents including the ion exchange resins used in the

process of ion exchange process. This process employs a range of oxidizing agents, such as the Fenton's Reagent (mixture of hydrogen peroxide and iron catalyst in appropriate proportion), ozone, hydrogen peroxide, activated sodium persulfate, which can degrade the organic molecules present in the effluent. Although the process is comparable with the process of incineration, but this process can efficiently take place in lower temperatures. So wet oxidation is advantageous over incineration (IAEA, 1992; Twissel and Holt, 1996).

3. Acid Digestion: This is another oxidative degradation process, which utilizes a mixture of nitric acid and phosphoric acid under conditions of extremely high temperatures of around 200°C and atmospheric pressures or moderate pressures of about 20 psig. The major organic constituents of the waste are first degraded, followed by mineralization by the strong acids, leading to generation of inorganic species. The inorganic end products can be removed via immobilization in ceramic or glass and the gaseous end products can be rendered safe by passing them through an off-gas scrubbing system, thus converting the oxides of nitrogen into nitric acid, which can be again recycled for the process (IAEA, 1992; IAEA, 2001).
4. Distillation: This is the most simple and cheap method, as the solvent can be recycled for subsequent runs. Distillation is another method that is used for the reduction or scaling down of the volumes of the radioactive effluents, thereby aiding in waste management and disposal (IAEA, 2001).

8.4.1.4 TREATMENT OF RADIOACTIVE SOLID WASTE

1. Decontamination: This process employs the techniques of washing, mechanical cleaning, electrochemical, and chemical removal or by other methods. to free up radioactive contamination from different sites. In the process of chemical decontamination, dilute or concentrated chemical reagents are used to dissolve the layers of radioactive contamination from surfaces, while in case of mechanical decontamination technique, a range of techniques can be employed to remove contamination, which includes cleaning by scrubbing, washing, wiping or seeping, and removal by scarifying, grit blasting, spalling or drilling. However, wet abrasive method involves combination of mechanical methods along with certain abrasive chemicals and compressed air in the presence of water. Wet abrasive methods

are rather preferred over dry methods, as in case of the dry method, huge amount of dust is generated leading to environmental pollution (NEA, 1999; Chang, 2001).

2. Compaction: This method is chiefly employed for solid waste volume reduction and for the concentration of the radionuclides present in the wastes. Contaminated plastic, cloth, paper, and any form of absorbent substance containing radioactive wastes are treated by the conventional compactors, while metal items, such as pipes and valves, wooden materials are to be treated using super compactors. This device generates a huge compaction force upon the contaminated materials, thereby reducing their volumes by a great extent. This aids in the process of handling and disposal of these contaminants (IAEA, 2002).
3. Cutting, crushing, and shredding: These are some of the simple and cost-effective methods used for the size reduction of large radioactive solid wastes into smaller forms for easy handling.
4. Incineration: This process is applicable not only to liquid effluents, but plays an equally important role in the combustion of solid radioactive wastes. However, this technique demands effective reduction in size and volume of the wastes, followed by mixing and blending in an appropriate manner to accelerate the process of incineration (Chang, 2001; IAEA, 1992, 2001, 2006; NEA, 1999).

8.4.2 BIOLOGICAL TREATMENTS

Several technologies pertaining to the cleaning up of the radioactive wastes from the environment have been developed, which is associated with the huge potential of microbes and plants to remediate toxins. Mitigation of environmental problems with plants and microbes is a rather cheap and effective method, with almost zero negative impact on the surrounding environment. Bioremediation is the term used to describe the role of these living agents in the cleaning up strategy (Barkay and Schaefer, 2001; Dushenkov, 2003; Ibeanusi et al., 2004).

8.4.2.1 PHYTOREMEDIATION OF RADIOACTIVE WASTES

The phytoremediation technology, often termed as green remediation is rather a subset of the bioremediation process, which employs plants to remove the toxins from the environment (Eapen et al., 2007; Olson and Fletcher, 2000). This technology is a rather new and illustrates the potential role of plants to remediate soils contaminated with radionuclides, thereby rendering the environment safe for life. The plants have the potential to remove and immobilize the radioactive species from the contaminated soil water or sludge within their system. Usually, plants by virtue of their roots absorbs the radionuclides and accumulates these within their system by translocation of these radioactive metals through the stem and accumulation of the leaves and stems, or can cause adsorption of these materials on the root surface, followed by precipitation or reduction of the harmful materials within the rhizosphere, rendering these radioactive species almost inert and effectively remediating the affected areas (Pavel and Gavrilescu, 2008). However, this process is quite time-consuming and is effective for sites with lower levels of contamination. In the literature, plants have been reported to carry out the uptake and accumulation of a range of radioactive elements including Uranium (238), Radium (226), Thorium (232), etc. However, there are several factors that modulate this uptake process in plants. According to the properties of the soil like pH, temperature, water, and organic matter content, the rate of uptake of these radioactive elements varies. Moreover, the type of the species of the plants and also the concentration of the radioactive effluent in the plant growth media or soil will affect the uptake to an appreciable amount.

The process of phytoremediation in plants are carried out in six different ways:

1. Phytoextraction: Plants can degrade the toxins present in soil by accumulating these pollutants in the harvestable parts of the plants like roots, stems, and leaves (Kumar et al., 1995).
2. Phytovolatilization: It is one of the most distinctive processes carried out by plants, wherein the plants mediate the transformation of contaminants in the soil into volatile molecules. Phytovolatilization mainly involves the uptake followed by the volatilization of these pollutants into the surrounding air. However, this process is associated with great concern, as it remediates the soil but pollutes the surrounding air, which in turn has serious effects on the environment. The volatile form of the pollutants has greater mobility in the air than its soil bound forms and can travel longer distances posing serious threats to life (Burken and Schnoor, 1998; Burken and Schnoor, 1999).

3. Phytostabilization: It is the process of reducing the rate of migration of pollutants in the soil by precipitation or immobilization of these contaminants. This prevention of migration in turn reduces the bioavailability of the contaminants present in the soil, thus reducing its potential adverse effects on the environment (Smith and Bradshaw, 1979).
4. Phytodegradation: Certain plants have the potential to express certain metabolic products and a range of hydrolytic enzymes, which can degrade the contaminants present in the growth media or soils. This process is more common for breakdown of organic pollutants present in the soil, thus very specific to certain types of pollutants only (Burken and Schnoor, 1998).
5. Rhizodegradation: It involves both the role of plant roots and the microbes growing in the associated rhizosphere in the process of breakdown of the pollutants in the soil. By virtue of a mutual interaction between the plant and its associated microbes, certain metabolites and enzymes are extruded into the rhizosphere, which carries out the biotransformation and destruction of pollutants in the soil (Zhou et al., 2011; Rugh, 2001; Sors et al., 2005).
6. Rhizofiltration: This process demonstrates the role of the plant root system to bring about the uptake and accumulation or precipitation of toxic components in the soil and waters exclusively within the plant roots. The roots of the plants serve as a reservoir of the pollutants, thus remediating the surrounding contaminated soil (Dushenkov et al., 1995; Vera Prasad and de Oliveira Freitas, 2003).

Moreover, plant roots can actively uptake the radionuclides present in the soil and cause intracellular metabolism of these elements to render them inert and safe. Depending on the type of plant species, the mode of remediation will vary. However, some plants tend to modulate between these phytoremediation strategies, depending on the type of radionuclide exposed to its roots.

Sometimes, in order to enhance the rate of uptake of the radionuclides by the roots, the bioavailability of these elements near the roots must to be increased. Several chelating agents and micronutrients play an immense role in the increased bioavailability (Ehlken and Kirchner, 2002). Moreover, the plants employed to remediate the radioactively polluted sites must have the level of persistence to thrive in such a hostile environment. The prevailing climatic factors and soil properties also play a crucial role to modulate the uptake of radionuclides, explosives, heavy metals, and even contaminated organic materials like crude oil by the roots, and efficiently carry out the detoxification process (Efe and Elenwo, 2014; Sinha et al., 2009). Hyperaccumulator plant species must fulfil an essential criterion, which describes that the radionuclide element concentration within the plant system should be greater than the concentration present in the soil. This criterion is called transfer factor (TF) and can be defined as the ratio of the concentration of the radioactive element within the plant to the concentration of that element in the tailings containing the radioactive contamination (Baker and Brooks, 1989). The TF will determine the rate of uptake of the elements from the contaminated tailings and the extent of accumulation of the contaminating element within the plant. The value of TF can vary between tissues within the same plant, depending on the rate of metabolism of the radionuclides within the respective tissues. Moreover, the age of the plant and surface area of the roots and shoots or the plant biomass will govern the metabolic pathways as well. Thus, by examining the values of TF, one can infer about the plants or rather the plant tissue types which have the potential to remove and degrade radioactive elements like uranium, thorium, and radium from contaminated sites (Soudek et al., 2007; Lauria et al., 2009; Soudek et al., 2011) (Figure 8.3).

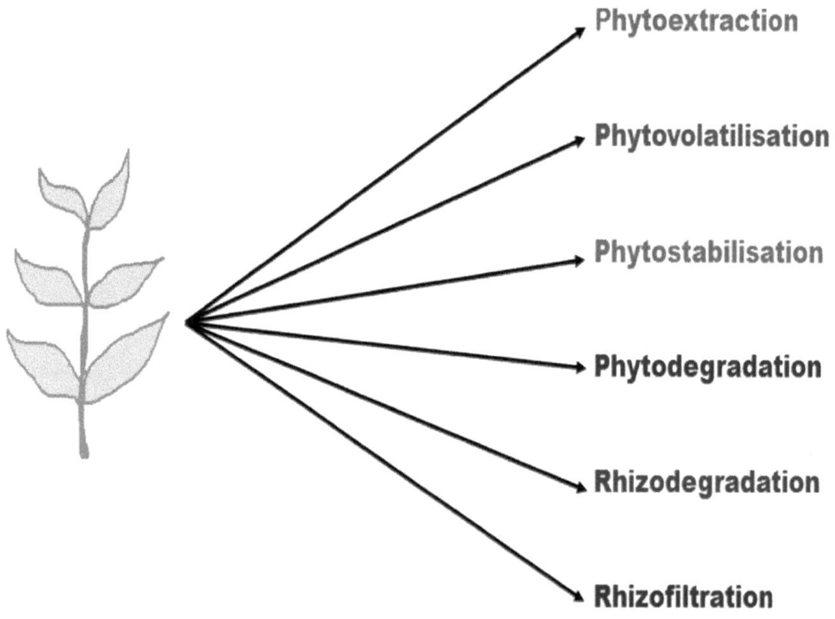

FIGURE 8.3 Strategies for phytoremediation.

Yet, another factor, called the phytoremediation factor (PF) was devised for better understanding of the potential plants that can be employed for the sake of radionuclide remediation. This factor is more specific in governing the efficacy of a plant to remove contamination from tailings. The PF is defined as the concentration of the target radionuclide within the plant shoot only to the concentration of the radionuclides in the contaminated tailings (Li et al., 2011). Basically, PF deals with all the plant parts above the ground, that is, the leaves, stem, seeds, and stalk but not the roots. Therefore, PF is referred as an index to define the capability of a particular plant species to remediate the contaminants from the sites where the plant grows. So, plants having relatively higher biomass, qualified the criteria of PF and can be administered for their probable roles in bioremediation of radionuclides. Depending on the PF, *Phragmites australis* and *Macleaya cordata* were found to be potential players in the phytoremediation of uranium contaminated sites (Hu et al., 2014a, 2014b). *Azolla imbircata* also had immense potential to degrade uranium contaminations in water and sludge (Hu et al., 2012, 2014a, 2014b). On the other hand, *Phragmites australis* has also role in phytoremediation of thorium contamination in soils, while another species *Phragmites multifida* demonstrated roles in cleaning up of Radium (^{226} Ra) species from soils (Hu et al., 2012, 2014a, 2014b). Therefore, the PF serves as an important parameter in the characterization of potential plant species for the remediation of different radionuclides.

8.4.2.2 MICROBIAL TREATMENT OF RADIOACTIVE WASTES

During the past few decades, there has been a lot more emphasis on the role of different microbes in the detoxification of sites with radioactive contamination (Kimber et al., 2011). This process is often referred to as bioremediation, which is a rather cheap and environment-friendly method, hence has gained much of the attention of the public. The greatest advantage of this technology is that the process can be administered in situ, that is, at the site of the contamination (Lloyd and Renshaw, 2005; Vidali, 2001).

8.4.2.2.1 Microbial Treatment of Uranium

Uranium 238 is the maximally abundant isotope of uranium found in the environment. This isotope of uranium has a long half-life of 4.47×10^{9} years and is a potent emitter of alpha particles (Perkins and Thomas, 1980).

Upon exposure to certain microorganisms, the U (VI) is biotransformed to U (IV), leading to generation of uraninite mineral, which is insoluble in nature and is rather minimally potent in causing adverse effects on the environment as compared with uranium. A range of bacterial species have been reported to carry out this reduction step, essential for bioremediation of uranium contaminations. *Geobacter metallireducens* and *Shewanella oneidensis* are usually associated with the reduction of Fe (III) ores, can also reduce U (VI) and conserve the energy generated for its anaerobic growth (Lovley et al., 1991). Some of the sulfate-reducing species of *Desulfovibrio*, such as *Desulfovibrio desulfuricans* and *Desulfovibrio vulgaris*, and few species of anaerobic *Clostridium* have the potential to reduce U (VI), but is unable to conserve the energy generated for the purpose of their growth (Lovley and Phillips, 1992; Francis, 2003). The family of enzymes carrying out this biotransformation was first characterized in *Desulfovibrio vulgaris.* It has been demonstrated by in vitro experiments, that purified form of Cytochrome c3 has the power of reducing U (VI), in the presence of an electron donor like hydrogenase. Initially, it was considered that only the outer surface localized cytochrome type-c of *Geobacter sulfurreducens* has a role in the reduction of U (VI), but now it has been shown that the surface of the pili, by virtue of its electrical conductivity also plays a role in the reduction process (Beliaev et al., 2005; Orellana et al., 2013). Also, in *Shewanella oneidensis* MR1, a number of c- type cytochromes have been reported to have an appreciable role in uranium reduction. Few of these reducing cytochromes include the plasma membrane-associated Tetraheme cytochrome (CymA), outer membrane-associated Decaheme cytochromes like MtrC and OmcA, the periplasm-associated Decaheme cytochrome (MtrA), and the transmembrane MtrB subunit of the outer membrane-associated complex MtrABC protein (Shi et al., 2006; Hartshorne et al., 2009; Schicklberger et al., 2011). The most widely accepted method employed for the sake of uranium precipitation and reduction is the application of Fe (III) and sulfate-reducing bacterial species stimulated in the presence of acetate in the form of an electron donor. Moreover, *Geobacter* species have also been administered for the reduction and precipitation of uranium contaminations, to render them harmless to the environment. In the past few years, these bacterial species have been widely used for the remediation of uranium contaminated groundwater, with acetate being used as a potential electron donor in the reduction process.

8.4.2.2.2 Microbial Treatment of Neptunium

Neptunium is one of the most dreaded radioactive contaminants, produced in huge amounts by nuclear reactors. This element also has a long half-life like uranium, ranging to almost 2.14×10^{6} years. Neptunium is a potential alpha emitter and displays higher solubility under aerobic conditions. Therefore, the removal of Np (IV) is preferentially carried out in reducing environments, as hydrolysis and subsequent removal by interaction with the surfaces become more favorable (Dozol and Hagemann., 2009; Kaszuba and Runde, 1999).

Several microorganisms including *Pseudomonas aeruginosa*, *Scenedesmus obliquus*, *Streptomyces viridochromogenes*, *Micrococcus luteus*, and few marine algal species have been reported to uptake considerable amounts of Np from contaminated sites (Fisher et al., 1983; Strandberg and Arnold, 1988). Another species of Pseudomonads, *Pseudomonas fluorescens* can efficiently absorb sufficient quantities of Np at neutral pH (pH 7) (Songkasiri et al., 2002). NpO_2^+ has the tendency to be reduced into Np (IV) in the absence of oxygen or anoxic conditions, however, certain intermediates generated during the conversion process can hinder the precipitation and separation of the contaminant (Rittmann et al., 2002). Also, addition of certain complexing ligands might be necessary to prevent the oxidation and solubilization of the element, preventing its removal. The bioconversion of Np (V) to Np (IV) by *Shewanella putrefaciens* followed by the precipitation of the reduced product in the form of phosphates under the action of *Serratia* sp. (has inherent phosphatase activity), is one of the most efficient methods for Np bioremediation. Similar attributes of reducing Np (V) to Np (IV) have also been displayed by sulfate-reducing bacterial members like *Desulfovibrio desulfuricans*.

8.4.2.2.3 Microbial Treatment of Plutonium

Plutonium has a range of isotopes. By virtue of its dynamic redox chemistry, dealing with Plutonium contamination is rather more complex as compared with other elements. The most abundant oxidation state of Plutonium is Pu (IV), but other forms including Pu (VI), Pu (V), and Pu (III) are quite stable. Due to range of oxidation states being available in nature, studies with different microbes on the bioremediation of Plutonium isotopes have been tried with small concentrations of the element, except in cases of in situ bioremediation procedures for the removal of Plutonium-239 in the form of carbonates, considerably higher concentrations of the element have been treated. However, some additional mechanism can also be operative along with the bioremediation strategy in contaminated natural sediments. Spontaneous reversal into the previous form in case of Plutonium isotopes is the greatest challenge in treating Plutonium contamination in nature (Fisher et al., 1983). Several Fe (III) reducing bacterial species have the power of reducing Pu (IV) to Pu (III), but Pu (III) spontaneously gets re-oxidized to its precursor. Certain microorganisms can generate certain species-specific ligands that can specifically bind to the trivalent species generated due to reduction of Pu (IV), thus leading to its sequestration and detoxification. *Shewanella oneidensis, Shewanella putrefaciens*, and *Geobacter metallireducens* expresses enzymes that can specifically carry out the reduction of Pu (VI) and Pu (V) into Pu (IV). Microbes are often administered to convert all forms of Pu available in nature into the most abundant form of Plutonium in nature, that is, Pu (IV), which can be further treated altogether. Moreover, application of certain sequestering molecules from outside or certain modification in the pH and other properties of the contaminated media can cause alterations in the redox state and solubility of the contaminant, thereby increasing the ease of removal and detoxification (John et al., 2001). The next step would involve the conversion of Pu (IV) into a more mobile and less potent form, that is, Pu (III), by the intervention of another set of microbes. However, this reductive step is rather more difficult to achieve in natural sediments and thus restricted to laboratories mostly.

8.4.2.2.4 Microbial Treatment of Americium

Americium is another alpha particle emitter, and upon emission, it gets converted into Neptunium-237. It has a relatively shorter half-life of 432.2 years, as compared with other elements discussed previously. Americium-241 exists in a range of oxidation states, starting from II to VII. Am (III) is the most abundant form available in nature and the divalent form is the only solid form of Americium found in nature. Several fungal species, such as *Rhizopus arrhizus*, *Candida utilis*, and bacterial members like *Escherichia coli* have the power of altering the solubility, followed by the absorption of Am (III) from the contaminated sites (Fisher et al., 1983; Wurtz et al., 1986; Watson and Ellwood, 1994; Dhami et al., 1998). Moreover, other microorganisms, such as *Pseudomonas gladioli, Flavobacterium* spp., *Ochrobactrum anthropic*, and *Chryseobacterium indologenes* also demonstrate considerable influence on the associated microbial consortium that actively participates in this bioremediation process (de Pa'dua Ferreira et al., 2011; Marumo et al., 2008).

8.4.2.2.5 Microbial Treatment of Technetium

Technetium serves as a potential component of most nuclear wastes generated from nuclear power plants. It is an emitter of beta particles and has a half-life of 2.13×10^5 years. Technetium has a wide range of solubility, depending on its redox state. In toxic environments, technetium exists in the form of pertechnetate ion (TcO_4^-) having an oxidation state of VII. This form of Technetium has the highest mobility and has the tendency to get weakly adsorbed on the surfaces of minerals in the environment (Bondietti and Francis, 1979). However, Tc (VII) cannot establish stable complexes with chemical ligands and escaping removal, which is usually employed for the sequestration and separation of other forms of radionuclides. On the other hand, reduced forms of Technetium are insoluble and several microorganisms exists in nature that carry out the reduction of Tc (VII) to Tc (IV), followed by subsequent precipitation in the form of oxides. *Shewanella putrefaciens, Rhodobacter sphaeroides, Geobacter metallireducens, Escherichia coli, Paracoccus denitrificans,* some Pseudomonads and a number of sulfate-reducing bacterial species have been found to carry out this reduction step (Llyod et al., 1997, 2001, 2002). Usually, under anaerobic conditions *E. coli* can catalyze the reduction of Tc (VII), under the action of the formate hydrogen lyase (FHL) complex (hydrogenase part of the complex). Some strains of *Desulfovibrio desulfuricans* can also carry out this reduction step using formate as the source of electrons. These strains have a periplasm-associated Formate dedydrogenase component coupled via cytochrome to hydrogenase component (Llyod et al., 1999). In another sulfate-reducing species of *Desulfovibrio (Desulfovibrio fructosovorans),* a periplasm-associated nickel–iron hydrogenase component plays an immense role in the reduction of Tc (VII). It can be inferred that the aerial oxidation of iron drives the reduction and subsequent mobilization of Tc-99 in contaminated effluents. However, the reduced form of Tc (IV) is rather recalcitrant to mobilization and tend to remain adsorbed on sediments, posing a challenge to the complete removal of this contaminant from nature.

8.5 CONCLUSION AND FUTURE PERSPECTIVES

The identification and characterization of the different forms pf radioactive waste and their sources is the key step for proper management and disposal of these harmful wastes. Due to the extensive use of radioactive species in industries, nuclear power plants, and research labs, there has been a steady hike in the concentrations of the radioactive end products generated from these sectors. This in turn has led to decline in the quality of life on this earth, posing severe health hazards to humans and other living organisms. Therefore, it is essential for the generators of these wastes to be aware of the proper ways of handling and mitigation of these harmful wastes. By virtue of the long half-lives of the radionuclides, they tend to decay for extended periods of time and imposes severe adverse effects on life. Therefore, storage of these harmful species becomes the most challenging part of the entire management strategy. Firstly, demarcation of the various forms of radioactive wastes must be carried out on the basis of their half-lives, ranging from few days to enormous number of years. After the demarcation is made, it is necessary to come up with cost-effective, yet safe methods for the treatment of these wastes. According to the nature of the waste, viz., solid, liquid or gas, the nature of treatment and disposal will vary among other forms of radioactive wastes.

A series of physiochemical techniques for rendering radionuclides harmless or less potent have been employed for the treatment of these wastes. Although, a series of physical and chemical treatment techniques are available, still microbial and plant-mediated remediation of radionuclides from contaminated sites have gained much more popularity these days. Microbes and plants have the potential to modulate the solubility of these species and also carry out the biotransformation of these molecules into rather harmless or lesser damage causing end products. These days extensive research is going on to chalk out the biogeochemical signaling pathways in plants and microorganisms that are necessary for the control on radioactive wastes. However, the knowledge of the necessary enzymes and other interacting molecules that will carry out the remediation step, still remains partially addressed. The information about uptake, accumulation, and further treatments on the wastes still remains in a rather gray area and demands extensive research for proper understanding of the phenomenon. Concrete information about the nature of interactions between the radionuclide and the plant or microbe and the signaling pathways associated with the remediation process, might help in better understanding of the entire process. Therefore, this area demands more intervention by the scientific world for improvisation of better wide-scale methods for efficient detoxification of the environment from radioactive wastes.

ACKNOWLEDGMENTS

Financial assistance from the Council of Scientific and Industrial Research (CSIR), Government of India, through the research grant [38(1387)/14/EMR-II], Science and Engineering Research Board, Government of India through the grant [EMR/2016/004799] and Department of Higher Education, Science and Technology and Biotechnology, Government of West Bengal, through the grant [264(Sanc.)/ST/P/S&T/1G-80/2017] to Dr. Aryadeep Roychoudhury is gratefully acknowledged.

KEYWORDS

- **nuclear reactors**
- **nuclear fission**
- **radioactive wastes**
- **irradiation**

REFERENCES

Baker, A. J.; Brooks, R. Terrestrial Higher Plants Which Hyperaccumulate Metallic Elements. A Review of Their Distribution, Ecology and Phytochemistry. *Biorecovery* **1989,** *1*, 81–126.

Barkay, T.; Schaefer, J. Metal and Radionuclide Bioremediation: Issues, Considerations and Potentials. *Curr. Opin. Microbiol.* **2001,** *4*, 318–323.

Beliaev, A. S.; Klingeman, D. M.; Klappenbach, J. A.; Wu, L.; Romine, M. F.; Tiedje, J. M.; Nealson, K. H.; Fredrickson, J. K.; Zhou, J. Global Transcriptome Analysis of *Shewanella oneidensis* MR-1 Exposed to Different Terminal Electron Acceptors. *J. Bacteriol.* **2005,** *187*, 7138–7145.

Bondietti, E. A.; Francis, C. W. Geologic Migration Potentials of Technetium-99 and Neptunium-237. *Science* **1979,** *203*, 1337–1340.

Burken, J. G.; Schnoor, J. L. Distribution and Volatilization of Organic Compounds Following Uptake by Hybrid Poplars. *Int. J. Phytoremediation* **1999,** *1*, 139–151.

Burken, J. G.; Schnoor, J. L. Predictive Relationships for Uptake of Organic Contaminants by Hybrid Poplar Trees. *Environ. Sci. Technol.* **1998,** *32*, 3379–3385.

Chang, H. O. *Hazardous and Radioactive Waste Treatment Technologies*; Handbook CRP Press: USA, 2001.

de Pa'dua Ferreira, R. V.; Sakata, S. K.; Isiki, V. L. K.; Miyamoto, H.; Bellini, M. H.; de Lima, L. F. C. P.; Marumo, J. T. Influence of Americium-241 on the Microbial Population and Biodegradation of Organic Waste. *Environ. Chem. Lett.* **2011,** *9*, 209–216.

Dhami, P. S.; Gopalakrishnan, V.; Kannan, R.; Ramanujam, A.; Salvi, N.; Udupa, S. I. Biosorption of radionuclides by *Rhizopus arrhizus*. *Biotechnol. Lett.* **1998,** *20*, 225–228.

Dozol, M.; Hagemann, R. Radionuclide Migration in Groundwaters: Review of the Behaviour of Actinides (Technical Report). *Pure Appl. Chem.* **2009,** *65*, 1082–1102.

Dushenkov, S. Trends in Phytoremediation of Radionuclides. *Plant Soil* **2003,** *249*, 167–175.

Dushenkov, V.; Kumar, P. N.; Motto, H.; Raskin, I. Rhizofiltration: The Use of Plants to Remove Heavy Metals from Aqueous Streams. *Environ. Sci. Technol.* **1995,** *29*, 1239–1245.

Eapen, S.; Singh, S; D'Souza, S. F. Phytoremediation of Metals and Radionuclides. In *Environmental Bioremediation Technologies*; Singh, S. N., Tripathi, R. D., Eds.; Springer-Verlag: Berlin Heidelberg, 2007.

Efe, S. I.; Elenwo, E. I. Phytoremediation of Crude Oil Contaminated Soil with *Axonopus compressus* in the Niger Delta Region of Nigeria. *Nat. Resour.* **2014,** *5*, 59.

Ehlken, S.; Kirchner, G. Environmental Processes Affecting Plant Root Uptake of Radioactive Trace Elements and Variability of Transfer Factor Data: A Review. *J. Environ. Radioact.* **2002,** *58*, 97–112.

Fisher, N. S.; Bjerregaard, P.; Huynh-Ngoc, L.; Harvey, G. R. Interactions of Marine Plankton with Transuranic Elements. II. Influence of Dissolved Organic Compounds on Americium and Plutonium Accumulation in a Diatom. *Marine Chem.* **1983,** *13*, 45–56.

Francis, A. J. Microbial Transformations of Radioactive Wastes and Environmental Restoration through Bioremediation. *J. Alloys Compd.* **2003,** *213*, 226–231.

Hartshorne, R. S.; Reardon, C. L.; Ross, D.; Nuester, J.; Clarke, T. A.; Gates, A. J.; Mills, P. C.; Fredrickson, J. K.; Zachara, J. M.; Shi, L.; Beliaev, A. S.; Marshall, M. J.; Tien, M.; Brantley, S.; Butt, J. N.; Richardson, D. J. Characterization of an Electron Conduit between Bacteria and the Extracellular Environment. *PNAS* **2009,** *106*, 22169–22174.

Hox, X. L.; Roos, P. Critical Comparison of Radiometric and Mass Spectrometric Method for the Determination of Radionuclides in Environment, Biological and Nuclear Waste Samples. *Analyt. Chim. Acta* **2008,** *608* (2), 105–139.

Hu, N.; Ding, D.; Li, G. Natural Plant Selection for Radioactive Waste Remediation. In *Radionuclide Contamination and Remediation through Plants*; Springer: Cham, 2014; pp 33–53.

Hu, N.; Ding, D.; Li, G.; Wang, Y.; Li, L.; Zheng, J. Uranium Removal from Water by Five Aquatic Plants. In *Progress Report on Nuclear Science and Technology in China (Vol. 2). Proceedings of Academic Annual Meeting of China Nuclear Society in 2011, No. 2-Uranium Mining and Metallurgy Sub-Volume*, 2012.

Hu, N.; Ding, D.; Li, G.; Zheng, J.; Li, L.; Zhao, W.; Wang, Y. Vegetation Composition and ^{226}Ra Uptake by Native Plant Species at a Uranium Mill Tailings Impoundment in South China. *J. Environ. Radioact.* **2014,** *129*, 100–106.

IAEA. Application of Ion Exchange Processes for the Treatment of Radioactive Waste and Management of Spent Ion Exchanger. Technical Reports Series No. 408. International Atomic Energy Agency (IAEA) Vienna, 2002.

IAEA. Application of Thermal Technologies for Processing of Radioactive Waste, IAEA-TECDOC-1527, International Atomic Energy Agency (IAEA) Vienna, 2006.

IAEA. Decontamination and Decommissioning of Nuclear Facilities. Technical Reports Series No. 401. International Atomic Energy Agency (IAEA) Vienna, 2001.

IAEA. Treatment and Conditioning of Radioactive Organic Liquids. IAEA-TECDOC-656 International Atomic Energy Agency (IAEA) Vienna, 1992.

Ibeanusi, V. M.; Grab. D. A.; Jensen, L.; Ostrodka, S. Radionuclide Biological Remediation Resource Guide. US Environmental Protection Agency, Region 5, Superfund Division, 2004.

John, S. G.; Ruggiero, C. E.; Hersman, L. E.; Tung, C. S.; Neu, M. P. Siderophore Mediated Plutonium Accumulation by *Microbacterium flavescens* (JG-9). *Environ. Sci. Technol.* **2001,** *35*, 2942–2948.

Kaszuba, J. P.; Runde, W. H. The Aqueous Geochemistry of Neptunium: Dynamic Control of Soluble Concentrations with Applications to Nuclear Waste Disposal. *Environ. Sci. Technol.* **1999,** *33*, 4427–4433.

Khelurkar, N.; Shah, S.; Jeswani, H. A Review of Radioactive Waste Management. In *International Conference on Technologies for Sustainable Development*; 2015; pp 4–6.

Kimber, R.; Livens, F. R.; Lloyd, J. R. Management of Land Contaminated by the Nuclear Legacy. *R. Soc. Chem.* **2011,** *32*, 82.

Kumar, P. B. A. N.; Dushenkov, V.; Ensley, B. D.; Chet, I.; Raskin, I. Phytoremediation: A Novel Strategy for the Removal of Toxic Metals from Environment Using Plants. *Biotechnology* **1995,** *13*, 1232–1238.

Lauria, D. C.; Ribeiro, F. C.; Conti, C. C.; Loureiro, F. A. Radium and Uranium Levels in Vegetables Grown Using Different Farming Management Systems. *J. Environ. Radioact.* **2009,** *100*, 176–183.

Lee, M. H.; Yang, W.; Chae, N.; Choi, S. Performance Assessment of HEPA Filter against Radioactive Aerosols from Metal Cutting During Nuclear Decommissioning. *Nucl. Eng. Technol.* **2020,** *52*, 1043–1050.

Li, G. Y.; Hu, N.; Ding, D. X.; Zheng, J. F.; Liu, Y. L.; Wang, Y. D.; Nie, X. Screening of Plant Species for Phytoremediation of Uranium, Thorium, Barium, Nickel, Strontium and Lead Contaminated Soils from a Uranium Mill Tailings Repository in South China. *Bull. Environ. Contam. Toxicol.* **2011,** *86*, 646–652.

Liu, J.; Wentai, D. Overview of Nuclear Waste Treatment and Management. *E3S Web of Conferences* **2019,** *118*, 04037.

Lloyd, J. R.; Renshaw, J. C. Bioremediation of Radioactive Waste: Radionuclide- Microbe Interactions in Laboratory and Field- Scale Studies. *Curr. Opin. Biotechnol.* **2005,** *16*, 254–260.

Lloyd, J. R.; Cole, J. A.; Macaskie, L. E. Reduction and Removal of Heptavalent Technetium from Solution by *Escherichia coli*. *J. Bacteriol.* **1997,** *179*, 2014.

Lloyd, J. R.; Mabbett, A.; Williams, D. R.; Macaskie, L. E. Metal Reduction by Sulphate-Reducing Bacteria: Physiological Diversity and Metal Specificity. *Hydrometallurgy* **2001,** *59*, 327–337.

Lloyd, J. R.; Ridley, J.; Khizniak, T.; Lyalikova, N. N.; Macaskie, L. E. Reduction of Technetium by *Desulfovibrio desulfuricans*: Biocatalyst Characterization and Use in a Flowthrough Bioreactor. *Appl. Environ. Microbiol.* **1999,** *65*, 2691–2696.

Llyod, J. R.; Chesnes, J.; Glasauer, S; Bunke, D. J.; Livens, F. R.; Lovely, D. R. Reduction of Actinides and Fission Products by Fe (III)-Reducing Bacteria. *Geomicrobiol. J.* **2002,** *19*, 103–120.

Lovley, D. R.; Phillips, E. J. P.; Gorby, Y. A.; Landa, E. R. Microbial Reduction of Uranium. *Nature* **1991,** *350*, 413–416.

Lovley, D. R.; Phillips, E. J. Reduction of Uranium by *Desulfovibrio desulfuricans*. *Appl. Environ. Microbiol.* **1992,** *58*, 850–856.

Marumo, J. T.; Isiki, V. L. K.; Miyamoto, H.; Ferreira, R. V. P.; Bellini, M. H.; de Lima, L. F. C. P. Investigation of the Radiation Risk Due to Environmental Contamination by ^{241} am from Lightning Rods Disposed at Uncontrolled Garbage Dumps. *Radiat. Environ. Biophys.* **2008,** *47*, 131–137.

NEA. Decontamination Techniques Used in Decommissioning Activities. A Report Nuclear Energy Agency (NEA) Task Group on Decontamination 1999.

Olson, P. E.; Fletcher, J. S. Ecological Recovery of Vegetation at a Former Industrial Sludge Basin and Its Implications to Phytoremediation. *Environ. Sci. Pollut. Res.* **2000,** *7*, 95–204.

Orellana, R.; Leavitt, J. J.; Comolli, L. R.; Csencsits, R.; Janot, N.; Flanagan, K. A.; Gray, A. S.; Leang, C.; Izallalen, M.; Esterc, T.; Lovley, D. R. U (VI) Reduction by Diverse Outer Surface c-Type Cytochromes of *Geobacter sulfurreducens*. *Appl. Environ. Microbiol.* **2013,** *79*, 6369.

Pavel, L. V.; Gavrilescu, M. Overview of Ex Situ Decontamination Techniques for Soil Cleanup. *Environ. Eng. Manage. J. (EEMJ)* **2008,** *7*.

Perkins, R. W.; Thomas, C. W. US Department of Energy, Pacific Northwest National Laboratory, *Technical Report PNNL* **1980,** 13443.

Rittmann, B. E.; Banaszak, J. E.; Reed, D. T. Reduction of Np(V) and Precipitation of Np(IV) by an Anaerobic Microbial Consortium. *Biodegradation* **2002,** *13*, 329–342.

Rugh, C. L. Mercury Detoxification with Transgenic Plants and Other Biotechnological Breakthroughs for Phytoremediation. *In Vitro Cell. Dev. Biol. Plant* **2001,** *37*, 321.

Schicklberger, M.; Bücking, C.; Schuetz, B.; Heide, H.; Gescher, J. Involvement of the *Shewanella oneidensis* Decaheme Cytochrome MtrA in the Periplasmic Stability of the β-Barrel Protein MtrB. *Appl. Environ. Microbiol.* **2011,** *77*, 1520.

Shi, L.; Chen, B.; Wang, Z.; Elias, D. A.; Mayer, M. U.; Gorby, Y. A.; Ni, S.; Lower, B. H.; Kennedy, D. W.; Wunschel, D. S.; Mottaz, H. M.; Marshall, M. J.; Hill, E. A.; Beliaev, A. S.; Zachara, J. M.; Fredrickson, J. K.; Squier, T. C. Isolation of a High-Affinity Functional Protein Complex between OmcA and MtrC: Two Outer Membrane Decaheme *c*-Type Cytochromes of *Shewanella oneidensis* MR-1. *J. Bacteriol.* **2006,** *188*, 4705.

Sinha, R. K.; Valani, D.; Sinha, S.; Singh, S.; Herat, S. Bioremediation of Contaminated Sites: A Low-Cost Nature's Biotechnology for Environmental Cleanup by Versatile Microbes, Plants and Earthworms. *Solid Waste Manage. Environ. Remediation* **2009,** *978*, 1.

Smith, R. A.; Bradshaw, A. D. The Use of Metal Tolerant Plant Populations for the Reclamation of Metalliferous Wastes. *J. Appl. Ecol.* **1979,** 595–612.

Songkasiri, W.; Reed, D. T.; Rittmann, B. E. Bio-Sorption of Neptunium (V) by *Pseudomonas fluorescens*. *Radiochim. Acta* **2002,** *90*, 785.

Sors, T. G.; Ellis, D. R.; Salt, D. E. Selenium Uptake, Translocation, Assimilation and Metabolic Fate in Plants. *Photosynth. Res.* **2005,** *86*, 373–389.

Soudek, P.; Petrová, Š.; Benešová, D.; Tykva, R.; Vaňková, R.; Vaněk, T. Comparison of ^{226}Ra Nuclide from Soil by Three Woody Species *Betula pendula*, *Sambucus nigra* and *Alnus glutinosa* during the Vegetation Period. *J. Environ. Radioact.* **2007,** *97*, 76–82.

Soudek. P.; Petrová, Š.; Benešová, D.; Dvořáková, M.; Vaněk, T. Uranium Uptake by Hydroponically Cultivated Crop Plants. *J. Environ. Radioact.* **2011,** *102*, 598–604.

Strandberg, G.; Arnold, W. D. Microbial Accumulation of Neptunium. *J. Ind. Microbiol.* **1988,** *3*, 329–331.

Twissel, M. A.; Holt, N. S. Wet Oxidation of Organic Radioactive Waste. European Commission Nuclear Science and Technology. EUR 16776, Luxembourg. European Communities 1996.

Vara Prasad, M. N.; de Oliveira Freitas, H. M. Metal Hyperaccumulation in Plants: Biodiversity Prospecting for Phytoremediation Technology. *Electr. J. Biotechnol.* **2003,** *6*, 285–321.
Vidali, M. Bioremediation: An Overview. *Pure Appl. Chem.* **2001,** *73*, 1163–1172.
Wang, S. A.; Alekseev, V.; Ling, J.; Liu, G. K. Polarity and Chirality in Uranyl Borates: Insights into Understanding the Vitrification of Nuclear Waste and the Development if Nonlinear Optical Materials. *Chem. Mater.* **2010,** *22* (6), 2155–2163.
Watson, J. H. P.; Ellwood, D. C. Biomagnetic Separation and Extraction Process for Heavy Metals from Solution. *Miner. Eng.* **1994,** *7*, 1017–1028.
Weber, W. J.; Navrotsky, A.; Stefanovsky, S.; Vans, E. R.; Vernaz, E. Materials Science of High- Level Nuclear Waste Immobilisation. *Mrs Bull.* **2009,** *34* (1), 46–53.
Wurtz, E. A.; Sibley, T. H.; Schell, W. R. Health Physics. *Radiat. Safety J.* **1986,** 79–88.
Zhou, Q.; Cai, Z.; Zhang, Z.; Liu, W. Ecological Remediation of Hydrocarbon Contaminated Soils with Weed Plant. *J. Resour. Ecol.* **2011,** *2*, 97–105.

CHAPTER 9

Health Care Waste: Pollution Potential from Generation to Disposal, Management, and Treatment

SAMI ULLAH QADIR[1*], VASEEM RAJA[2], and NASEER AHMAD DAR[3]

[1]*Department of Environmental Sciences, Govt. Degree College, Kokernag 192221, Jammu and Kashmir 192221, India*

[2]*Department of Botany, Govt. Degree College for Women's Pulwama, Jammu and Kashmir 192301, India*

[3]*Department of Environmental Sciences, Govt. Degree College Shopian, Jammu and Kashmir 192303, India*

[*]*Corresponding author. E-mail: samievs84@gmail.com*

ABSTRACT

Establishment of health care units (HCUs) in both developing as well as in developed countries is of fundamental necessity. Keeping in view the increased incidence of human diseases in developing countries, these HCUs need to be well-equipped in order to treat the patients. While treating patients in these HCUs, large amounts of food materials, equipment's, medicines, and instruments are used for the treatment of patients which in turn produce large amounts of discarded materials commonly called as health care waste (HCW) or biomedical waste (BMW). The wastes released from patients suffering from contagious diseases spread this infection to human directly or indirectly through different environmental segments if not properly handled. These infections become widespread due to the haphazard disposal. Most HCUs in India has no proper treatment facilities for disposal and management

Waste Problems and Management in Developing Countries. Umair Riaz, Shazia Iqbal, & Moazzam Jamil (Eds.)

of HCW. Indian hospitals approximately generate about 1.45–2.0 kg of waste per day per patient, where it is much higher in western countries. From Indian HCUs, about 0.33 million tons of BMW is annually generated, the major portion of about 30.35% comprises bandages and linen infectious waste, 0.3–0.5% comprises disposable syringes, 9–10% comprises plastics, 40–45% includes food waste and the rest (3–5%) comprises glass waste. In developed western countries, nearly 15–20% of the total waste generated is hazardous and infectious in nature. Whereas, this proportion is much higher in Indian HCUs due to lack of proper waste segregation conveniences excluding some large ones.

Keeping this serious issue in mind it calls for global attention for proper installation of waste management facilities in these HCUs. In India, health care waste management received serious attention after the enactment of BMW (management and Handling Rules 1998). This legislation impelled the concerned authorities at HCUs for installation of proper waste treatment facilities to meet the increased requirements. For installation of these facilities various problems are being faced by various HCUs at various places in India. To overcome these problems, the current study will analyze those issues and current deficiencies. Further, the authors will try to suggest a plan to develop the present system of waste management. Further, the current study will analyze the issues like generation, composition, management handling, and disposal of HCW.

9.1 INTRODUCTION

The mushrooming growth of health care units/establishments (HCUs) in India owing to high disease incidence has occasioned in the generation of large quantities of health care waste (HCW). These wastes generated from HCUs including hospitals, nursing homes, pathological labs, clinics, research centers, vet-health care establishments, etc. are impending sources of health risks and creates a frightening situation for regional and local administration. The waste thus generated from these HCUs is commonly called as HCW or biomedical waste (BMW). In literature, waste generated by these health care facilities is recognized by terms like clinical waste, infectious waste, health care facility waste, and medical waste (Yazie et al., 2019). The waste generated from these HCUs while delivering health care services include chemicals, sharps, pharmaceuticals, body parts and fluids, radioactive elements and equipment's used during and or for the treatment (Muduli and Barve, 2012; Odonkor and Mahami, 2020). Although, very less amount of waste is spawned from these HCUs and has a very huge potential to endanger human life by causing injuries and transmission of infectious diseases (Yazie et al., 2019) due to presence of pathogens in adequate concentrations (Muduli and Barve, 2012) and needs special treatment and disposal. HCW generated from these HCUs is generally of two types: general waste or nonhazardous and Hazardous waste. Out of the total waste produced in an HCUs, 10–15% is regarded as infectious, hazardous, toxic, or radioactive nature with potential human and environmental risks (Odonkor and Mahami, 2020; Yazie et al., 2019) and the remaining 85% consists of nonhazardous or general HCW (Emmanuel et al., 2014) (Figure 9.1), nonhazardous general waste generally comes out of administrative, kitchen, and other housekeeping units at HCUs and mostly consist of packaging and demolition constructional waste (Emmanuel et al., 2014). If both of these categories are mixed with each other, then the entire waste stream becomes unsafe and hazardous (Mmereki et al., 2015). However, the infectious portion ranges somewhat between 16% and75% depending upon the nature of the country. World over around 30 hazardous blood-borne pathogens get transmitted by inadequate, insufficient, and indecorous management of HCW which exposes over 2 million health care professionals to various contagious diseases (Odonkor and Mahami, 2020; Yazie et al., 2019). The highest risk of HCW is due to infectious and sharp components, including syringes, needles, and blades, etc. Health care professionals and people dealing with HCW disposal often get injuries due to manifestation of contaminated sharp components present in the waste and can even get contacted to health threatening diseases like AIDS/HIV and other types of harmful infections like Hepatitis B and C as has been evident from the Gujrat state of India, where at least 240 persons get infected with Hepatitis B due to the use of contaminated equipments supplied for medical services (Harhay et al., 2009; Solberg, 2009). Thus, for management and disposal of HCW, great attention is being paid. Management and disposal of HCW in transition and developing countries are worsened by factors like lack of economic and technological capacities, social snags, lack of training to medical staff dealing with HCW, absenteeism of treatment amenities, policy or action plans for proper supervision to reduce, recover and reuse noninfectious or common waste to lessen the disposal expenditures, and to treat chemical waste originating from HCUs, coupled with derisory organizational setup requirement to manage BMW, insufficient measures for assessment of this waste, lack of adequate management plans and accountabilities of general people and various agencies dealing with management of HCW (Harhay et al., 2009).

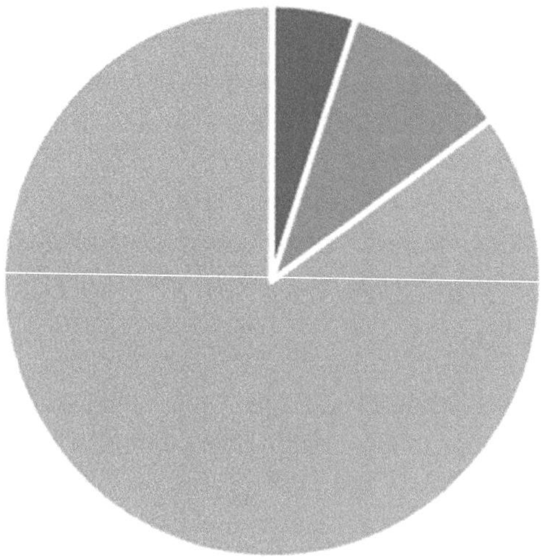

FIGURE 9.1 Waste composition in health care units.

Globally, management of HCW is a serious and challenging issue with potential implications for environment and human health (Chudasama et al., 2014; Mane et al., 2016; Singh et al., 2014) for example, globally around 16 billion injections are annually administered to patients suffering from several diseases, the HCW in the form of needles and syringes arising out of this medical service is not adequately managed (Odonkor and Mahami, 2020), thus disposed improperly in different environmental segments. These improperly managed needles and syringes, like other types of HCWs endangers human health by spreading infections as they are contaminated with potentially hazardous microbes which can infect health care management personals, and local people (Oli et al., 2015).

The HCW contaminated with microbes are reported to possess multidrug resistance and get mixed with various water bodies in the environment during cleaning and disposal on land. The main risk involved is the transfer of resistant genes to humans from environmental bacteria. The release of different antibiotics during cleaning of HCUs into water ways leads to the resistance in different bacterial strains to these antibiotics through the transfer of genes horizontally and selective pressures. Selective pressure manifested with antibiotic resistance occurs due to the higher perseverance or growth of bacterial strains transporting resistant genes that lead to the blockage of defensive cellular mechanism by the antibiotic complexes. These resistant genes are very often positioned on those plasmids susceptible to gene transmission horizontally, responsible for the central corridor by which those genes get transferred which are resistant to different antibiotics (Odonkor and Mahami, 2020).

9.2 EMERGENCE OF HEALTH CARE WASTE

Between 990 and 2000 waste production increased by tenfold and is foreseeable to double again by 2025 which means that waste generation is faster than other environmental waste product including greenhouse gases (GHGs). In different states of India, large number of HCUs are both public and private institutions. In addition to these large HCUs, there are also large number of dispensaries and small clinics for outpatient care only. In rural areas, the health care services provided to urban poor are carried out by government-established community health care centers. The next category of these HCUs consists of those functioning at divisional and district levels, hospitals of medical colleges, and those established within public or private domain. The recent initiative of Indian Government to establish multicare speciality infirmaries in different cities is also growing rapidly. The usual specialty bed in all these HCUs in India is between 30 and 100 beds or even more (Patil and Shekdar, 2001). The substantial quantity of leftover is generated within these HCUs which needs to be managed within these establishments themselves. However, the waste generated within rural HCUs is much lesser than those generated in urban health care establishments. As per the health information statistics, rural areas account for 20% of the total bed strength and 80% of beds are present within urban HCUs. As per the reports, the total amount of HCW generated in India increased from 288 tons per day (TPD) in year 2007 to 530 TPD in 2018, generated from about 168,869 HCUs (Datta et al., 2020), out of which, only 447 TPD is treated, and 37 TPD remains untreated. The common BMW treatment facilities (CBMWTFs) across the country increased from 93 in 2008 up to 198 in 2017 and about 28 treatment facilities are in the process of construction. The HCUs with CBMWTFs facilities were 131,837 and about 21,870 HCUs are having their own treatment handling facilities for HCW on-site (Datta et al., 2020). The causes accountable for this huge generation of HCW in India include burgeoning growth of population, increasing affluence, increasing suburbanization, changes in methods of procuring, packaging of materials, increased use of plastics, life style diseases, and expanding growth of medical tourism.

9.3 QUANTIFICATION AND COMPOSITION OF HCW IN INDIA

The mushrooming growth of health care sector around the world especially in India over the last few decades coupled with indiscriminate use of single-use disposable items has led to the generation of large quantities of HCW in India (Patwary et al., 2009). As per the reports of WHO, 85% of the waste belongs to the nonhazardous category out of the total waste generated, and is considered to be general domestic waste, 10% of HCW is of infectious type and 5% belongs to the category of noninfectious type of waste (Emmanuel et al., 2014; ENVIS Center 1 CPCB, 2014). Sharps make about 1% of the waste and are responsible for the transmission of major diseases in humans and animals (Abdulla et al., 2008), 3% of the waste generated from HCUs consists of the waste of pharmaceuticals and chemicals, whereas radioactive substances, genotoxic waste, and heavy metal constitute 1% of the total waste generated (Emmanuel et al., 2014). As per the suggested literature, the HCW generated in different countries of the world is presented in (Table 9.1). As per the findings of Komilis et al. (2010), the rate of waste generation ranges from 2.1 to 4.2 kg/bed/day in outsized HCUs and 0.5–1.8 kg/bed/day in District HCUs. According to the studies conducted in different states of India, the average waste generated in Indian states are presented in Table 9.2. The average generation of HCW in India ranges between 0.3 and 2.0 kg/bed/day (Patil and Shekdar, 2001). Of the total waste collected, 50% belongs to contagious waste of category I and II (ENVIS Center 1 CPCB, 2014), solid waste like dressings and linen infectious waste (30–35%), throwaway syringes (0.3–0.5%), plastics (9–10%), food waste (40–45%), and glass waste (3–5%). In developed western countries 3–10 kg/day/patient HCW is generated on an average, of which, roughly, 15–20% of the total waste is found to be dangerous and disease-ridden (Mandal and Dutta, 2009), whereas this percentage is much higher (50–75%) in Indian HCUs (Mandal and Dutta, 2009; Patil and Shekdar, 2001). The HCW generation depends on several factors, the most responsible factors include the infrastructure and location, capacity, methods of waste management, instrumentation facility, proportion of reusable and the use of disposable items in delivery of health services, application of hospital and national waste supervision policy

(Mandal and Dutta, 2009), specialization of HCUs, waste segregation, and the total number of persons treated per day (Thakur and Ramesh, 2017). The literature points out that the generation of HCW also depends upon local legislation (Askarian et al., 2004; Bdour et al., 2007; Mohee, 2005).

TABLE 9.1 Country Wise Generation of HCW.

S. no.	Country	Generation rates	Source
1	India	0.25–2.31 kg/bed/day	(Gupta and Boojh, 2006; Katoch and Kumar, 2008; Patil and Pokhrel, 2005)
2	Korea	0.14–0.49 kg/bed/day	(Jang et al., 2006)
3	Jordan	1.9–3.5 kg/bed/day	(Abdulla et al., 2008)
4	Turkey	0.63–3.83 kg/bed/day	(Birpinar et al., 2009)
5	Bangladesh	0.25 kg/bed/day	(Patwary et al., 2009)
6	Iran	0.4–1.91 kg/bed/day	(Taghipour and Mosaferi, 2008)
7	North Greece	0.58 kg/bed/day	(Sanida et al., 2010)
8	Egypt	0.23–2.07 kg/bed/day	(Abd El-salam, 2010)
9	South Greece	1.4 kg/bed/day	(Komilis et al., 2010; Sanida et al., 2010)
10	Turkey	2.11–3.83 kg/bed/day	(Eker and Bilgil, 2011)
11	Netherlands	2.7 kg/bed/day	(ENVIS Center 1 CPCB, 2014)
12	France	2.5 kg/bed/day	(ENVIS Center 1 CPCB, 2014)
13	USA	4.5 kg/bed/day	(ENVIS Center 1 CPCB, 2014)
14	Latin America	2.63–3.8 kg/bed/day	(ENVIS Center 1 CPCB, 2014)

TABLE 9.2 HCW Generated in Different States of India.

State	Type of HCU	Waste generated kg/bed/day	Composition
Kolkata	Govt. Private, nursing homes (large hospitals)	1.044–1.38 kg	20–30% infectious 50–75% general
New Delhi	Govt. and private hospitals	1.5–1.8 kg	45% infectious
Mumbai	Tertiary HCU	1.13 kg	46% infectious
Jaipur	Large tertiary HCU	1.5 kg	25–35% infectious
Manipal	Large tertiary HCU	0.775 kg	16–26% infectious
Punjab	Large tertiary HCU	1.05–1.50 kg	15–30% infectious

Source: Mandal and Dutta (2009)

The effective segregation and management of this anatomical and human waste will decrease the magnitude of hazardous contagious waste in HCUs demanding burning/deep burial (Thakur and Ramesh, 2017).

9.4 HEALTH CARE WASTE CLASSIFICATION

The prime responsibility for the management of HCW lies with health care establishments within which it is generated. The health care establishments are accountable for separation, gathering, conveyance, pretreatment, and waste storing before collection by some common BMW treatment facility institution or operator within or outside these HCUs (Sharma et al., 2020). Thus, for proper management of HCW, the technical knowhow during HCW handling, segregation, treatment and disposal need to be comprehensively understood (Sharma et al., 2020; Thakur and Ramesh, 2017) and practiced by concerned staff in compliance with the BMW Management and Handling Rules 2016 (CPCB, 2016). Out of total waste produced from the health care establishments 75–90% of the waste belongs to the general category or nonhazardous waste (Chauhan and Singh, 2016; Mandal and Dutta, 2009) analogous to the waste generated from domestic units, due to its non-risky nature. Such wastes require no special methods of handling, transportation, and disposal but are managed according to the rules specified for the management of general municipality waste (Pruss et al., 1999; Taghipour and Mosaferi, 2008). This type of waste is also labeled as nonregulated medical waste (Lee et al., 2004). The outstanding 10–25% of generated waste within these HCUs belongs to the category of infectious or special waste or hazardous waste (Chauhan and Singh, 2016; Yang et al., 2021) as specified by the definitions proposed by WHO and US Environmental Protection Agency. This class of waste comprises infectious waste, human pathological waste, genotoxic, pharmaceuticals, chemicals, hassled ampules, and wastes of radioactive nature (Table 9.3), most of which are lethal, damaging, oncogenic, and liable to cause infections (CPCB, 2016; Emmanuel et al., 2014; ENVIS Center 1 CPCB, 2014; Taghipour and Mosaferi, 2008). Thus this type of waste needs extraordinary techniques of management (handling, treatment, and disposal), therefore, most of the hazardous or special wastes are preserved as controlled health waste (Pruss et al., 1999).

The waste generated within HCUs are commonly categorized into three types.

- Biomedical waste
- General waste
- Other waste

TABLE 9.3 Classification of HCW.

Sl. No	Class	Waste type
1	Human anatomical waste	Human tissues, organs, body parts; fluids; fetuses; unused blood products
2	Animal wastes	Animal tissues, organs, body parts, carcasses, fluid, blood; experimental animal used in research, waste generated by veterinary hospitals
3	Microbiology and biotechnology wastes	Waste from laboratory cultures, stocks or specimens of microorganisms, live or attenuated vaccines, human and animal cell cultures used in research, infectious agents from research and industrial laboratories, wastes from production of biological, toxins, dishes and devices used to transfer of cultures
4	Waste sharps	This category includes both used and unused sharps (e.g., hypodermic; intravenous or other needles; autodisable syringes; syringes with attached needles; infusion sets; scalpels, blades; broken-glasses, etc.) capable of causing punctures and cuts
5	Discarded medicines and cytotoxic drugs	Waste comprising outdated, contaminated and discarded drugs and medicines; items contaminated by or containing pharmaceuticals; Cytotoxic waste containing substances with genotoxic properties (e.g., waste containing cytostatic drugs–often used in cancer therapies; genotoxic chemicals)
6	Soiled wastes	Items contaminated with blood fluids including cotton, dressings, soiled plaster casts, linens, bedding
7	Solid wastes	Disposable items other than the waste sharps, tubing, catheters, IV sets, etc.
8	Liquid wastes	Waste generated from laboratories, washing, cleaning, housekeeping and disinfection activities
9	Incineration ash	Ash from incineration of any medical wastes
10	Chemical wastes	Waste containing chemical substances (e.g., laboratory reagents; film developer; disinfectants; that are expired and no longer needed; solvents; waste with high content of heavy metals, for example, batteries; broken thermometers and blood pressure gauges); Chemicals used in production of biologicals, disinfection, insecticides, etc.
11	Radioactive waste	Waste containing radioactive substances (e.g., unused liquids from radiotherapy or laboratory research; contaminated glassware; packages; or absorbent paper; urine and excreta from patients treated or tested with unsealed radionuclides; sealed sources
12	Nonhazardous or general waste	Waste that does not pose any particular biological, chemical, radioactive, or physical hazard

9.4.1 BIOMEDICAL WASTE

Any waste that is spawned within HCUs through treatment, analysis, vaccination, or pathological laboratories (Caniato et al., 2015; Gupta and Boojh, 2006; Joshi et al., 2015; Patil and Pokhrel, 2005) is termed biomedical waste. Out of the total waste spawned from different HCUs, 10–15% of waste belongs to the category of BMW (Muduli and Barve, 2012; Taghipour and Mosaferi, 2008; Yazie et al., 2019). The infectious waste generated within these health care establishments if not properly managed is liable to endanger human and environment because of its infectious nature (Pruss et al., 1999). This portion of the waste is assumed to comprise pathogens including microbes, viruses, pests in ampule application or amounts liable to cause various human and animal diseases (Emmanuel et al., 2014). This type of BMW generated from such medical facilities is managed as per the BMW management and handling Rules 2016. Ministry of Environment and Forests (MoEF) while issuing statement on the BMW (Management and Handling) Rules in 1998 for regulating the management of health care waste. The organization however instituting the permissible structure, familiarized the term BMW for all waste categories produced from health care establishments and vet-hospitals. HCWs were classified into different classes as listed below. In order to protect the human health, these rules also make the health care waste generator accountable for separation, stuffing, storing, conveyance, and dumping (Patil and Shekdar, 2001) of the BMW. Based on the segregation pathways, color coding (CPCB, 2016) and type of ampule for various classes of HCW and their analogous handling and discarding approaches were also framed as per the BMWM Rules 2016. The ampules carrying or storing HCW should be symbolled accurately with suitable biohazard or cytotoxic signs.

9.4.2 PATHOLOGICAL WASTE

Considered as subclass of infectious BMW, frequently categorized individually chiefly when superior methods of handling, treatment, and disposal are in operation (Emmanuel et al., 2014; Taghipour and Mosaferi, 2008). This class of waste comprises human body parts, tissues, organs, body parts, fluids, and other waste generated from operation and postmortems on patients with contagious ailments. This class of waste also includes human fetuses and polluted animal cadavers (Patil and Shekdar, 2001).

Recognizable body parts of human or animals are sometimes christened as anatomical waste. Wastes of pathological nature may also include vigorous parts of human body that have been detached in medical techniques or arise during operations of medical research (Emmanuel et al., 2014; Pruss et al., 1999).

9.4.3 PHARMACEUTICAL WASTE

This category of waste includes all the unwanted, unexploited, expired, fragmented and polluted products and medicines, vaccines, and drugs that are no longer in use and need to be discarded properly owing to their chemical and biological properties (Emmanuel et al., 2014). This category of waste also comprises rejected products and substances used in the treatment and shipping of drugs which involves decanters, packaging material contaminated with drug remains, handbags, covers, pipes and vials contaminated with drug remains (Pruss et al., 1999).

9.4.4 GENOTOXIC WASTE

It includes the category of wastes, highly hazardous in nature endangering human and animal life by causing genetic mutations (mutagenic), birth defects (teratogens), or cancerous growth (carcinogenic). The term genotoxic is commonly applied to all those discarded substances which causes defects in the human genetic material particularly to the DNA. This class of waste includes all those wastes that are likely to endanger cells (cytotoxic) or suppressing the growth and multiplication of cells (cytostatic) or by inhibiting the normal growth and development of abnormal tissues (antineoplastic), and the chemicals used in cancer therapies (chemotherapeutics) (Emmanuel et al., 2014; Pruss et al., 1999). The specialized departments of different HCUs like that of oncology and radiotherapy sections make the wider use of cytotoxic drugs for the treatment of cancer. Keeping in mind the wider incidence of this dreaded disease, cytotoxic drugs are nowadays used in other hospitals and different clinics in diverse parts of the globe. Out of the total waste generated from specialized HCUs, genotoxic waste comprises only 1% together with cytostatic and radioactive constituents (Emmanuel et al., 2014; Mandal and Dutta, 2009).

9.4.5 CHEMICAL WASTE

This category of HCWs includes all those discarded solids, liquids, and gaseous chemicals arising from diagnosis, experimental, housekeeping, cleaning, and disinfecting processes. These wastes are labeled as hazardous chemical waste if they are liable to endanger human and animal life because of their toxicity, corrosivity (pH < 2 for acids; and >12 for bases) flammability, and reactivity characteristics. The hazardous chemicals which are mostly extensively employed in HCUs and most likely originate in the waste include formaldehyde, photographic chemicals including fixers (hydroquinone 5–10%, potassium hydroxide 1–5% and silver 1%), acetic acid, solvents (methylene chloride, chloroform, trichloroethylene, refrigerants), and compounds of nonhalogenated nature (Xylene, methanol, acetone, isopropanol, toluene, ethyl acetate), chemicals of organic nature (phenols, perchlorethylene, vacuum pump oils, pesticides and rodenticides), inorganic chemicals including acids and alkalis (e.g., H_2SO_4, HCl, HNO_3, NaOH), ammonia solutions, $KMnO_4$, $K_2Cr_2O_7$, $NaHSO_3$, and Na_2SO_3. All other wastes which do not possess these characteristics are termed as nonhazardous chemical wastes. This category of wastes includes all those sugars, amino acids, and inorganic and organic salts widely used in transfusion liquids. Of the total waste generated from HCUs 3% of the waste consists of chemical and pharmaceutical wastes (Mandal and Dutta, 2009).

9.4.6 NONHAZARDOUS GENERAL WASTE

All wastes excluding BMW and which do not endanger human health any way or which has not been in contact with contagious, harmful, and substances of radioactive nature (CPCB, 2016) come under he category of nonhazardous general waste. A large portion of about 85% of the waste generated (Taghipour and Mosaferi, 2008) from the HCUs consists of nonhazardous waste and is comparable with the domestic and municipal waste characteristics (Yang et al., 2021). More than 50% of the nonhazardous waste generated from HCUs consists of packaging material (cardboards), office paper (15%), plastic (10%), rags (15%), food remains and sweepings from hospital premises (53.5%) (Mandal and Dutta, 2009), bags, containers, cans, metals, glass, textiles, and wood, etc. (Emmanuel et al., 2014; Mmereki et al., 2015; Pruss et al., 1999). General nonhazardous waste generated from different HCUs are additionally categorized into dehydrated and wet nature, and need to be managed individually. Almost all nonhazardous or municipal waste is disposed either through open dumping or in landfills or through incineration. The handling, treatment, and disposal of all nonhazardous wastes come under the preview of Solid Waste Management and Handling Rules 2016.

9.4.7 *OTHER WASTES*

The other wastes generated within HCUs include the electronic gadgets, such as computer parts, discarded batteries, printers, cartridges, photocopying ink and toners (Emmanuel et al., 2014; Pruss et al., 1999), radioactive materials, and all those wastes which do not come under the preview of HCWs. However, the management, handling, and dumping of all such wastes are carried out according to the requirements placed down under the Electronic Waste (Management) Rules 2016, Batteries (Management and Handling) Rules 2011 and rules framed under Atomic Energy Act, 1962 (CPCB, 2016).

9.5 WASTE: PHYSICOCHEMICAL CHARACTERIZATION

The important aspect of waste characterization is the physicochemical composition of the HCW. This information is indispensable in evolving plans for waste reduction and laid down an important salvaging programs that necessitates a thoughtful understanding of the general waste. Creating prerequisite conditions for facilities like treatment technologies are essential for the management of infectious portion of the waste, for example, microwave and steam treatment trust on least amount of wetness present in the stream of waste, organic load, and water content are the main factors affecting the chemical composition of the HCW, while the amount of incombustible portion, calorific value, and dampness of the waste affect the process of incineration.

For the estimation of storage, transport, and treatment and chamber capacity, the influence of physical properties like bulk density is an essential factor. These physicochemical characteristics like bulk density of HCW vary from nation to nation and waste to waste within the HCUs in different countries. The bulk density of 594 kg/m^3 was reported from the HCWs generated from the urban hospitals of Tanzania; 218 kg/m^3 for total amount of generated waste, 211 kg/m^3 for nonhazardous general waste, 226 kg/m^3 for contaminated infectious portion of waste and 110 kg/m^3 with boxes and 100 kg/m^3 excluding these boxes.

Wide variation in the average moisture content between departmental waste and countries of different HCW was reported. It was reported 26.76% in Italy, more than 50% of the moisture content was noticed in HCWs generated from the departments like obstetrics, dialysis, and pediatrics. However, low moisture content of 14.15% was reported from the HCW generated within the hospitals of Turkey. Additionally, these hospitals had an average of 8% of noncombustible waste materials (Altin et al., 2003).

Residual percentage from BMW was found to be temperature-dependent as it was observed 66% at 110C, 15% at 550°C and 14% at 1000°C as reported from the health care system of Italy. The estimated composition of the HCW was found to contain 37% carbon, 18% oxygen, and 4.6% hydrogen coupled with other elements that are found in bulk quantities (Liberti et al., 1994). Health care waste contain toxic metals that are readily emitted in to the atmosphere during combustion. These toxic metals include Pb, Hg, Cd, As, Cr, and Zn. Furthermore, persistent organic pollutants (POPs), such as polychlorinated dioxins and furans cannot be anticipated constantly from elementary essential composition. These dioxins and furans are extremely lethal at lower applications. The reduction in the concentrations of halogenated plastics leads to the decrease in the amounts of HCl together with other halogenated contaminants. In modern hospitals, as much as 40% of the waste is composed of chlorine-based plastics. Internationally accepted symbol and number system is recognized for common plastics in order to facilitate the process of recycling, for example, 1 – for polyethylene terephthalate, 2 – for high-density polyethylene, 3 – for polyvinyl chloride, 4 – for low-density polyethylene, 5 – for polypropylene, and 6 – polystyrene. However, no labeling system was established for many polyvinyl chloride products, such as blood bags, gloves, external feeding substances, wrapping films, etc.

9.6 HEALTH CARE WASTE AND ASSOCIATED ISSUES

Globally, every year, huge amounts of contagious and dangerous BMWs are produced from different health establishment units. Lack of financial and other limitations (Policy and regulatory) mostly in developing and middle-income countries (Caniato et al., 2015) leads to inefficient and inadequate management of these HCWs coupled with lack of proper education among the staff members about the familiarity of procedures. Adequate treatment and management is one of the severe issues about these HCWs (Birpinar et al., 2009). In poorly developed countries, the supervision of HCW is commonly substituted either to illiterate or ill-educated persons especially those of laborers mostly associated with cleaning processes, who perform most of the activities dealing with collection and packing of HCW without any guidance and proper protection. For the control of infections within the HCUs, efficient management of wastes within these establishments is of critical importance, which subsequently has a great bearing on the care quality and occupational health of workers (Diaz, 2003).

9.6.1 TRAUMA AND INFECTIOUS RISKS

This hazardous and infectious HCW causes serious environmental and health care issues (Jang et al., 2006) due to inadequate treatment and inefficient supervision. Lack of proper disposal practices cause health risks to people openly convoluted in the collection and disposal of these trashes. Thus, it requires adequate treatment and management prior to its disposal (Muduli and Barve, 2012). Keeping in view, the per capita HCW production within the health care facilities is highly dangerous because of its contamination by pathogenic agents. The pathogenic organisms present in the HCW are resistant to treatment and thus possess high pathogenicity (Abdulla et al., 2008). Inadequate supervision of HCW leads to ecological contamination by toxic hazardous chemicals (Jang et al., 2006), unpleasant odors, multiplication and growth of harmful insects, flies, cockroaches, vermin, rodents (Abd El-salam, 2010), and pathogens that may lead to pollution of underground water sources (Abd El-salam, 2010) by unregulated disposal of HCW in landfills and transmission of contagious diseases, such as AIDS, hepatitis, cholera, typhoid fever, tuberculosis, cancers (Abd El-salam, 2010; Abdulla et al., 2008; Gupta and Boojh, 2006; Mandal and Dutta, 2009), hemorrhagic fever, herpes, measles, shigellosis, salmonellosis, pneumonia, septicemia, bacteremia, anthrax, helminthic infections, candidiasis, and others caused due to the contamination of wastes by human anatomical, microbiological, biotechnological, soiled waste components, and through injuries caused by blood contaminated sharps (Mandal and Dutta, 2009; Muduli and Barve, 2012). The detailed account of infections with mode of transmission (Figure 9.2) and responsible organisms are outlined in (Table 9.4).

In the world, over 5.2 million individuals comprising 4 million kids die every year owing to waste-related diseases. Thus, proper waste management is an important factor for decreasing environmental pollution and human health risks (Askarian et al., 2004). As a result, awareness increased among medical staff and common masses about the infectious diseases caused by

TABLE 9.4 Infections Caused by Exposure to HCW, Causative Agents and Transmission Routes.

Infection type	Causative agents	Route of transmission
Gastro enteric infections	Enterobacteria, for example, *Salmonella*, *Shigella* spp., *Vibrio cholerae*, *Clostridium difficile*, helminths	Feces and/or puke
Respiratory infections	*Mycobacterium tuberculosis*, measles virus, *Streptococcus pneumoniae*, severe acute respiratory syndrome (SARS)	Breathe in excretions, spit
Ocular infection	Herpesvirus	Eye discharges
Genital infections	*Neisseria gonorrhoeae*, herpesvirus	Genital discharges
Skin infections	*Streptococcus* spp.	Pus
Anthrax	*Bacillus anthracis*	Skin discharges
Meningitis	*Neisseria meningitidis*	Cerebrospinal fluid
Acquired immunodeficiency syndrome (AIDS)	Human immunodeficiency virus (HIV)	Plasma, sexual discharges, body discharges
Hemorrhagic fevers	Junin, Lassa, Ebola and Marburg viruses	All bloody produces and excretions
Septicemia	*Staphylococcus* spp.	Plasma
Bacteremia	Coagulase-negative *Staphylococcus* spp. (including methicillin-resistant *S. aureus*), *Enterobacter*, *Enterococcus*, *Klebsiella* and *Streptococcus* spp.	Nasal discharge, skin interaction
Candidemia	*Candida albicans*	Plasma
Viral hepatitis A	Hepatitis A virus	Feces
Viral hepatitis B and C	Hepatitis B and C viruses	Plasma and body liquids

Source: Emmanuel et al. (2014).

HCW like that of HIV, hepatitis B and C (Adu et al., 2020). It is worthy to note that HCW can lead to the transmission of more than 30 hazardous blood-borne pathogenic organisms (Odonkor and Mahami, 2020). Annually, more than 2 million health care professionals get exposed to percutaneous wounds due to the presence of sharps in HCW. In 2000, it was reported that globally, sharp injuries to occupational health workers have affected about 66,000 Hepatitis B (HBV), 16,000 Hepatitis C (HCV), and 200–5000 HIV infections due to their contact to percutaneous damages and the fractions of these injuries account to about 37%, 39%, and 4%, respectively (Pruss-ustun et al., 2005). Animal waste released from different veterinary hospitals and other related sources leads to spread of infections like rabies and anthrax, etc.

Wastes such as sharps, cytotoxic, radioactive substances (Borowy, 2020), and the wastes released form incineration of BMW cause different kinds of injuries. These wastes are likely to cause diseases, such as dermatitis, conjunctivitis, bronchitis, and pulmonary diseases are caused by exposure to air contaminated bioaerosols at refuse disposal site (ICRC, 2011). Diseases, such as cancers, genetic mutations, and poisonings are also caused by wastes contaminated by radioactive, cytotoxic, pharmaceutical, and chemical liquid wastes in different health care facilities. These are responsible for these dangerous diseases because of their hazardous properties (toxic, carcinogenic, mutagenic, corrosive, and explosive).

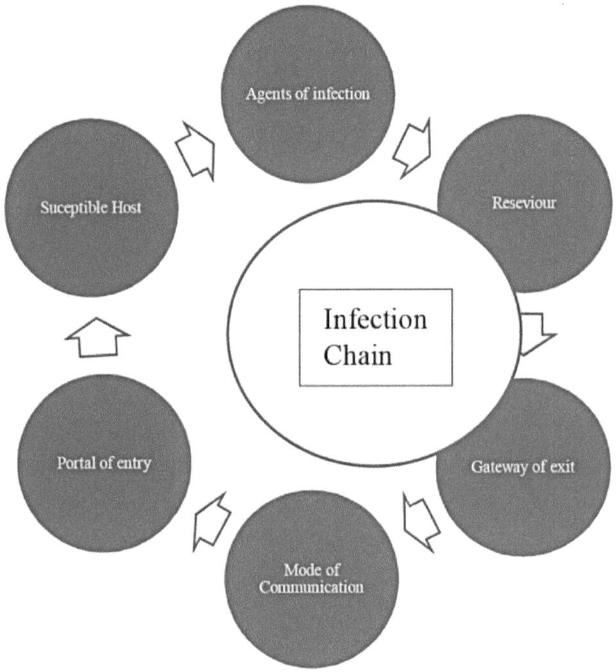

FIGURE 9.2 Sequence of infection.

9.6.2 ENVIRONMENTAL RISKS

Improper segregation of HCW and other waste components from the point of origin to disposal can trigger a *domino-* like effect on different environmental components that incur dangers to water, animals or soil, and surface and ground water resources. Improper segregation and disposal of HCW has the potential to contaminate water sources as different pollutants get mixed with ground water sources leached from waste disposal sites or landfills. It was reported that heavy metals released from the landfill leachate were above the permissible limit of drinking water standards. The concentration of different elements, viz. aluminum (Al), vanadium (V), Chromium (Cr), Manganese (Mn), cobalt (CO), Nickel (Ni), barium (Br), lead (Pb), and iron (Fe) were 2.050, 0.9775, 0.503, 0.128, 0.773, 0.8575, and 39.25 mg/L (Manzoor and Sharma, 2019). While conducting a comprehensive study on ground water quality in the neighborhood of waste dumping sites in logos Nigeria, it was reported that all the water physiochemical parameters were above the permissible level recommended for drinking water supply (Oluseyi et al., 2014). The incinerated BMW results in the discharge of pollutants, such as toxic metals, polycyclic aromatic hydrocarbons, and some hazardous waste materials that result in the contamination of shallow and underground water resources. In order to check the discharge of these harmful chemicals into the surroundings, it is necessary to remove the toxicity of the incinerated ash prior to its dumping in landfills (Heera et al., 2014). From the environmental analysis of biomedical ash, it was observed that it displayed augmented hardness of (1320 mg/L) and chloride (8500 mg/L) content in leachate when compared with the WHO and EPA drinking water recommendations. Metals like mercury (Hg) when released into the environment from HCW evaporates readily and remain suspended in the atmosphere for a year or more. Mercury (Hg) when released into the aquatic environments accumulates into the sediments and gets converted into more toxic organic derivative called methyl mercury. Mercury (Hg) is commonly used in thermometers, dental alloys, etc. These health care establishments are also accountable for the mercury pollution of superficial water sources, which subsequently get accumulated into fish tissues, and subsequently affects humans through food chains (Emmanuel et al., 2014; ICRC, 2011).

Poor and unregulated sewage and waste water management generated from the HCUs results in pollution of soil and water sources with pathogenic and deadly chemicals. Releasing of chemicals and pharmaceutical waste water can damage the working of biological sewage treatment plants and septic tanks which results in the contamination of water sources and ecosystems. Excreted antibiotics, drugs and associated metabolites with excreta of patients under medical supervision end up in sewage. Domestic waste water generated from HCUs showed 2–10 times more bacterial strains resistant to different antibiotics than inland waste water, a sensation which subsidizes to the occurrence and proliferation of microorganisms such as methicillin-resistant *Staphylococcus aureus* (MRSA) (ICRC, 2011).

Soil pollution due to HCW is carried out by infectious waste, discarded and expired medicines, chemicals and polluted water released from the HCUs. Ash and other harmful metals and metalloids are released during treatment processes. The metals, such as Cd, Pb, and Hg, etc. are reported to cause soil pollution, these metals when existing in waste water, will become available to plants from metal-contaminated soils and enter the food chain. In addition to metals, large quantities of nitrates and phosphates are also reported from the landfill leachates. Excessive release of metals and other elements from the HCW water in soils are harmful to plants, animals, and human beings (Babu et al., 2009). Heavy metals released from the dumping of BMW were found to be responsible for deteriorating soil quality and loss of vegetation cover. Further, a significant effect was also observed in soil physiochemical parameters near a landfill site. Soils at this landfill site showed high pH, total dissolved solids (TDS), and electrical conductivity (EC) and metals, such as lead (Pb), copper (Cu), nickel (Ni), chromium (Cr), and Zinc (Zn) (Ali et al., 2014). Thus, the principal ecological risk connected with the waste dump locations is the probable danger postured to topsoil. Since unregulated disposal of waste onto superficial soils, a large amount of pollutants infiltrate freely and ultimately pollute the top soil and disturb the plants in the vicinity of waste landfills.

The major ideology of health care training is *"Do no harm"* but paradoxically, the health care exercise itself results in noteworthy contamination. The medical sector is also blamable for significant amounts of nationwide pollution discharges and effects comprising acid rain (12%), emission of GHGs (10%), formation of smog (10%), criteria air pollutants (9%), depletion of ozone (1%), carcinogenic and noncarcinogenic (1–2%) (Eckelman and Sherman, 2016). Combustion of HCW also releases different kinds of harmful pollutants into the ambient air, notable among them includes the dust, black carbon, ammonia, particulates of sulfates and nitrates. In addition to these, harmful hazardous organic compounds, such as diethyl phthalate, decane, dodecane, nonane, methenamine, cyclobutane, carbon disulfide and acetone diperoxide (Manzoor and Sharma, 2019) are also released. Incineration of HCW releases organic and inorganic contaminants from stack emissions, incinerated dust, and fly ash. The organic emissions include polychlorinated dibenzo-*p*-dioxin/dibenzofuran (Adu et al., 2020) and polycyclic aromatic hydrocarbons (Borowy, 2020), while the inorganic emission from BMW incineration includes the fumes and ash containing lethal metals and metalloids. Inorganic acidic gases that get released into the atmosphere include hydrogen chloride (HCl), hydrogen fluoride (HF), hydrogen bromide (HBr), sulfur oxides (So_x), and nitrogen oxides (NO_x)

(Singh and Prakash, 2007). The toxic metals associated with incinerator emissions include volatile metals, such as Hg, Cd, Cr, As, and Be responsible for neurological, lung, and kidney damages (Lee et al., 2004; Segura-Munoz et al., 2004). Different forms of these metals, at various levels and through different media and pathways of exposure, have shown to cause a wide range of carcinogenic and noncarcinogenic health hazards (Rushton, 2003).

9.7 HIERARCHY OF HEALTH CARE WASTE MANAGEMENT

Management of waste help in the protection of environment and human health is accomplished by a range of approaches. These methods when outlined in preferential order is called "waste hierarchy." In this hierarchy, the most appropriate methods are placed at the top and the least desirable at the bottom. The desirability of every method is judged by its environmental and health impacts, public acceptability, and economic affordability. This waste hierarchy is basically based on the concept of sustainable use of resources integrating the perception of "3R" that is *reduce*, *reuse*, and *recycle*. Eco-friendly management of waste will help us to avoid the wasteful use of resources or to recuperate as much as possible from the waste within or nearby these HCUs moderately than its disposal by scorching or landfilling. The most efficient method helps us to avoid the production of waste thereby help us in the minimizing the waste quantity. The second most preferable method of waste hierarchy is the recovery of waste materials for secondary use. Non-recoverable waste products then must be dealt with the least desirable methods such as treatment or landfilling in order to minimize the impact on environment and public health.

9.7.1 HEALTH CARE WASTE MINIMIZATION

Good HCW management desires that the waste should be reduced and minimized by its *reuse* and *recycling*. The use of disposable items should therefore be minimized, and the use of products with long life reusable items like syringes with proper sterilization should be promoted except where their use becomes absolutely essential in the performance of essential measures. Reduction in waste could ultimately be eco-friendly in nature. Waste minimization is fundamentally applied at the source of its generation. To achieve the waste minimization, the principal focus should be on health care employed staff to adopt scientific approaches with efficient and reduced use of material, these professionals can also take necessary measures in the reduced production of wastes by changing strategies of stock control and purchasing (Emmanuel et al., 2014).

Municipalities, households, private companies including some public sectors are a large array of institutions where process of *recycling* is largely practiced. When considering *recycling*, it is a less efficient process than *reusing* waste items, because of frequent and substantial energy input and transportation to recycling facilities (Emmanuel et al., 2014; Pruss-ustun et al., 2005; Pruss et al., 1999).

Most commonly, the process of recovery is defined in more than one way. Most simple definition states "*energy recovery*" whereby discarded material is transformed to fuel for the generation of electricity. This generated electricity may subsequently be used in health care establishments. Alternatively, the term "*waste recovery*" encompasses the process of converting recyclable waste products into new commodities and production of compost or soil conditioner from composting of organic waste products (Emmanuel et al., 2014).

In large HCUs, the process of *recycling* is essentially practiced for large portion of nonhazardous waste materials which help in considerable reduction in financial costs and thereby help in waste minimization.

9.8 SEGREGATION, STORAGE, AND TRANSPORT OF HEALTH CARE WASTE

Waste segregation is considered to be one of the most appropriate and effective methods for waste management and minimization. The various types of HCWs are separated into diverse classes, and collected in different waste bins with different colors, so as to aid in the separation of hazardous and nonhazardous wastes, involving different processes of treatments called as waste segregation. The accurate arrangement for assortment of diverse waste substances in diverse colored waste bins is to be done by numerous infirmaries, and it is the responsibility of the sanatorium clinicians and harbors to confirm accurate discarding of diverse waste stuffs in their corresponding dust bins or ampules. The scheme of using different decorated bins and bags to accumulate different forms of solid HCWs is acknowledged as *Color coding*. Such a scheme ultimately aids in independently amassing the nonhazardous medicinal waste substancesthat includes the unadulterated stuffing constituents, such as plastics, cardboard, food and kitchen waste, garden waste. as given in Table 9.5, such nonhazardous wastes can then effortlessly and carefully be predisposed of by *landfilling* but only if it is permitted of BMW that would cause adversative effects on human health and atmosphere.

TABLE 9.5 Color Coding Segregation Scheme Suggested by WHO.

Color	Type of ampule	Waste type	Method of treatment
Yellow	Plastic bag	Human body Waste/animal waste, Microbiology, Biotechnology and soiled waste	Incineration/deep burial
Red	Disinfected ampule/plastic bag	Microbiology, Biotechnology and soiled waste	Autoclave/microwave/chemical treatment
Blue white/Translucent	Plastic bag/Perforation proof ampule	Sharps and solid waste	Autoclave/microwave/Chemical treatment/shredding/destruction
Black	Plastic bag	Rejected medicines, cytotoxic drugs, incineration ash, and chemical waste	Secured landfill disposal

9.8.1 LABELING FOR IDENTIFICATION OF HCW

Classification is indispensable for the accurate documentation and benign organization of HCW. Labeling makes the identification and handling of different types of HCWs easier. It will also help to warn the staffs, patients, and community about the survival of the wastes and their potential health hazards (Emmanuel et al., 2014; Pruss-ustun et al., 2005).

All labeling and mark symbols should be properly followed according to the global symbols and color coding. All the infested, dirty, compulsive, human and sharp trashes should be labeled with *biohazard symbol* in black color, representing *cytotoxic wastes*. The sign of cytotoxic hazard, that is, *C* in a triangle is used for this labeling. At the point of waste generation, labeling should be in tagged form or paste sticker properly devoted to the particular collection container or bag prior to its collection by scrubbing staff. This *tagging system* of waste will allow *waste audits* steered at dealing/disposal site, to recognize those areas that are in compliance or noncompliance with the required HCW management practices. If needed, a corrective action can then be easily initiated. It would be superior, if the carrying bags and containers delivered are previously branded with suitable hazard sign. Some of the symbols are shown in Figure 9.3.

Biohazard **Radiation Symbol (Old)** **Radiation Symbol (new)**

FIGURE 9.3 Symbol representing biohazard, radiation, and chemical hazard.

9.8.2 STORAGE OF HEALTH CARE WASTE

According to the rules, the HCW should be properly kept in designated plastic bags or ampules in a separate room or apartment within these care establishments or any other health care facility generating the waste depending on the amount of generated waste and frequency of collection. (Pruss et al., 1999). The infectious waste cannot be stowed for more than 48 h in summer time and 72 h in winter season under temperate climatic conditions. Under warm climatic conditions the storage timing should not exceed 48 h during the cool season and 24 h during hot season (Emmanuel et al., 2014). However, the HCW needs to be stowed, if instantaneous handling and clearance cannot be carried out. These wastes should be cooled *on-site* to prevent decaying and evolution of unpleasant odors. These wastes may then be heaved and organized in *off-site* conveniences of storage that is constructed in R.C.C in acquainted zones, away from the common public. These storing facilities must be correctly planned to allow proper outflow of waste wash down and rain water.

9.8.3 TRANSPORTATION OF HCW TO TREATMENT AND DISPOSAL SITES

The BMW should be regularly elevated from proper labeled areas from hospitals and other research facilities in wards or rooms. For waste collection from individual bins located in hospital storage rooms often open trolleys are often employed. While transportation of waste, each bin or container be properly washed and disinfected it with 1% solution of hypochlorite (bleaching powder) once in a week at least (Pruss-ustun et al., 2005). These trolleys and waste carts should not be used for any other purposes except for waste storage and transportation. Automated transport system and waste handling should be preferred than manual methods. The lorries employed for the conveyance of the waste should not be used for any other purpose. Such vehicles should be properly labeled and designated for the purpose by the concerned hospital authorities. Waste tags should be placed in tact at the end of transportation (Emmanuel et al., 2014; Pruss et al., 1999).

9.9 TREATMENT AND DISPOSAL METHODS FOR HEALTH CARE WASTE

All the disposable plastic items in HCW are susceptible for salvaging without appropriate disinfection methods, if not completely demolished. Several cases of repacking, sale, and supply of contaminated syringes, catheters, and other medical equipment's after low temperature autoclaving or by just washing from unauthorized factories in different parts of country including the capital city of Delhi has been testified recently (Harhay et al., 2009). This practice of using contaminated medical equipment's is a serious health issues for the patients to whom these were used.. According to the reports of Indian Clinical Epidemiology Network, greater than 30% of 2–6 billion vaccinations administered annually were performed with contaminated and repacked medical syringes. This practice of using repacked and reused medical equipment's were nationally reported from 10% of medical health services. Used needles and syringes were sold off to waste pickers (Solberg, 2009). It is now a national obligation to shred or maim all reusable medical objects at the source of its generation, keeping in view the health risks associated (Adu et al., 2020) with the use of reusable and repacked medical items. All the HCW should be managed in an eco-friendly manner.

The various methods employed at various health care establishments for disposing of hazardous HCW includes chemical and thermal methods. The chemical method of treatment involves *chemical disinfection* using some appropriate chemical disinfectant. The thermal method depends on the heat for destruction of pathogens. Thermal processes are the most widely used processes globally (Emmanuel et al., 2014) and include autoclaving, hydro-claving, microwaving, and Incineration. In addition to these methods, for disposing of some individual type of HCW, *landfilling* is also followed some times. However, it should be kept in mind that the landfill site should always be away from the residential areas, because of their environment and human health risks. Due to these constraints, HCWs are not generally taken for burial in landfills, but are usually disposed of by using any of the thermal processes outlined above. The brief account of these methods is given as under.

9.9.1 CHEMICAL DISINFECTION

In chemical practice, the stream of waste is initially disinfected with some appropriate chemical disinfectant like chlorine compounds like bleaching powder. Other chemicals, such as alcohol, iodine, phenolic compounds, hexachlorophene, lime solution, peracetic acid (Emmanuel et al., 2014), and formaldehyde may sometimes be used for the process of disinfection. However, in most of the instances, 1% hypochlorite solution (Bleaching powder) is used. All these disinfectants are used in the form of aqueous solutions in which wastes are kept submerged for some time. Before disinfection many of the waste items need shredding to provide sufficient time of contact between wastes and the disinfectant. The disinfected waste is then usually shredded. The shredded wet and chemical disinfected laden waste will be now available for final disposal (Garg, 2017). The chemical disinfectant method is used for the disposal of waste sharps, gloves, urine bags, catheters, tubes, blood bags, and syringes. (Emmanuel et al., 2014). These waste products after shredding and disinfection can be buried and if liquid in nature they are discharged into sewers after proper disinfection.

9.9.2 AUTOCLAVING

This is a low heat thermal practice or steam sterilization whereby a waste is conveyed in a direct connection with a steam for a sufficient time duration in an organized fashion to sterilize the streams of waste (Emmanuel, 2001). This can be done with the help of three types of autoclaves.

9.9.2.1 GRAVITY TYPE

This includes those autoclaves in which evacuation of air from the treatment chambers is carried out with the help of a stem pressure alone. Such autoclaves operate with a steam temperature of about 121°C, for full steam penetration into the densely packed stream of waste, these autoclaves require a cycling time of nearly 60–90 min (Garg, 2017).

9.9.2.2 PRE-VACUUM TYPE OR HIGH VACUUM TYPE

In these types of autoclaves, the air from the treatment chambers is evacuated with the help of using vacuum pumps. Which enables them to reduce the cycle time by about 30–60 min, as the time required to heat the air within the chamber is reduced. These systems require an operating temperature of about 132°C.

9.9.2.3 RETORT TYPE AUTOCLAVES

The retort type autoclaves require a large volume of treatment chambers for much higher steam temperatures and pressures, and hence their cycle times are reduced substantially than those of above discussed systems.

After low heat treatment, the wastes are shredded and suitably disposed. Wastes disposed of in this manner are indicated in Table 9.5.

9.9.3 HYDROCLAVING

Hydroclaving is the advanced type of autoclaving. In hydroclaving, subsidiary heating is carried out by directing the steam into the external jacket of a double-walled ampule and within the inside container is rotated by an appropriate machinery. This causes the waste fragmentation and tumbling against the hot walls of the vessel. The moisture content present in the waste gets converted to steam and pressurizes the vessel. In the absence of enough moisture to pressurize the vessel, small quantity of steam is supplied externally till suitable pressure is achieved. The time of treatment required for sterilization is 15 min. at 132°C or 30 min at 121°C temperature. In hydroclaving process, enough of sterilization is carried out, and the resultant waste stream gets fragmented and dehydrated with reduction in weight and volume (Garg, 2017).

9.9.4 MICROWAVE TREATMENT TECHNOLOGY

It is essentially a steam-based waste treatment method in which treatment of waste is carried out with the help of moist heat and steam generated with the help of microwave energy. The existing moisture in the waste is rapidly heated with the assistance of microwave-based energy carried at a frequency and a wavelength of 240 MHz and 12.24 cm. Thus, common microwave system of treatment consists of chamber or treatment area in which microwave-based energy is directed from a magnetron (microwave generator), especially for each 1.2 Kw output at least two to six magnetrons are required. In some instances, microwave structures are semicontinuous and some are available in batch processing (Emmanuel, 2001; Emmanuel and Stringer, 2007). The categories of wastes usually treated in microwave structures are alike to those treated in autoclaves (e.g.,sharps, cultures and stocks, blood contaminated tools, and laboratory wastes). some microwave systems are also tested for the treatment of animal waste and can possibly be used in the treatment of pathological and tissues wastes (Devine et al., 2007).

9.9.5 INCINERATION

In most developing countries, incineration is one of the prime and most accepted method of hazardous HCW treatment before its final disposal (Yazie et al., 2019). Incineration is a dry oxidation waste combustion process, carried out at high temperature under controlled circumstances, to change transmissible and pathological medical waste materials to gases and inert inorganic remains (Garg, 2017) or mineral incombustible matter, which results in the substantial decrease in size and mass of the wastes (Emmanuel et al., 2014). Incineration requires a temperature from about 200°C to 1000°C. This process involves physical and chemical break down of organic waste materials through combustion, pyrolysis, and gasification. Incinerators are generally used for the treatment of hazardous liquid and solid wastes like those of hospitals and industries.. Liquid waste incinerator and Rotary kiln type incinerators are two main types of incinerators normally used in the treatment of HCWs. Almost 90% of all hazardous waste incinerator installations belong to these two types of incinerators, and only 10% belongs to fluidized-bed type and starved air/pyrolysis kind of incinerator, primarily used for the treatment of sludge waste, sewage waste, and municipal refuse, respectively.

Liquid waste incinerators are a type of incinerators primarily used in the treatment of hazardous liquid wastes. In such incinerators, liquid wastes are injected at 350–700 kPa through a special type of nozzle called an atomizing nozzle into the furnace or incineration chamber. The ejected liquid waste in the form of fine droplets gets volatized in the stream of hot gas which helps in gas oxidization. For effective annihilation of liquid hazardous wastes, the process involves diminishing unevaporated drops and unreacted vapor. In order to increase the destructive waste efficiency, the time main factors kept in mind are the residence time, temperature, and turbulence of the incinerator.

The rotary kiln incinerators are the most versatile of all the incinerators, as they can be used to burn all forms (solid, liquid, gaseous, and sludge) of wastes. Globally, these incinerators are used for the combustion of nonliquid hazardous HCWs than any other type of incinerator.

Some of the disadvantages associated with this type of treatment are the removal of residual ash and combustion by-products. The incineration of HCW mainly produces gaseous emissions, such as steam, CO_2, NOx, and wide range of volatile substances including toxic metals, halogen acids, particulate matters and solid remains especially those of incinerated ashes (Emmanuel et al., 2014; Pruss et al., 1999).

9.10 IRRADIATIVE PROCESSES

Radiation-based equipment involves the bombardment with electron beams like that of Cobalt-60, or irradiation with highly ionizing ultraviolet rays (UV rays). These technologies necessitate the proper protection of the working staff to avert any work-related contacts. Electron beam irradiation involves showering of high-energy electrons to destroy pathogenic microorganisms present in the streams of waste by causing chemical separation and disagreement of cell walls. The pathogen-destruction effectiveness depends on the dose immersed into the stream and mass of the waste, which, in turn, is dependent on waste density and energy of the electron beams. As an enhancement to other handling machineries, germicidal ultraviolet radiation (UV-C) has also been used. Irradiation does not change the stream of waste tangibly and would need a chopper or shredder to condense the waste to be unrecognizable (Emmanuel, 2001).

9.11 INTERNATIONAL CONVENTIONS AND AGREEMENTS

Three internationally accepted agreements and conventions which deal with the HCW management, sustainable development, and environmental protection, which should be comprehensively understood and kept in consideration while formulating policies of waste management. These internationally adopted agreements are the Basel Convention on hazardous waste, Stockholm Convention on POPs, and convention related to mercury waste pollution commonly known as Minamata Convention.

Basel Convention on hazardous waste is one of the most inclusive among different environmental treaties on hazardous and other wastes. The objectives of Basel Convention are the protection of atmosphere and public health from the hostile effects rising from the generation, management, and disposal of hazardous waste particularly HCWs from different health care establishments.

Stockholm Convention on POPs is another internationally accepted treaty dealing with the protection of public health and environment from persistent organic pollutants (Furans and Dioxins). POPs are toxic contaminants released into the environment due to the incineration of HCWs which accumulate in the adipose tissues of organisms and cause large number of diseases. Comprehensive guidelines dealing with the Best Available Techniques and Best Environmental Practices (BEP) were formulated in 2006. These BEPs include reduction in waste generation, waste separation, resource recovery, recovering, training, appropriate assemblage, and methods of transportation.

Universal treaty to safeguard environment and social health from harmful effects of mercury was signed by more than 90 countries in Japan on 10 of October 2014. This treaty is commonly called as Minamata Convention. This treaty deals with the phasing out of various mercury containing health equipments in health care establishments.

9.12 CONCLUSION

With respect to present and future life on the planet, environmental pollution has become one of the significant concerns of the environment. In this scenario, health care facilities also play a very prominent role toward this environmental problem. It is the accountability of every health care establishment to warrant safe disposal and treatment of the waste they generate, as to protect environmental and human health. The administrative authorities in the government or Central Pollution Control Boards (CPCB) are also responsible to guarantee the suitable execution of the legal environmental laws and the statues, concerning such vital pollution issues.

KEYWORDS

- **biomedical waste**
- **health hazards**
- **environmental risks**
- **collection**
- **treatment**
- **environmental legislation**

REFERENCES

Abd El-salam, M. M. Hospital Waste Management in El-Beheira Governorate. *Egypt. J. Environ. Manage.* **2010,** *91*, 618–629. DOI: 10.1016/j.jenvman.2009.08.012

Abdulla, F.; Qdais, A. H.; Rabi, A. Site Investigation on Medical Waste Management Practices in Northern Jordan. *Waste Manage.* **2008,** *28*, 450–458. DOI: 10.1016/j.wasman.2007.02.035

Adu, R. O.; Gyasi, S. F.; Essumang, D. K.; Otabil, K. B. Medical Waste-Sorting and Management Practices in Five Hospitals in Ghana. *Hindawi J. Environ. Public Health* **2020,** 1–14.

Ali, S. M.; Pervaiz, A.; Afzal, B.; Hamid, N.; Yasmin, A. Open Dumping of Municipal Solid Waste and Its Hazardous Impacts on Soil and Vegetation Diversity at Waste Dumping Sites of Islamabad City. *J. King Saud Univ.—Sci.* **2014,** *26*, 59–65. DOI: 10.1016/j.jksus.2013.08.003

Altin, S.; Altin, A.; Elevli, B.; Cerit, O. Determination of Hospital Waste Composition and Disposal Methods: a Case Study. *Polish J. Environ. Stud.* **2003,** *12*, 251–255.

Askarian, M.; Vakili, M.; Kabir, G. Results of a Hospital Waste Survey in Private Hospitals in Fars Province, Iran. *Waste Manage.* **2004,** *24*, 347–352. DOI: 10.1016/j.wasman.2003.09.008

Babu, B. R.; Parande, A. K.; Rajalakshmi, R.; Suriyakala, P.; Volga, M. Management of Biomedical Waste in India and Other Countries: A Review. *J. Int. Environ. App. Sci.* **2009,** *4*, 65–78.

Bdour, A.; Altrabsheh, B.; Hadadin, N.; Al-sharief, M. Assessment of Medical Wastes Management Practice: A Case Study of the Northern Part of Jordan. *Waste Manage.* **2007,** *27*, 746–759. DOI: 10.1016/j.wasman.2006.03.004

Birpinar, M. E.; Bilgili, M. S.; Erdogan, T. Medical Waste Management in Turkey: A Case Study of Istanbul. *Waste Manage.* **2009,** *29,* 445–448. DOI: 10.1016/j.wasman.2008.03.015.

Borowy, Iris. Medical Waste: The Darkside of Healthcare. *História, Ciências, Saúde—Manguinhos*, Rio de Janeiro, *Supl Set* **2020,** *27*, 231–251.

Caniato, M.; Tudor, T.; Vaccari, M. International Governance Structures for Health-Care Waste Management: A Systematic Review of Scientific Literature. *J. Environ. Manage.* **2015,** *153*, 93–107. DOI: 10.1016/j.jenvman.2015.01.039

Chudasama, R.; Rangoonwala, M.; Sheth, A.; Misra, S.; Kadri, A.; Patel, U. V. Biomedical Waste Management: A Study of Knowledge, Attitude and Practice among Health Care Personnel at Tertiary Care Hospital in Rajkot. *J. Res. Med. Dental Sci.* **2014**.

Chauhan, A.; Singh, A. Healthcare Waste Management: A State-of-the-Art Literature Review Ankur Chauhan and Amol Singh. *Int. J. Environ. Waste Manage.* **2016,** *18*, 120–144.

CPCB, 2016. Guidelines for Management of Healthcare Waste as Per Biomedical Waste Management Rules, Directorate General of Health Services Ministry of Health & Family Welfare **2016,** 1–74.

Datta, P.; Mohi, G. K.;Chander, J. Biomedical Waste Management in India : Critical Appraisal. *J. Lab. Phys.* **2020,** *10*, 6–14. DOI: 10.4103/JLP.JLP

Devine, A. A.; Grunden, A. M.; Krisiunas, E.; Davis, D. K.; Rosario, G.; Scott, S.; Faision, S.; Cosby, W. M. Testing the Efficacy of a Combination of Microwave and Steam Heat for Log Reduction of the Microbial Load Following a Simulated Poultry Mass Mortality Event. *Appl. Biosaftey* **2007,** *12*, 79–84.

Diaz, L. Metro Manila Solid Waste (TA 3848-PHI) Final Report no:11 *Medical Waste Management*. Republic of Philippines, 2003.

Eckelman, M. J.; Sherman, J. Environmental Impacts of the U. S. Health Care System and Effects on Public Health. *Plos One* **2016,** *11*, 1–14. DOI: 10.1371/journal.pone.0157014

Eker, H. H.; Bilgil, M. S.; Statistical Analysis of Waste Generation in Healthcare Services: A Case Study. *Waste Manage. Res.* **2011,** *29*, 791–796. DOI: 10.1177/0734242X10396755

Emmanuel, J. Non-Incineration Medical Waste Treatment technologies. A Resource for Hospital Administrators, Facility Managers, Health Care Professionals, Environmental Advocates, and Community Members, 2001.

Emmanuel, J.; Pieper, U.; Pruss, A.; Rushbrook, P.; Stringer, R.; Townend, W.; Wilburn, S.; Zghondi, R. *Safe Management of Wastes from Health-Care Activities*, 2nd ed.; WHO: Geneva, 2014.

Emmanuel, J.; Stringer, R. For Proper Disposal Hospital Waste: A Global Inventory of Alternative Medical Waste Treatment Technologies, 2007.

ENVIS Centre 1 CPCB. Bio-Medical Waste management-An Overview Control of Pollution Water, Air & Noise **2014**, 1–24.

Garg, S. K. *Sewage Disposal and Air Pollution Engineering*, 37th. ed.; Khanna Publishers: New-Delhi, 2017.

Gupta, S.; Boojh, R. Report : Biomedical Waste Management Practices at Balrampur Hospital, Lucknow, India. *Waste Manage. Res.* **2006,** *24*, 584–591. DOI: 10.1177/0734242X 06068342

Harhay, M. O.; Halpern, S. D.; Harhay, J. S.; Olliaro, P. L. Health Care Waste Management : A Neglected and Growing Public Health Problem Worldwide. *Trop. Med. Int. Health* **2009,** *14*, 1414–1417. DOI: 10.1111/j.1365-3156.2009.02386.x

Heera, S.; Kunal, Rajor, A. Bacterial Treatment and Metal Characterization of Biomedical Waste Ash. *Hindwai J. Waste Manage.* **2014,** 1–6.

International Committee of the Red Cross (ICRC). *Medical Waste Management*; Geneva, Switzerland, 2011; pp 1–159.

Jang, Y.; Lee, C.; Yoon, O.; Kim, H. Medical Waste Management in Korea. *J. Environ. Manage.* **2006,** *80*, 107–115. DOI: 10.1016/j.jenvman.2005.08.018.

Joshi, S. C.; Diwan, V.; Tamhankar, A. J.; Joshi, R.; Shah, H.; Sharma, M.; Pathak, A.; Macaden, R.; Lundborg, cecilia S. Staff Perception on Biomedical or Health Care Waste Management: A Qualitative Study in a Rural Tertiary Care Hospital. *Plos One* **2015,** *10*, 1–15. DOI: 10.1371/journal.pone.0128383.

Katoch, S. S.; Kumar, V. Modelling Seasonal Variation in Biomedical Waste Generation at Healthcare Facilities. *Waste Manage. Res.* **2008,** *26*, 241–246. DOI: 10.1177/0734242X 08093433.

Komilis, D.; Nikolaos, K.; Vassilopoulos, P. Hazardous Medical Waste Generation in Greece: Case Studies from Medical Facilities in Attica and from a Small Insular Hospital. *Waste Manage. Res.* **2010,** *29*, 807–816. DOI: 10.1177/0734242X10388684.

Lee, B.; Ellenbecker, M. J.; Moure-ersaso, R. Alternatives for Treatment and Disposal Cost Reduction of Regulated Medical Wastes. *Waste Manage.* **2004,** *24*, 143–151. DOI: 10.1016/j.wasman.2003.10.008.

Liberti, L.; Tursi, A.; Costantino, N.; Ferrara, L.; Nuzzo, G. Optimization of Infectious Hospital Waste Management in Italy: Part 1- Wastes Production and Characterization Study. *Waste Manage. Res.* **1994**, *12*, 373–385.

Mandal, S. K.; Dutta, J. Integrated Bio-Medical Waste Management Plan for Patna City. *Inst. Town Planners, India J.* **2009,** 1–25.

Mane, V.; Nimbannavar, S. M.; Yuvaraj, B. Y. Knowledge, Attitude and Practices on Biomedical Waste and Its Management among Health Care Workers at a Tertiary Care Hospital in Koppal, Karnataka, India. *Int. J. Community Med. Public Health* **2016,** *3*, 2953–2957.

Manzoor, J.; Sharma, M. Impact of Biomedical Waste on Environment and Human Health. *Environ. Claims J.* **2019,** *31*, 311–334. DOI: 10.1080/10406026.2019.1619265

Mmereki, D.; Baldwin, A.; Li, B.; Liu, M. Healthcare Waste Management in Botswana: Storage, Collection, Treatment and Disposal System. *J. Mater. Cycl. Waste Manage.* **2015,** 1–15. DOI: 10.1007/s10163-015-0429-0

Mohee, R. Medical Wastes Characterisation in Healthcare Institutions in Mauritius. *Waste Manage.* **2005,** *25*, 575–581. DOI: 10.1016/j.wasman.2004.10.003

Muduli, K.; Barve, A. Barriers to Green Practices in Health Care Waste Sector: An Indian Perspective. *Int. J. Environ. Sci. Dev.* **2012,** *3*, 393–399.

Odonkor, S. T.; Mahami, T. Healthcare Waste Management in Ghanaian Hospitals: Associated Public Health and Environmental Challenges. *Waste Manage. Res.* **2020,** 1–9. DOI: 10.1177/0734242X20914748

Oli, A. N.; Ekejindu, C. C.; Adje, U. D.; Ezeobi, I.; Ejiofor, S. O.; Ibeh, C. C.; Ubajaka, F. C. *Asian Pacific J. Trop. Biomed.* **2015,** *17*, 1–22. DOI: 10.1016/j.apjtb.2015.09.019

Oluseyi, T.; Adetunde, O.; Amadi, E. Impact Asseeement of Dumpsites on Quality of Near by Soil and Underground Water: A Case Study of an Abandoned and a Functional Dumpsite in Logos, Nigeria. *Int. J. Sci. Environ.* **2014,** *3*, 1004–1015.

Patil, A. D.; Shekdar, A. V. Health-Care Waste Management in India. *J. Environ. Manage.* **2001,** *63*, 211–220. DOI: 10.1006/jema.2001.0453

Patil, G. V.; Pokhrel, K. Biomedical Solid Waste Management in an Indian Hospital: A Case Study. *Waste Manage.* **2005,** *25*, 592–599. DOI: 10.1016/j.wasman.2004.07.011

Patwary, M. A.; O'Hare, W.; Thomas, S. G.; Elahi, K. M.; Hossain, S. S.; Sarker, M. H. Quantitative Assessment of Medical Waste Generation in the Capital City of Bangladesh. *Waste Manage.* **2009,** *29*, 2392–2397. DOI: 10.1016/j.wasman.2009.03.021

Pruss-ustun, A.; Rapiti, E.; Hutin, Y. Estimation of the Global Burden of Disease Attributable to Contaminated Sharps Injuries Among Health-Care Workers. *Am. J. Ind. Med.* **2005,** 482–490. DOI: 10.1002/ajim.20230.

Pruss, A.; Giroult, E.; Rushbrook, P. *Safe Management of Wastes from Health-Care Activities*; World Health Organisation: Geneva, 1999.

Rushton, L. Health Hazards and Waste Management. *Br. Med. Bull.* **2003,** *68*, 183–197. DOI: 10.1093/bmb/ldg034

Sanida, G.; Karagiannidis, A.; Mavidou, F.; Vartzopoulos, D.; Moussiopoulos, N.; Chatzopoulos, S. Assessing Generated Quantities of Infectious Medical Wastes: A Case Study for a Health Region Administration in Central Macedonia, Greece. *Waste Manage.* **2010,** *30*, 532–538. DOI: 10.1016/j.wasman.2008.11.019

Segura-Munoz, S.; Takayanagui, A. M.; Trevilato, T. M.; Santos, C. Trace Metal Distribution in Surface Soil in the Area of a Municipal Solid waste Landfill and a Medical Waste Incinerator. *Bull. Environ. Contamin. Toxicol.* **2004,** *72*, 157–164.

Sharma, H. B.; Vanapalli, R. K.; Cheela, Shankar, V. R.; Ranjan Prakash, V.; Jaglan, K. A.; Dubey, B.; Goel, S.; Bhattacharyya, J. Challenges, Opportunities, and Innovations for Effective Solid Waste Management during and Post COVID-19 Pandemic. *Resour. Conserv. Recycl.* **2020**, *162*, 105052. DOI: 10.1016/j.resconrec.2020.105052

Singh, A.; Kumari, R.; Wakhlu, A.; Srivastava, K.; Wakhlu, A.; Kumar, S. Assessment of Bio-Medical Waste Management in a Government Healthcare. *Int. J. Health Sci. Res.* **2014,** *4*, 203–208.

Singh, S.; Prakash, V. Toxic Environmental Releases from Medical Waste Incineration: A Review. *Environ. Minitor. Asses.* **2007,** 67–81. DOI: 10.1007/s10661-006-9503-3

Solberg, K. E. Trade in Medical Waste Causes Deaths in India A Burgeoning Trade in Used Medical Equipment Has Caused a Deadly Outbreak of Hepatitis B. *Lancet* **2009,** *373*, 1067. DOI: 10.1016/S0140-6736(09)60632-2.

Taghipour, H.; Mosaferi, M. Characterization of Medical Waste from Hospitals in Tabriz, Iran. *Sci. Total Environ.* **2008,** *407*, 1527–1535. DOI: 10.1016/j.scitotenv.2008.11.032

Thakur, V.; Ramesh, A. Analyzing Composition and Generation Rates of Biomedical Waste in Selected Hospitals of Uttarakhand, India. *J. Mater. Cycles Waste Manage*. **2017**. DOI: 10.1007/s10163-017-0648-7

Yang, L.; Yu, X.; Wu, X.; Wang, J.; Yan, X.; Jiang, S.; Chen, Z. Emergency Response to the Explosive Growth of Health Care Wastes during COVID-19 Pandemic in Wuhan, China. *Resour. Conserv. Recycl.* **2021**, *164*, 105074. DOI: 10.1016/j.resconrec.2020.105074

Yazie, T. D.;Tebeje, M. G.; Chufa, K. A. Healthcare Waste Management Current Status and Potential Challenges in Ethiopia: A Systematic Review. *BMC Res. Notes* **2019,** *12*, 1–7 DOI: 10.1186/s13104-019-4316-y

CHAPTER 10

Food Loss and Waste, Types, Sources, Pollution Potential, and Country-Wise Comparison

SAMINA KHALID[1*], MUHAMMAD IRFAN ULLAH[2], and AMAN ULLAH MALIK[3]

[1]*Department of Environmental Sciences, COMSATS University Islamabad, Vehari Campus, Punjab, Pakistan*

[2]*Department of Entomology; College of Agriculture, University of Sargodha, Punjab, Pakistan*

[3]*Postharvest Research and Training Center, Institute of Horticultural Sciences, University of Agriculture Faisalabad, Punjab, Pakistan*

[*]*Corresponding author. E-mail: saminakhalid@cuivehari.edu.pk*

ABSTRACT

Food loss and waste (FLW) are becoming increasingly relevant today due to their detrimental effects on sustainability, natural resources (water, soil, minerals, biodiversity, and energy depletion), environment (emission of greenhouse gasses (GHG), global warming, and depletion of ozone), economy (high production prices), and human health (malnutrition and toxic emission from rotting food waste). According to the Food and Agricultural Organization (FAO) assessment, almost one-third of the food provided for human intake is lost and wasted annually in the whole agri-food supply chain. This chapter provides a concise overview of FLW, its description given by different organizations, various types/classifications of food waste, drivers of food waste at several points from farm to fork, pollution caused by food waste, and its country-wise comparison.

Waste Problems and Management in Developing Countries. Umair Riaz, Shazia Iqbal, & Moazzam Jamil (Eds.)

10.1 INTRODUCTION

In 2019, about 690 million individuals were starving, which is around 8.9% of the global population. The COVID-19 pandemic will significantly worsen this situation, adding about 82–132 million undernourished people by the end of this year (Food and Agricultural Organization (FAO) and UNICEF, 2020). On the other side, one-third of the food grown for human intake is lost or discarded worldwide (Gustavsson et al., 2011), which is unjustifiable. The FLW has a substantial adverse effect on the climate, the economy, culture, and public being (Abiad and Meho, 2018). The above 2 billion people worldwide are suffering from micronutrient deficiency (Serafini et al., 2015). During 2019, 144 million kids below 5 years of age were stunted, with more than 9 out of 10 stunted children were living in Asia and Africa (FAO and UNICEF). The main factor for increased hunger and starvation worldwide is poor people's failure to consume nutritious diets (FAO and UNICEF). Reducing food waste can improve supply chain productivity by lowering food costs and increasing food affordability to marginal families(Gustavsson et al., 2011). The FLW also has economic implications, as the value of wasted food can be utilized in other developmental works. Annual spending on food waste in Britain and America was assessed at £680 and USD 1600, respectively (Lipinski et al., 2013); in Finland, EUR 460 (Silvennoinen et al., 2014); in Sweden, EUR 585 (Gjerris and Gaiani, 2013); and in Australia, $616 (Baker et al., 2009) per household. According to Statistics Canada (2010), Canadians waste food worth $27 billion per annum, and 50% of that waste was comprised of domestic waste (Gooch et al., 2010). Decreasing food waste can increase supply chain productivity the efficiency of value chains by lowering food prices and increasing food access to marginal households (Gustavsson et al., 2011) and hence relieving food insecurity through redistribution.

Food loss and waste (FLW) had high water, carbon, energy, nutrients (nitrogen and phosphorous), and ecological footprints (Song et al., 2015). The consumption of petroleum products (irrigation, pesticide, herbicide and fertilizer manufacturing, food processing, storage, and transportation) for the production of wasted food adversely affects the environment due to the release of greenhouse gasses (GHG) responsible for global warming and climate change (Woods et al., 2010). Production of wasted food also causes erosion, depletion of habitat, wastage of water supplies, and pollution due to the indiscriminate application of fertilizers and pesticides (Food and Nations, 2013). When food is degraded in the disposal area, methane gas is released 25 times more effective in global warming than carbon dioxide (CO_2) (CHANGE, 2007). As a result, if FLW mitigate reliance on natural resources, poverty and malnutrition will be minimized, and climate change challenges will be tackled.

Considering the importance of FLW in sustainability and elimination of hunger, governmental, nongovernmental organizations and academics have engaged in FLW quantification and management research projects in many countries in recent decades. For example, in 1997, The United States Department of Agriculture Economic Research Service (USDA-ERS) established the Loss-Adjusted Food Availability Data Series, which includes FLW data in terms of amount, values, and calories of ~200 commodities from three different points of the agri-food chain, that is, farm to retail, during retail and on final ingestion stage. In the United Kingdom, an agency known as the Waste and Resources Action Program (WRAP) has been established to lessen FLW, and since 2007 has delivered a series of studies on FLW that occurs in the agri-food chain (WRAP, 2008, 2009). The FAO has delivered numerous reports on FLW globally (Gustavsson et al., 2011; Food, 2014). A European Commission (EC) Framework Program 7 funded project Food Use for Social Innovation by Optimizing Waste Prevention Strategies (FUSIONS) (2012–2016) is moving for a extra resource-proficient Europe by substantially decreasing food waste through social innovations (Östergren et al., 2014; FUSIONS, 2016). In 2015, another EC funded project was recognized as Resource Efficient Food and dRink for the Entire Supply cHain (2015–2019) focused on eliminating preventable food waste and enhancing food supplies recovery, which has 26 project allies from 12 European countries and China. In 2016, the first global standard for quantifying FLW was jointly released by major international organizations such as World Resources Institute (WRI), World Business Council for Sustainable Development, FAO, United Nations Environment Program, and WRAP (Hanson et al., 2016). Last year, the United Nations General Assembly, in its 74th session, affirmed 29 September as the International Day for the recognition of FLW.

As food waste is a global problem, several countries focus on this problem at the community, institutional, and academic levels. Due to the absence of a precise global description of FLW and adequate measurement methodologies, FLW management is difficult.

10.2 FLW DEFINITIONS

"Food" can be described as any substance—whether cooked or processed, semicooked or semiprocessed, or raw—that is planned to be eaten or consumed by humans. **Food waste or food loss** is the food that is not consumed. However, the concept of FLW is challenging as there is an ambiguous concept of FLW in the literature. One cause of misperception is various terms (Schneider, 2013) like food loss, bio waste, food waste, kitchen waste, postharvest waste, and food and beverage waste. Its inconsistent terms are used among authors. In literature, various organizations report the difference among food loss and food waste (FAO, 2011; Parfitt et al., 2010). However, their meaning varies in association to food loss and food waste. The difference in definitions used by several organizations is due to their different goals for quantifying FLW. For instance, if some entity or organization aims to increase food safety, FLW can only be specified in only inedible parts. On the other hand, if any agency or organization's goal is to minimize GHG emissions from a waste management perspective, FLW could be described as edible and inedible parts.

Owing to the absence of commonly accepted notions of food loss and food waste, it becomes difficult to measure the precise loss of food and waste, conduct generalized research, and establish policy goals. Several descriptions of FLW are being used by various organizations listed in Table 10.1.

This chapter follows the FLW definitions (Figure 10.1) described by (HLPE, 2014).

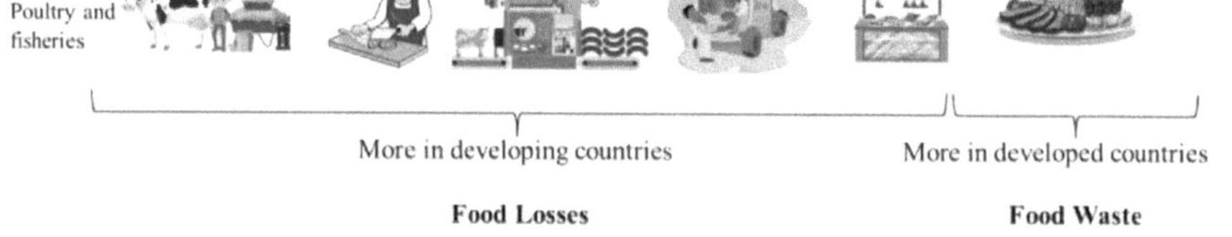

FIGURE 10.1 The FLW in the agri-food chain.

Table 10.1 revealed that the given definitions are alike in stating the decline in the quality or quantity of food directed for human ingestion. On the other hand, these are distinct when discussing external factors and describing the association among FLW. The FLW happens anywhere alongside the agri-food chains (FAO, 2011). According to the earlier FAO concept (1981), all food items proposed for human consumption are thrown out, destroyed, disposed of, or eaten by pests at any point of the agri-food chain in their place

TABLE 10.1 Similarities and Differences in FLW Definitions by Various Organizations.

Food type	Entity/ organization	According to the definition	
		Included in FLW	**Not included in FLW**
Edible parts	FAO	Edible Food planned for human ingestion but is not consumed Food used as feed and bioenergy is also considered FLW (Food, 2014). Food loss: reduction in quantity and quality of food or food losses happens from production to processing phases in the FSC, Food losses happening from harvest/slaughter/catch up stages of FSC (FAO, 2019) Food waste: An element of food loss that has been left to rot or perish due to negligence by the consumer or happens at the last step of FSC in retail and distribution and final ingestion (FAO, 2019)	Inedible parts and food used for animal feed are not included as food waste (FAO, 2011). Food used as feed, industrial use, and other like fertilizers and inedible components of food is not recognized as food waste (FAO, 2019).
	USDA	Food waste is the subcomponent of food loss and happens when eatable produce go unconsumed	
	HELP	Food loss: A reduction in the mass of food at all phases of FSC before consumer level Food waste: Food fit for human ingestion being thrown away or left to rot at the consumer level	
Edible and inedible parts	USEPA	Unconsumed food and food scraps produced during preparation from dwellings, commercial areas, and institutions are called food waste	
	CalRecycle	Any rejected food, due to overproduction, unsold Food (for instance, onion scraps or carrot tips), as well as any leftovers in dishes is called food waste	
	FUSION	**Food and its inedible parts** removed from FSC is called FLW Food waste: Crops not harvested, composted, used for bioenergy production, incineration, dispose of in landfills, sewers, or discarded in the sea is called food waste. Food that is recovered is also called food waste. Food loss: unharvested crops -pre-slaughter losses of livestock or losses of milk due to cow sickness and mastitis.	Food used for animal feeding and biomaterial production

TABLE 10.1 *(Continued)*

Food type	Entity/ organization	According to the definition	
		Included in FLW	**Not included in FLW**
	EU commission	**Food and its inedible** parts lost from FSC	Food used for animal feed, bio-based product formation and sent for redistributionInedible crop and animal parts
	EU	**Food and its uneatableparts** lost from the agri-food chain Food loss: loss before, during, and after meal preparation or loss during manufacturing, processing, delivery, marketing, and food consumption activities	Animal feed, bio-based products, and food sent for redistribution
	WRI	**Food and its uneatable components** detached from FSC	
	WRAP	Food and drink are thrown away	

are referred to as food waste. In a 2011 FAO report, food loss is described as a decline in quality or quantity of food suitable for human ingestion in the initial to middle phases. Food waste refers to later steps of agri-food chain, that is, in marketing and ingestion phases. Food waste or loss is quantified only for food items proposed for human ingestion, eliminating feed, and inedible produce components. According to another set of definitions, FAO describes food loss as the decrease in nutritional value (quality) or quantity (weight or dry matter) of food, which was initially formed for human intake and food waste, as food that is fit for human ingestion but thrown away, either subsequently left it to rot or held past its expiration date (Food and Nations, 2013). FAO in 2014 describes food waste as a component of food loss as food that is used as feed, and bioenergy is also considered FLW, whereas inedible components of the food are not measured as FLW (Food, 2014). In 2019, FAO modified its definition and considered food used as feed, industrial use, and other uses like fertilizer are not recognized as food waste. Agreeing to this interpretation, food waste is owing to supplier and end-user activities.

In a study conducted by FUSIONS, EU food waste is characterized as "Any food and its uneatable components, separated from the agri-food chain to be predisposed of (comprising of crops tilled in without harvesting, composted, anaerobic digestion, cogeneration, bio-energy formation, burning, sewage disposal, dumping ground, or thrown away to sea), or recuperated (Östergren et al., 2014). Food and its inedible parts as animal feed or for the manufacture of biomaterials are not considered food waste. According to FUSIONS EU, all losses are stated as food waste, and no explanation of food loss is used.

High-Level Panel of Experts (HLPE) describes food loss as a reduction during the initial four points of the agri-food chain and food waste during the final ingestion step of the agri-food chain. Their terminology of food waste is attributed purely to consumer behavior. HLPE defined the loss of food as "a reduction, at all steps of the FSC before the consumer level, in the quantity of food that was initially proposed for human ingestion, irrespective of the source." Similarly, food waste "food suitable for human ingesting but thrown away or left to rot at the consumer level—irrespective of the cause" (HLPE, 2014).

According to USDA definition, food waste is a subcomponent of food loss and happens when an eatable food is not consumed. The food that is still fit for human intake at the moment of disposal is known as food waste" (Buzby et al., 2014).

10.3 TYPES OF FOOD WASTE

Food waste types or classifications are based on the elements of the analysis to be categorized.

Food waste is classified on the following bases:

- Phases of agri-food chain.
- Food commodity groups.
- Food edibility and avoidability.

10.3.1 PHASES OF AGRI-FOOD CHAIN

In the agri-food chain, FLW can arise at each step (WRAP, 2015). Reliant on the agri-food chain phases, FLW is classified into the following five types (Gustavsson et al., 2011).

10.3.1.1 FARM LOSSES/WASTE

These losses and wastes that occur during crop production may include plant and animal losses and waste. These losses include spillage, mechanical injury, and inadequate harvesting equipment, for example, fruit picking or threshing and on-farm postharvest crop sorting. Crops and animals are rejected as they cannot meet high-quality criteria or are not economical to harvest (Lipinski et al., 2013). For livestock, poultry, and fisheries, livestock deaths during breeding, fish disposal during fishing, and decreased milk production due to animal sickness are included. Climate change and environmental factors increase pests, weeds, and diseases in plants and animal breeds. Animal death occurred during breeding on the farm.

10.3.1.2 POSTHARVEST LOSSES/WASTE

These losses and waste include losses after crop or animal product is harvested before reaching the processing stage. These losses and waste happen to owe to spillage and deprivation during various postharvest handling steps from farm to distribution. Animal death occurs during transport to the abattoir, milk spillage in transition from farm to market, and fish losses during icing, packaging, storage, and transportation (FAO, 2011). Losses due to disease and pest attack after harvesting and storage.

10.3.1.3 PROCESSING LOSSES AND WASTE

Processing losses can occur in both industrial and domestic processing. Losses occur due to spillage in processing activities like sorting, washing, unpeeling, cutting and boiling, etc. For livestock, poultry, and fisheries losses during fish canning, animal slaughtering, and trimming, spillage during milk handling and milk is processing (Lipinski et al., 2013).

10.3.1.4 DISTRIBUTION RETAIL AND MARKETING LOSSES AND WASTE

FLW arises during transportation and distribution and in the wholesale and retail market.

10.3.1.5 CONSUMER WASTE

This waste occurs both in the home and out of the home in the food service industry.

10.3.2 FOOD COMMODITY GROUPS

In (FAO, 2011) classified food losses and waste into the following seven commodity groups:

10.3.2.1 CEREALS

Cereals (excluding beer) include barley, maize, millet, sorghum, rice (milled), rye, oats, and wheat.

10.3.2.2 TUBERS AND ROOTS

Root tubers include sweet potatoes, potatoes, cassava, and yams.

10.3.2.3 PULSES AND OILSEEDS

Oilseed and pulses include soybeans, mustard seed, groundnuts (shelled), cottonseed, sunflower seeds, coconuts (incl. copra), rape, and other oil crops. Nuts are also part of this group.

10.3.2.4 FRUIT AND VEGETABLES

All fruits and vegetables, including bananas.

10.3.2.5 MEAT

Include meat of pig, poultry, goat, and bovine meat.

10.3.2.6 FISH AND SEAFOOD

It includes freshwater and marine and other marine crustaceans' aquatic plants and animals.

10.3.2.7 DAIRY PRODUCTS

Dairy products include milk and milk products.

In 2019, FAO classified food losses and waste into five food commodities: cereal and pulses, roots and tubers and oilseed crops, fruits and vegetables, animal and animal products, and fish and fish products.

10.3.3 FOOD EDIBILITY AND AVOIDABILITY

Food losses and waste may be classified as edible (avoidable) or inedible (unavoidable) food waste (Beretta et al., 2013; Quested and Johnson, 2009). According to their avoidability, domestic food and beverage waste are clustered into three categories: avoidable, possibly avoidable, and unavoidable (Quested and Johnson, 2009) (Figure 10.2).

10.3.3.1 EDIBLE OR TYPICALLY EDIBLE OR AVOIDABLE FLW

These are defined as any substance at one point edible and not usually considered inedible. These foods can be easily prohibited from going to waste. In the processing industry, avoidable food waste comprises smashed stocks and foodstuffs that have not been used. This food waste happens due to bulk purchasing, poor preparation, insufficient storage, and bigger serving sizes. *Examples* include an apple or half a pack of cheese. An entire tray of lasagna is left over every day at a buffet—a slice of bread, plate residues, pizza, liquid coffee, meat, and banana without the peel.

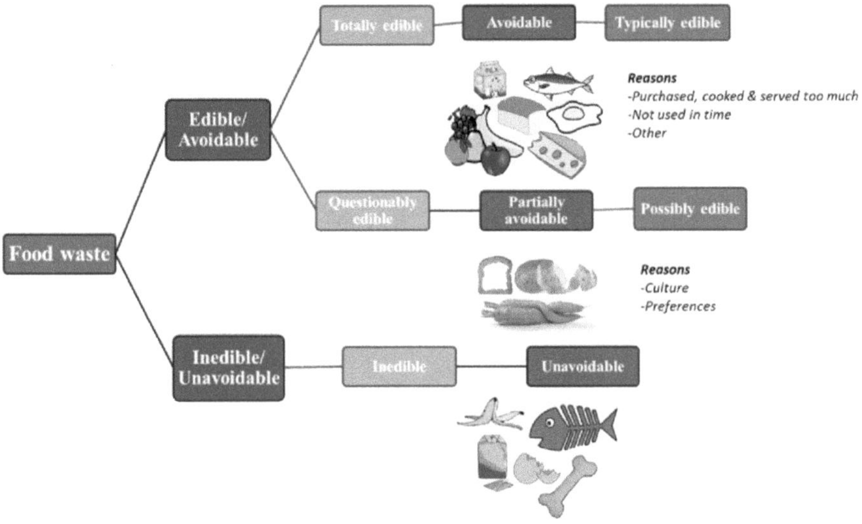

FIGURE 10.2 Food waste types according to food edibility.

10.3.3.2 PARTIALLY EDIBLE OR POSSIBLY AVOIDABLE OR QUESTIONABLY EDIBLE FLW

These are the food waste that can safely be eaten but is not be considered edible by some people due to culture or liking (e.g., potato skins, carrot tops, and beet green) and that can be eaten if considerable skill or effort are used to make this part of food eatable. *Examples* include citrus rinds, potato, pumpkin and apple skins, bent carrots, and beet tops that can be cooked similarly to collard greens or spinach instead of thrown away. Also, marginally stale bread can be consumed for crumbs or croutons.

10.3.3.3 INEDIBLE OR UNAVOIDABLE FLW

These refer to any substances that are not consumed under usual conditions. This includes banana peels, apple cores, pineapple skin, eggshells, bones, coffee grounds, tea bags, and uneatable butchery waste.

10.4 DRIVERS/SOURCES/CAUSES OF FLW

The food value chain FLW can happen at various stages, from agricultural production to final intake in the household and food service industry (Figure 10.3). In less-developed nations food is generally lost in the agricultural production-to-processing phase whereas, in industrialized countries, food is mostly thrown away or, to a greater extent, wasted, even if food is quiet fit for human consumption in later stages agri-food chain (HLPE, 2014).

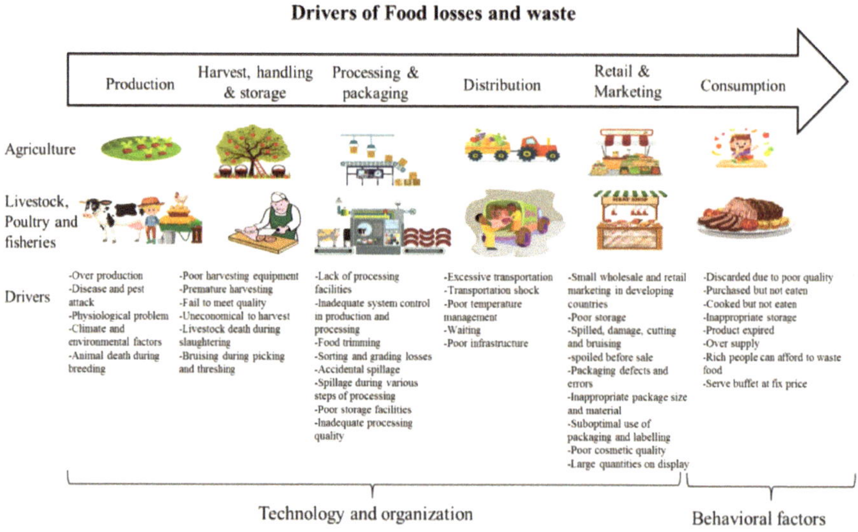

FIGURE 10.3 Drivers of FLW in the agri-food chain.

10.4.1 PRODUCTION

Food is often lost in developed countries when production surpasses demand (Kantor et al., 1997). Farmers ensure the delivery of the contracted amount, whereas, forestalling unpredictable bad weather or pest attacks, deliberately cultivating more than required to meet their agreements (Left-Out 2012). The surplus produce is sold to financially lower ended sectors like processing and animal feed. Edible produce is often eaten by pests and despoiled by fungus or disease, which renders produce inedible and results in food waste.

10.4.2 HARVESTING, HANDLING, AND STORAGE

Animal death occurs in transportation to butchery or is not approved for slaughter and fish fallen or despoiled after being caught. The crop that remains unharvested in the field due to less profit from selling the crop than the cost of harvesting (Johnson et al., 2018) or is deliberately discarded for failure to comply with quality standards (Kantor et al., 1997). In less developed and sometimes in industrialized nations, food can be lost owed to early harvesting. Marginal growers occasionally harvest immature crop due to financial constraints or shortage of food (Bond et al., 2013) thus crop lacks required nutritional and organoleptic qualities and are wasted as it did not fulfill the consumer requirements. Grain/crops are left behind in the field due to inadequate harvesting equipment or severe falls in prices (Food, 2014). Fruits injured during picking or cereals damage during threshing. In developing countries, inadequate storage facilities and proper infrastructure produce more after harvest food losses (FAO, 2011). In warm areas owing to deficiency of appropriate arrangement for cooling, storage, shipping, and marketing, fresh produce of plant and animal origin can be spoilt just after moving from the farm or after the catch (Stuart, 2009). In low-income nations, the lack of cold storage facilities is the main reason for FLW (Lipinski et al., 2013) as storage facilities are economical for a few crops and are not used year-round (FAO, 2011). Lack of technical knowledge regarding storage techniques also results in more food losses and waste. For example, chilling sensitive produce is stored at low temperatures, mixed product storage like meat, fruit and vegetables, and milk in a single room produces waste. Many diseases and insects pest spoil the stored grains. Also, diseases occurred due to inadequate drying of produce.

10.4.3 PROCESSING AND PACKAGING

In developing nations, inadequate processing facilities can cause high FLW. In some instances, the processing industry failed to process and preserve the fresh farm products to fulfill the market demand. Food shortages and waste are also attributed to inaccurate order predicting (Goriwondo et al., 2011) and factory operation inefficiencies (HLPE, 2014). Processed foods may be lost or discarded due to lousy order forecasting and unreliable production methods (Lipinski et al., 2013). Poor hygienic conditions at the processing level contaminate the animal products. Sometimes eatable fruit or cereals are sorted out as these are not appropriate for processing (Lipinski et al., 2013). Milk dropped in processing and pasteurization activities. Fish dropped or spoiled during the canning or smoking process. Excessive food trimming to achieve proper size and shape (e.g., lettuce, cabbage, and carrot) during processing can result in losses and waste of the edible product. Growers and processors use many chemicals regulated or unregulated to fill the market demand of superior crop quality and use inappropriate doses of these chemicals, which may cross a safe level and have harmful effects on consumers (Gustavsson et al., 2011). Lack of temperature control and poor hygienic handling ends up in unsafe food thrown away and ends up in waste.

10.4.4 TRANSPORT AND DISTRIBUTION

Failure of logistics and excessive, complicated, and costly products in the agri-food chain lead to more FLW(Jedermann et al., 2014). In less developed countries, poor infrastructure and logistics management also cause major food losses and waste due to transportation shock. Unskilled labor during the loading and offloading of produce results in higher FLW. In undeveloped countries, fresh produce is frequently transported in open trucks without using proper containers or packed in sacks and polyethylene bags, causing compression injuries during distribution (Kitinoja et al., 2010). Poor temperature management during distribution results in food losses and waste. Worldwide food trade enhances food miles in the agri-food chain, subsequently enhancing food losses and waste. Shipment rejections and delays due to failing the sanitary and phytosanitary requirements at custom checking end up in food waste.

10.4.5 RETAIL AND MARKETING

Supermarket's demand for higher aesthetic standards causes rejection at the farm gate level (Aschemann-Witzel et al., 2017). Therefore, a more prominent crop share never leaves the farm or is sold to lower-end markets like processing and animal feed and not used for human consumption (Stuart, 2009). Supermarkets show large quantities and a wide variety of items on display to attract clients and thereby wasted (HLPE, 2014). Enormous quantities of items on display raise the chances of a few of them attaining their "sell by" date earlier than being sold and end up in waste (Stuart, 2009). In developing countries, insufficient market systems may lead to more food losses. To reduce these losses, the crops grown by farmers need to reach consumers efficiently. Some supermarkets, wholesale and retail are giving appropriate storage facilities and sales environment for food commodities. In developing countries, wholesale and retail markets are mostly small, unhygienic, congested, and want low-temperature facilities. Eatable commodities are dropped or damaged in the market or expired before being purchased. Confusion in interpreting food labels, for instance, "use by" and "best before" can lead to consumers to discard flawlessly good food (Richter and Bokelmann, 2017). Mixed expiry date products are kept on display, and buyers avoid close to expiry date products, which end up in food waste (Stuart, 2009). Retail stores offer large packages and higher discounts on food items or buy one free offer to encourage shoppers to buy more food than their requirement (Gjerris and Gaiani, 2013).

10.4.6 CONSUMPTION

The consumption phase involves the consumption of food in the household and food service sectors. At the consumption stage, FLW are mainly due to the abundance of food and consumer attitudes, which end up in food waste. Consumer-level food losses and waste are more advanced as people can afford food waste in developing countries. Household size influences food losses and waste (Jörissen et al., 2015; Tucker and Farrelly, 2016), but the results are mostly inconsistent in different studies. Females tend to create higher food waste than males (Silvennoinen et al., 2014; Koivupuro et al., 2012). A family with kids has tremendous food waste (Parfitt et al., 2010, Tucker and Farrelly, 2016) as mothers mostly avoid the use of leftovers for their kids and select to feed "fresh" food (Evans, 2011). Moreover, a high amount of work from young parents decreases food care (Jörissen et al., 2015). Families with higher qualified persons are inclined to generate higher food waste (Secondi et al., 2015). Rural households generate less food waste than those living in urban areas (Marangon et al., 2014). Human psychology to provide fair and enough food results in more food waste (Porpino, 2016; Visschers et al., 2016).

Purchasing food without proper meal planning (making a shopping list with products and their quantities required) leads to more food waste (Principato 2018). Little or no knowledge of food previously kept at home (Gaiani et al., 2018) results in overbuying or buying wrong products, leading to food waste. Wrong package size leads to more food waste (Wohner et al., 2019, Charbel et al., 2016). Food items are purchased but not eaten or cooked due to lack of time or discarded due to poor quality and bad taste or smell.

Improper storage at the wrong temperature and too long duration results in spoilage of food products(van Geffen et al., 2017). Poor cooking skills like; cooking tasteless food, inappropriate quantity, and damage (burning) also influence food waste (van Geffen et al., 2017; Evans, 2011). Cooked items are not eaten in time and stored as a leftover but not consumed later due to any reason (Khalid et al., 2019) or leftover is too little to stand to food waste (Williams et al., 2020). Food items are purchased in bulk quantities or on special offers without considering their expiry dates and, sometimes, due to misinterpretation of expiry dates, end up in food waste (Graham-Rowe et al., 2014). Sometimes people throw away relatively healthy food due to food safety concerns (Farr-Wharton et al., 2014). Rich people or people living in developed countries afford to waste food.

Food service industry losses and waste include FLW at canteens, care centers, hospitals, in-flight catering, military institutions, restaurants, schools, and transport hubs. In the food service industry, lack of menu planning and over preparation/production are the key causes of food waste. Improper storage and demand forecasting are also enhancing FLW. Consumers abet to waste more due to large portion sizes and a wide variety of menu. Moreover, the food service industry buffet is served at a fixed price, which increases FLW.

10.5 POLLUTION POTENTIAL

The FLW has a considerable influence on the environment both directly and indirectly linked to the dumping of surplus food and energy inputs consumed to grow food that is eventually thrown away, respectively (Blakeney, 2019). The growing of food needs resources like land, water, pesticides, fertilizers, and energy. In addition to this, it causes the production of pollutants and affects the environment (Muth et al., 2019). Chemical resources likewise fertilizer (nitrogen and phosphorous) and energy (carbon) travel through several steps of the food value chain, endure chemical alterations, are discharged into the atmosphere in diverse forms, and end up in numerous impacts (Muth et al., 2019).

According to Rutten, 2013, in the agricultural segment, FLW are dissipated and produce significant inadequacies in input characteristics, such as labor, land, water, fertilizers, and energy. In Europe, Scherhaufer et al. (2018) reported that 15–16% of the food value chain's total environmental effects were caused by food waste.

The environmental influence of food discard by consumers both at households and the food service industry is significant. Some food categories, such as meat, cocoa beans, coffee, butter, eggs, aircraft-traded crops, oils and fats, fish and cheese, have the highest effect on the environment per kilogram of edible food waste. The waste of these foodstuffs should be given special care. Losses of fruits, vegetables, and potatoes have minor environmental impacts per kilogram, but their effects are significant because these products are discarded in massive amounts (Anonymous, 2019). A brief overview of the environmental impact studies associated with food waste was conducted in various countries is presented in Table 10.2.

10.5.1 GLOBAL WARMING

Food losses and waste had significant impacts on climate change and global warming. When food waste is dumped into landfills, a considerable portion of it, around 40–70% by dry volume, is transformed into methane, a GHG that has 25 times higher heat-trapping potential than CO_2 100-yeartime scale (CHANGE, 2007). In food supply chain, various activities like food production (rearing livestock, fertilizer and pesticide application and using agricultural machinery), land-use change, food distribution, food packaging, and food preparation and waste management at home and food service sector adversely impacts the environment and cause global warming (Tonini et al., 2018). Greenhouse gases emission allied with FLW change the radiative forcing of the earth and atmosphere and therefore causes global warming (Muth et al., 2019). NOx (nitrogen oxides) releases linked with FLW contribute to tropospheric ozone creation and contributing to warming (Muth et al., 2019). The influence of food waste emissions on global warming is nearly equal to 87% of worldwide road transport emissions (Footprint and Change, 2015).

10.5.2 GREENHOUSE GAS (GHG) EMISSIONS

The uneaten food is liable for adding 3.3 Gigatons of GHG into the earth's atmosphere during 2007, making up food waste as the third topmost emitter after the US and China (Food and Nations, 2013; FAO, 2011). Livestock rearing has higher environmental impacts, round about 4.6–7.1 billion tons of GHGs (Fiala, 2008) compared to agricultural production. Monogastric animals contribute to GHG emissions through enteric fermentation in feed digestion, and some methane is also contributed from manure management. Cereals are the significant contributors to the carbon footprint of food

TABLE 10.2 Environmental Impact Studies on Food Waste Performed in Various Countries Worldwide.

Country	Agri-food chain sector	Results	References
Australia	All stages from production to consumption	Water consumed in various production stages to produce 1 kg of wasted mango fruit is estimated to be 867 L.	Ridoutt et al. (2010)
China	Household-level	Estimated household food waste impacts are 2500 kg CO_2 eq/t	Song et al. (2015)
Greece	Household-level	Annually 5672.5 gigagrams of CO_2 equivalent was connected with food waste	Abeliotis et al. (2015)
UK	In the FSC from farm to fork	The global warming effect of edible food waste was calculated between 2000 and 3600 kg CO_2-equivalent per ton.	Tonini et al. (2018)
UK	Household-level	More than 20 Mt of CO_2 equivalent emissions	WRAP (2008, 2009)
New Zealand	Household and industrial food waste	Total food waste presented 4.2×10^6 tons of CO_2 equiv. of which household comprised of 2.3×10^6 tons of CO_2 equiv. and industry 1.9×10^6 tons of CO_2 equiv. Total 963 kg of CO_2 equivalent/person. The total exemplified water consumption is 4.7 billion m^3, water used in food waste at household comprised of 3.1×10^9 m^3 and industry 1.6 billion m^3. Total 1087 m^3 of water/individual/annum is wasted.	Reynolds et al. (2016)
UK		Nearly 0.9 kg CO_2 equivalent/person/day of water is used to produce wasted food. This amount of water is roughly equals to the value of water used for consumption per capita per day in UK. Environmental impacts of meat and fish were higher as compared to other food commodities	Cooper et al. (2018)
UK		Food waste consisted of 3% of domestic greenhouse gas (GHG) emissions.	Chapagain and James (2011)
Sweden	Household level	Food waste impact by household ranged between 800 and 1400 kg CO_2-eq/t	Schott and Andersson (2015)
US	Consumption stage	300 million acre of land and 780 10^6 pounds of pesticides were used to grow wasted food.	Conrad et al. (2018)

TABLE 10.2 *(Continued)*

Country	Agri-food chain sector	Results	References
US	From production to consumption	Annually 113 million metric tons of CO_2 or 2% of national emission is related with food waste	Venkat (2011)
US	From production to consumption	One fourth of fresh water and 300 million barrels or 4% of fossil fuel are used to grow wasted food	Hall et al. (2009)
Global		Specifically, food waste charges about 500 kg of CO_2 equivalent and 250 km^2 of water per year per person and consume 28% of cultivable land worldwide.	Food and Nations (2013).
Global		An average individual's daily food waste embedded environmental footprints are: 58 L of freshwater, 124 g of CO_2 equivalent, 0.36 m^2 of cropland and nutrients including nitrogen (2.90 g) and phosphorus (0.48 g). Of food items meat, Cereals, and sugar having significant environmental impacts.	Chen and Graedel (2016)

wastage (34% of total), while meat and vegetables contribute 21% each (Food and Nations, 2013).

In the US, GHG emissions by edible food waste during 2009 accounted for almost 2% of the net country's GHG emissions or 368 kg of CO_2 equivalent per capita (Venkat, 2011). Similarly, in England, the GHG productions related to edible and partially edible food and drink waste during 2007 reported for approximately 428 kg CO_2 equiv per capita or 25.7 M tons of CO_2 equiv (Liu et al., 2013; Wrap, 2011), which is about three percent of England's GHG emissions; whereas, Chinese households generates lesser amount, which is almost 40 kg CO_2 equivalent per person per year (Song et al., 2015) as compared with UK and US.

10.5.3 WASTED FOOD WASTED RESOURCES

The resources that are consumed in growing the food are also wasted with food waste. FLW embed 23–24% of the total world freshwater resource, cropland, and fertilizers to produce food (Kummu et al., 2012).

10.5.3.1 WATER

Water is mostly depleted by the processes in FSC, like food production and food preparation. Household sectors consumed much water as compared with other food sectors. By decreasing food losses and waste, with the goal that less food would be required, fresh water consumption could, therefore, be decreased (Lundqvist et al., 2008). It is appraised that irrigation water consumed to produce food that is not consumed, would be sufficient to fully fil the household water requirements of 9 billion people (FAO, 2011). In a study, Vanham et al. (2015) stated that the food waste footprint (27 L/capita/day) of freshwater is approximately equaled to the municipal freshwater consumption in Europe. US Hall et al., 2009 assessed that food growth, which is ultimately wasted, requires over 25% of the total freshwater used. Worldwide, blue water footprint (includes both surface and underground water) of uneaten food is nearly 250 km^3 (FAO, 2011), equivalent to the river Volga's water discharge.

It is evaluated that cereals and fruit wastage is the key contributor to the irrigation water footprint of food with 52% and 18% contribution, respectively (Food and Nations, 2013). Animal rearing also has a huge influence on water resources. For instance, a hundred times more water is needed to form the same quantity of beef than vegetable protein (Pimentel et al., 1997). Animal feed production activities create substantial negative impacts on freshwater resources versus animal rearing, which only needs 1.3% of total agricultural water (Pimentel et al., 1997).

Phosphorus, which is used in food production, depletes huge freshwater quantities during phosphate mining fertilizer production.

10.5.3.2 LAND USE

The consumption of land for the cultivation of wasted food remains a challenge to the environment. In 2007, nearly 1.4 Gaga hectares, equivalent to 30% global cultivable land, were consumed to produce uneaten food (FAO, 2011), representing a land area greater than China or Canada's size and lower than the Russian Federation. Higher land footprints are associated with livestock and beef production (FAO, 2019). The use of machinery in farming practices compacted the topsoil, which destroys the beneficial microbes living in the soil, resulting in barren soil (Horrigan et al., 2002). Land used to produce wasted food is almost between 500 and 100 m^3 (Birney et al., 2017).

10.5.3.3 FOSSIL FUELS

Agri-food chain stages like production, storage, processing, distribution, and food preparation rely on fossil fuels and other energy sources (Sage, 2013; Pradhan et al., 2013). In agri-food chain, the production and food preparation process consumes fossil fuels; however, waste management conserves fossil fuels when energy is recuperated in biofuel by consuming food waste in landfills. So in future, food-associated energy usage would increase with a surge in food demand. In addition, changing diets, particularly toward more fresh food that needs cold storage facilities, results in an expected increase in per capita energy usage for food (Pradhan et al., 2013). Therefore, decreasing FLW can reduce resource consumption (e.g., water, pesticides, and fertilizer) and decrease the food systems energy needs (e.g., energy consumed to pump underground water, energy consumed in chemical, and fertilizers production). A study (Hall et al., 2009) described that uneaten foodcs growth requisite the expenses of about 300 million barrels of oil in the US. The combustion of fossil fuel during various agri-food chain activities releases greenhouse gases in the atmosphere like CO_2, N_2O, and chlorofloro carbons.

10.5.3.4 FERTILIZERS

The FLW released the chemicals (N, P, and C) into the atmosphere in two ways: 1) chemicals that are physically present in FLW are mobilized into the environment when food waste is disposed of, and 2), chemicals released into the environment in the cultivation of wasted food. Nitrogen fertilizer is used during the production of several crops (Camargo et al., 2013). The movement of reactive nitrogen in the atmosphere negatively impacts all the ecosystem and human health (Galloway et al., 2003). Nitrous oxide (having 300 times higher global warming capabilities than CO_2) is released at higher intensities due to fertilizer application for crop production (Aneja et al., 2009). Some of the nitrogen presents in FLW is also converted into nitrous oxide by microbes in an anaerobic environment, but this is a minute quantity compared to nitrogen lost during the agri-food chain to produce wasted food. Phosphorus fertilizers are manufactured from limited mining resources of phosphate rock (Chen and Graedel, 2016) when lost in the water resources results in harmful algal blooms (Schindler et al., 2016). Moreover, mining phosphate for phosphorous fertilizer production consumes large quantities of fresh water resources (Tenneson, 2009).

Fertilizers are produced with fossil fuels, mainly petroleum, and harm greenhouse gas emissions. For instance, 2200 pounds of coal energy is required for manufacturing 5.5 pounds of nitrogen (Czarnezki and Prescott, 2013). Furthermore, during food production (both animal and plant origin) run off of nutrients like nitrogen and phosphorous from farmers' fields alter the ecosystem of water resources increase algal growth and deplete dissolved oxygen (a term known as eutrophication) in water bodies (The Dead Zone, 2002). N is most related to coastal eutrophication, while P is mostly linked with eutrophication of freshwater.

10.5.3.5 PESTICIDES

Indiscriminate use of pesticides for growing food that is eventually discarded by consumers adversely affects the environment through groundwater contamination, bird mortality, adverse effect on human health, and acquired pesticide resistance in pests. Inaccurate pesticide application methods result in higher pesticide loss and damage to the environment. For instance, by applying pesticides through airplane technology, lesser amounts are reached to the targeted area than spray bloom (Pimentel,

1995). Several pesticides are applied on crops without being carefully monitored for their toxicity, results in endocrine disruption and some types of cancers in humans (Horrigan et al., 2002). Pesticide applications in crop production also affect the pollinators, which maintain the ecosystems by pollinating plants based on many food chains. Pesticides also pollute fresh water resources.

10.5.4 TERRESTRIAL AND OCEAN ACIDIFICATION

Terrestrial acidification is mostly caused by food production, land-use change, food preparation at the food service and household sectors, whereas waste management activities reduce terrestrial acidification. CO_2 associated with FLW when dissolving with ocean water causes acidification, which leads to ocean acidification, negatively influencing ocean life (Anthony et al., 2008).

10.5.5 PHOTOCHEMICAL OZONE FORMATION AND OZONE DEPLETION

All points along the agri-food chain from processing until consumption positively influence photochemical ozone formation. Food distribution (transportation), food preparation (cooking), and waste management (burning of biogas in landfills and anaerobic digestion plants) also positively influences photochemical ozone formation. The combustion of fossil fuels during transportation and cooking produces NOx. Ozone and smog are produced when NOx combines with volatile organic compounds present in the environment. N_2O emissions related to FLW significantly contribute to stratospheric ozone destruction.

10.5.6 PARTICULATE MATTER

Particulate matter is released into the atmosphere during crop production and the agri-food chain's distribution stages, harmfully influencing human health (Vinikoor-Imler et al., 2011). Food production, land-use change, and food preparation activities increased particulate matter in the atmosphere.

10.5.7 ECO TOXICITY AND HUMAN TOXICITY

All sectors in the agri-food chain positively influence ecotoxicity and human toxicity. Among these sectors, food production activities are the main contributor mainly due to excessive use of herbicides and pesticides.

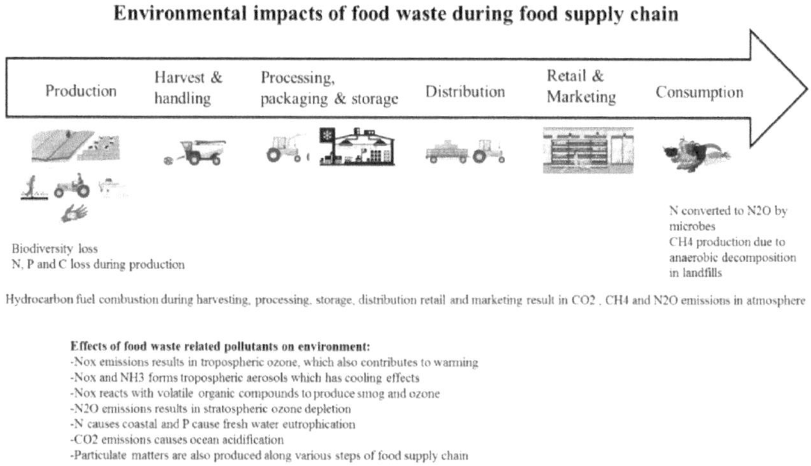

FIGURE 10.4 Environmental impacts of FLW during agri-food chain.

10.5.8 BIODIVERSITY LOSS

Benn (2010) defines biodiversity as the variation in the world's natural ecosystems. Biodiversity is influenced by both production and discarding of wasted food, yet the production stage has higher biodiversity impacts than the disposal stage (Roka, 2019; Roka et al., 2019). Most biodiversity loss causes can be ascribed to directly influence a species by overhunting, competing for invasive species, and overfishing or on related habitat by restraining resources through land-use change. Altering natural habitat for growing crops can cause soil deprivation, which leads to adverse effects on soil productivity, comprising declined soil structure, greater erosion, and diminished nutrient accessibility. During its decomposition in landfills, food also influences land biodiversity near the landfill, contaminating watercourses and underground water. Natural land used for agricultural farming causes habitat destruction, which leads to biodiversity loss (Chaudhary and Brooks, 2018). Food wastage has a higher biodiversity footprint. Agricultural expansion and intensification are a major threat to biodiversity loss globally. Biodiversity is mostly threatened by crop production (70%) rather than livestock raising (33%).

10.6 COUNTRY-WISE COMPARISON

The amount of food waste differs significantly amongst countries, being dependent on Gross National Income (GNI) (Adhikari et al., 2006), urbanization, and development levels (Chalak et al., 2019) of a country. In developing nations at the initial levels of the agri-food chain, that is, production to processing points about two-third of the food losses happen (FAO, 2011) approximately equals 44% of global food waste. This is generally ascribed to poor agricultural practices, technical, labor and monetary limitations, and insufficient infrastructure for processing, storage, and distribution (Gustavsson et al., 2011) (Gustavsson et al., 2011). Besides, in middle and high earnings nations, a considerable portion of food waste happens at consumer level, mostly determined by the consumers' attitudes, behaviors, and values (Gustavsson et al., 2011). It was appraised that industrialized countries like Europe and North America produce 95–115 kg/capita/annum FLW, which is significantly greater than that for unindustrialized nations like Sub-Saharan Africa and South/Southeast Asia, which produce 6–11 kg/capita/annum (Gustavsson et al., 2011). (Dung et al., 2014) described that food waste generation from developed nations (107 kg/per capita/year) is greater than that of developing nations (56 kg/individual/year), which was ascribed to the high living standards in developed nations (Lipinski et al., 2013). However, based on the high population and more developing nations, total food waste from both developing and industrialized countries is nearly alike, that is, 670 million tons in developed nations and 630 million tons in low-income nations (Food, 2014). On the bases of GNI, nations are divided into four groups; low income (GNI per person ≤$1035 or less in 2019); lower middle income (GNI per person between $1036 and $4045 in 2019); upper middle income (GNI per person between $ 4046 and $12,535 in 2019), and high income (GNI per person ≥ $12,535 in 2019) countries. Globally, every year nearly $940 billion financial losses happen from FLW (Footprint and Change). Sub-Sahara and Africa regions waste is roughly $4 billion worth of food (Mondiale, 2011), whereas in US an average family with four members' wastes up to $1500 per annum (Buzby et al., 2014) and in UK around £ 700 worth of food is discarded annually (WRAP, 2015). Food waste in developed nations can affect food insecure people worldwide by decreasing the accessibility of food human intake. Moreover, in developing countries, growing income and transition in food habits cause increased food waste.

Table 10.3 illustrates the global and country viz estimation of food losses and wastes in terms of quantity, value, and nutrition.

TABLE 10.3 Food Waste Generation in Terms of Quantity, Value, and Nutrition in Different Countries.

Country	FLW (quantity)	FLW (value)	FLW (nutrition)	References
Low-income countries (GNI per person ≤$1035 or less in 2019)				
Sub Sahara and Africa	120–170 kg/year/capita	Nearly USD 4 billion per year		FAO (2011), Mondiale (2011)
Lower middle income (GNI per person between $1036 and $4045 in 2019)				
Egypt	42% of respondents reported that they threw 25 g of food per week	21% of respondents reported that they waste more than the US $6 per month		Elmenofi et al. (2015)
Morocco	39% of respondents declared that their food waste per week is 250 g.	Their food waste value was more than the US $6 per month		Abouabdillah et al. (2015)
Nigeria		Nearly N 1499.67 worth of food was wasted per month		Akerele et al. (2017)
Pakistan	Average daily restaurant waste was 35.6 kg			Aamir et al. (2018)
Pakistan	About 35.01 g of cooked food, of which vegetables, fruits, cereals and dairy waste comprised of 11.62, 9.78, 2.45 and 1.98 g/person/day.		The overall quantity of household food waste denoted an assessed worth of energy, carbohydrates, protein, lipids, fiber, calcium, phosphorous, β-carotene and vitamin A 54.42 kcal, 10.58, 2.61, 2.21, 0.75 g, 22.49, 37.11 mg, 275.2 mcg and 96.83 RE respectively per person per day.	Khalid et al. (2019)
Vietnam- Thu Dau Mot	0.76 kg/household/day of household solid waste, of which organic waste comprised of 67%.			Trang et al. (2017)

TABLE 10.3 *(Continued)*

Country	FLW (quantity)	FLW (value)	FLW (nutrition)	References
Upper middle income countries (GNI per person between $4046 and $12,535 in 2019)				
Bangkok, Thailand	Estimated food waste production was 0.38–0.61 kg/capita/day			Liu et al. (2020)
Iran Tehran	27.6 kg/person/annum			Fami et al. (2019)
Lebanon	Food waste was nearly 0.2 kg/person/day in an urban area		Food waste contains energy (451.2 kcal), carbohydrates (37.5 g), protein (14.9 g), dietary fiber (2.9 g), calcium (165.2 mg), potassium (343.2 mg) and vitamin D (2.4 μg),	Chalak et al. (2019)
Lebanon	One kg per month per household		6 US $ per month per household	Charbel et al. (2016)
Malaysia	One-third of food is wasted in the out of home consumption stage	23% of the worth of procured food is wasted		Papargyropoulou et al. (2019)
South Africa		Food waste across the value chain R61.5 billion/year, which is almost US$7.7 billion)		Nahman and de Lange (2013)
South Africa		Domestic food waste R21.7 billion, which is almost US$2.7 billion/year		Nahman et al. (2012)
South Africa	Total estimated food waste was 177 kg/individual/year, and food waste at consumption level was 7 kg/individual/year.			Oelofse and Nahman (2013)

TABLE 10.3 *(Continued)*

Country	FLW (quantity)	FLW (value)	FLW (nutrition)	References
Turkey	Food waste per person was 318.8 g		215.7 kcal/person	Pekcan et al. (2006)
Turkey	26 million tons per year			Salihoglu et al. (2018)
High-income countries (GNI per person ≥ \$12,535 in 2019)				
Arab world	44% of food losses and waste occurred before consumption stage in food supply chain, while nearly 34% of food is lost at or in ingestion			HLPE (2014)
Australia	Total 7.3 10^6 tons of total food waste, out of which 4 10^6 tons represented municipal food waste and 3.2 10^6 tons was obtained from industry			Reynolds et al. (2016), Fry et al. (2016)
Australia	4.06 million tons/year			(Gault)
Belgium	Of the total food production annually 1011 kg or 2.1% is lost in the food processing industry			Dora et al. (2020)
EU-28	100 M tons/year			Timmermans (2015)
Helsinki, Finland	24 million kg/year or 23 kg/citizen			Silvennoinen and Korhonen (2013)
Japan	Annually 17 million M tons of food is wasted in Japan			Morisaki (2011)

TABLE 10.3 *(Continued)*

Country	FLW (quantity)	FLW (value)	FLW (nutrition)	References
Kingdom of Saudi Arabia	165–511 kg/capita/year			Baig et al. (2019)
Netherland	43–60 kg of food is wasted	€270–400 per individual per annum		Thönissen (2009)
New Zealand		Domestic food waste is equal to 750 million dollars per annum		Johnston and Davison (2011)
New Zealand	Domestic food waste 224,000 tons and industrial food waste 103,000 tons was wasted in 2011.	USD 292 million household food waste and $158 million in industrial food waste, which is approximately $131 per individual per year.	New Zealand food waste embodied 161×10^9 calories annually.	Reynolds et al. (2016)
New Zealand	148 kg/household/year			MINZ (2015)
Nordic countries	40,000–83500 M tons/year in each Nordic country			Stenmark (2011)
Qatar	In Qatar University, 329.5 kg of food is wasted per day or 757 g/sale in female and 980 g/sale in male housing complexes			Abdelaal et al. (2019)
Switzerland			48% of the total calories were wasted with food waste.	Beretta et al. (2013)
UK	In UK the domestic food and beverage waste comprised of 8.3 10^6 metric tons/year of food.	The worth of food waste was about $18.6 billion		Chapagain and James (2011)

TABLE 10.3 *(Continued)*

Country	FLW (quantity)	FLW (value)	FLW (nutrition)	References
UK	15% of avoidable food and beverage is wasted		16% avoidable calories were also wasted	Defra (2010)
UK	4.2 million tonnes/year			Council (2012)
UK			Many key nutrients are wasted by fruit, vegetable and salad waste. Fish and meat waste contributes to 14% of iron waste.	Cooper et al. (2018)
US	Food waste comprised of 29% of the available food supply			Buzby et al. (2011)
US	60 million tonnes		133 billion pounds	Buzby et al. (2014), Buzby et al. (2015)
US	Annually 40% food is wasted in US		1400 kcal/individual/day or 150 trillion kcal per annum	Hall et al. (2009)
US	34.69 million tons/year			USEPA (2014)
US	35.5 million tons/year or 0.615 pounds/person/day			Thyberg et al. (2015)
US		About US$90–100 billion/annum worth of FLW occurred, of which domestic food waste is worth US$48.3 billion per annum		Jones (2006)
US	Approximately 29% or 55 million M tons of food is squandered per annum, which is equal to 422 g/capita/day	Annually $198 billion food is wasted		Venkat (2011)

TABLE 10.3 *(Continued)*

Country	FLW (quantity)	FLW (value)	FLW (nutrition)	References
US			Wasted food account for 30% of daily calorie consumed	Conrad et al. (2018)
Global			1400 kcal/capita/day	Lundqvist et al. (2008)
Global	33% of total food production			Gustavsson et al. (2011)
Global	25% of total food production is wasted		614 kcal/individual/day	Kummu et al. (2012)
Global	Annually an individual food waste is nearly equal to 65 kg of which 25% comprised of vegetables, 24% cereals and 12% fruits.		The wasted food contained vitamins and minerals, which comprises 25–50% of daily recommended intake. Cereals, fruit and vegetables mostly contribute wasted nutrients. Meat dairy and contributes to wasted calcium, zinc and riboflavin.	Chen et al. (2020)

10.7 SUMMARY

Around one-third of food grown for human ingestion has been lost or wasted. The FLW has significant negative consequences on environment, economy, and human health. Many factors like lack of awareness, technological limitations, people attitude, and behaviors contribute to this waste. In the COVID 19 crisis, which aggravated the food waste problem by disintegrating the agri-food chain, food waste reduction is of utmost importance. The FLW vary with geographical locations and economic growth of a country.

KEYWORDS

- **agri-food chain**
- **environmental impacts**
- **food security**
- **food waste drivers**
- **global**

REFERENCES

Aamir, M.; Ahmad, H.; Javaid, Q.; Hasan, S. M. Waste Not, Want Not: A Case Study on Food Waste in Restaurants of Lahore, Pakistan. *J. Food Products Market.* **2018,** *24* (5), 591–610.

Abdelaal, A. H.; McKay, G.; Mackey, H. R. Food Waste from a University Campus in the Middle East: Drivers, Composition, and Resource Recovery Potential. *Waste Manage.* **2019,** *98*, 14–20.

Abeliotis, K.; Lasaridi, K.; Costarelli, V.; Chroni, C. The Implications of Food Waste Generation on Climate Change: The Case of Greece. *Sustain. Prod. Consump.* **2015,** *3*, 8–14.

Abiad, M. G.; Meho, L. I. Food Loss and Food Waste Research in the Arab World: A Systematic Review. *Food Sec.* **2018,** *10* (2), 311–322.

Abouabdillah, A.; Capone, R.; El Youssfi, L.; Debs, P.; Harraq, A.; El Bilali, H. et al. Household Food Waste in Morocco: An Exploratory Survey, 2015; pp 1353–1360.

Adhikari, B. K.; Barrington, S.; Martinez, J. Predicted Growth of World Urban Food Waste and Methane Production. *Waste Manage. Res.* **2006,** *24* (5), 421–433.

Akerele, D.; Afolayan, S.; Oyawole, F.; Sanusi, R. Socioeconomic Determinants of Food Waste among Households in Abeokuta, Ogun State, Nigeris. *Niger. J. Agric. Econ.* **2017,** *7* (2066-2018-1341), 25–35.

Aneja, V. P.; Schlesinger, W. H.; Erisman, J. W. Effects of agriculture upon the air quality and climate: research, policy, and regulations. ACS Publications, 2009.

Anthony, K. R.; Kline, D. I.; Diaz-Pulido, G.; Dove, S.; Hoegh-Guldberg, O. Ocean Acidification Causes Bleaching and Productivity Loss in Coral Reef Builders. *Proc. Natl. Acad. Sci.* **2008,** *105* (45), 17442–17446.

Aschemann-Witzel, J.; De Hooge, I. E.; Rohm, H.; Normann, A.; Bossle, M. B.; Grønhøj, A. et al. Key Characteristics and Success Factors of Supply Chain Initiatives Tackling Consumer-Related Food Waste–A Multiple Case Study. *J. Cleaner Prod.* **2017,** *155*, 33–45.

Baig, M. B.; Gorski, I.; Neff, R. A. Understanding and Addressing Waste of Food in the Kingdom of Saudi Arabia. *Saudi J. Biol. Sci.* **2019,** *26* (7), 1633–1648.

Baker, D.; Fear, J.; Denniss, R. What a Waste-An Analysis of Household Expenditure on Food. *Policy Brief No. 6*, 2009.

Beretta, C.; Stoessel, F.; Baier, U.; Hellweg, S. Quantifying Food Losses and the Potential for Reduction in Switzerland. *Waste Manage.* **2013,** *33* (3), 764–773.

Birney, C. I.; Franklin, K. F.; Davidson, F. T.; Webber, M. E. An Assessment of Individual Foodprints Attributed to Diets and Food Waste in the United States. *Environ. Res. Lett.* **2017,** *12* (10), 105008.

Blakeney, M. *Food Loss and Food Waste: Causes and Solutions*; Edward Elgar Publishing, 2019.

Bond, M.; Meacham, T.; Bhunnoo, R.; Benton, T. *Food Waste within Global Food Systems*; Global Food Security Swindon, UK, 2013.

Buzby, J. C.; Bentley, J. T.; Padera, B.; Ammon, C.; Campuzano, J. Estimated Fresh Produce Shrink and Food Loss in US Supermarkets. *Agriculture* **2015,** *5* (3), 626–648.

Buzby, J. C.; Farah-Wells, H.; Hyman, J. The Estimated Amount, Value, and Calories of Postharvest Food Losses at the Retail and Consumer Levels in the United States. *USDA-ERS Economic Information Bulletin* **2014,** 121.

Camargo, G. G.; Ryan, M. R.; Richard, T. L. Energy Use and Greenhouse Gas Emissions from Crop Production Using the Farm Energy Analysis Tool. *BioScience* **2013,** *63* (4), 263–273.

Chalak, A.; Abiad, M. G.; Diab, M.; Nasreddine, L. The Determinants of Household Food Waste Generation and Its Associated Caloric and Nutrient Losses: The Case of Lebanon. *Plos One* **2019,** *14* (12), e0225789.

CHANGE, O. C. *Intergovernmental Panel on Climate Change*; World Meteorological Organization, 2007.

Chapagain, A.; James, K. *The Water and Carbon Footprint of Household Food and Drink Waste in the UK*; Waste & Resources Action Programme (WRAP): Banbury, Oxon, UK; WWF, Godalming: Surrey, UK, 2011.

Charbel, L.; Capone, R.; Grizi, L.; Debs, P.; Khalife, D.; El Bilali, H. *et al.;* Preliminary Insights on Household Food Wastage in Lebanon. *J. Food Sec.* **2016,** *4* (6), 131–137.

Chaudhary, A.; Brooks, T. M. Land Use Intensity-Specific Global Characterization Factors to Assess Product Biodiversity Footprints. *Environ. Sci. Technol.* **2018,** *52* (9), 5094–5104.

Chen, M.; Graedel, T. A Half-century of Global Phosphorus Flows, Stocks, Production, Consumption, Recycling, and Environmental Impacts. *Global Environ. Change* **2016,** *36*, 139–152.

Conrad, Z.; Niles, M. T.; Neher, D. A.; Roy, E. D.; Tichenor, N. E.; Jahns, L. Relationship between Food Waste, Diet Quality, and Environmental Sustainability. *PloS One* **2018,** *13* (4), e0195405.

Cooper, K. A.; Quested, T. E.; Lanctuit, H.; Zimmermann, D.; Espinoza-Orias, N.; Roulin, A. Nutrition in the Bin: A Nutritional and Environmental Assessment of Food Wasted in the UK. *Front. Nutr.* **2018,** *5*, 19.

Council, C. C. Love Food Hate Waste, 2012.

Czarnezki, J. J.; Prescott, E. Food Court, 2013.

Defra. Household Food and Drink Waste Linked to Food and Drink Purchases, 2010.

Dora, M.; Wesana, J.; Gellynck, X.; Seth, N.; Dey, B.; De Steur, H. Importance of Sustainable Operations in Food Loss: Evidence from the Belgian Food Processing Industry. *Ann. Oper. Res.* **2020,** *290* (1), 47–72.

Dung, T. N. B.; Sen, B.; Chen, C.-C.; Kumar, G.; Lin, C.-Y. Food Waste to Bioenergy via Anaerobic Processes. *Energy Procedia* **2014,** *61*, 307–312.

Elmenofi, A.; Capone, R.; Waked, S.; Debs, P.; Bottalico, F.; El Bilali, H. An Exploratory Survey on Household Food Waste in Egypt, 2015; pp. 15–18.

Evans, D. Blaming the Consumer–Once Again: The Social and Material Contexts of Everyday Food Waste Practices in Some English Households. *Crit. Public Health* **2011,** *21* (4), 429–440.

Fami, H. S.; Aramyan, L. H.; Sijtsema, S. J.; Alambaigi, A. Determinants of Household Food Waste Behavior in Tehran City: A Structural Model. *Resour. Conserv. Recycl.* **2019,** *143*, 154–166.

FAO, G. Global Food Losses and Food Waste–Extent, Causes and Prevention. *SAVE FOOD: An Initiative on Food Loss and Waste Reduction*, 2011.

FAO, I.; UNICEF. WFP and WHO. *The State of Food Security and Nutrition in the World 2020. Transforming Food Systems for Affordable Healthy Diets*; Rome: FAO, 2020.

FAO, U. *The State of Food and Agriculture: Moving Forward on Food Loss and Waste Reduction*; Food Agriculture Organization Rome, 2019.

Farr-Wharton, G.; Foth, M.; Choi, J. H. J. Identifying Factors That Promote Consumer Behaviours Causing Expired Domestic Food Waste. *J. Consumer Behav.* **2014,** *13* (6), 393–402.

Fiala, N. Meeting the Demand: An Estimation of Potential Future Greenhouse Gas Emissions from Meat Production. *Ecol. Econ.* **2008,** *67* (3), 412–419.

Food and Nations, A.O.o.t.U. *Food Wastage Footprint: Impacts on Natural Resources*; FAO: Rome, Italy, 2013.

Food, F. S. Global Initiative on Food Loss and Waste Reduction. Definitional Framework of Food Loss; Working Paper, 2014.

Footprint, F. F. W.; Change, C. Food and Agriculture Organization, 2015.

Fry, J.; Lenzen, M.; Giurco, D.; Pauliuk, S. An Australian Multi-Regional Waste Supply-Use Framework. *J. Ind. Ecol.* **2016,** *20* (6), 1295–1305.

FUSIONS, E. *Food Waste Quantification Manual to Monitor Food Waste Amounts and Progression*; EU FUSIONS: Paris, 2016.

Gaiani, S.; Caldeira, S.; Adorno, V.; Segrè, A.; Vittuari, M. Food Wasters: Profiling Consumers' Attitude to Waste Food in Italy. *Waste Manage.* **2016,** *72*, 17–24.

Galloway, J. N.; Aber, J. D.; Erisman, J. W.; Seitzinger, S. P.; Howarth, R. W.; Cowling, E. B. et al. The Nitrogen Cascade. *Bioscience* **2003,** *53* (4), 341–356.

Gault, R. T. Environment Protection and Heritage Council.

Gjerris, M.; Gaiani, S. Household Food Waste in Nordic Countries: Estimations and Ethical Implications. *Etikk i praksis-Nordic J. Appl. Ethics* **2013,** *1*, 6–23.

Gooch, M.; Felfel, A.; Marenick, N. *Food Waste in Canada*; Value Chain Management Centre, George Morris Centre, November 2010.

Goriwondo, W. M.; Mhlanga, S.; Marecha, A. Use of the Value Stream Mapping Tool for Waste Reduction in Manufacturing. Case Study for Bread Manufacturing in Zimbabwe. Citeseer, 2011.

Graham-Rowe, E.; Jessop, D. C.; Sparks, P. Identifying Motivations and Barriers to Minimising Household Food Waste. *Resour. Conserv. Recycl.* **2014,** *84*, 15–23.
Gustavsson, J.; Cederberg, C.; Sonesson, U.; Van Otterdijk, R.; Meybeck, A. *Global Food Losses and Food Waste*; FAO: Rome, 2011.
Hall, K. D.; Guo, J.; Dore, M.; Chow, C. C. The Progressive Increase of Food Waste in America and Its Environmental Impact. *PloS One* **2009,** *4* (11), e7940.
Hanson, C.; Lipinski, B.; Robertson, K.; Dias, D.; Gavilan, I.; Grέverath, P. et al. Food Loss and Waste Accounting and Reporting Standard, 2016.
HLPE. Food Losses and Waste in the Context of Sustainable Food Systems. A Report by the High Level Panel of Experts on Food Security and Nutrition of the Committee on World Food Security. *HLPE Report*, 2014.
Horrigan, L.; Lawrence, R. S.; Walker, P. How Sustainable Agriculture Can Address the Environmental and Human Health Harms of Industrial Agriculture. *Environ. Health Persp.* **2002,** *110* (5), 445–456.
Jedermann, R.; Nicometo, M.; Uysal, I.; Lang, W. *Reducing Food Losses by Intelligent Food Logistics*; The Royal Society Publishing, 2014.
Johnson, L. K.; Dunning, R. D.; Bloom, J. D.; Gunter, C. C.; Boyette, M. D.; Creamer, N. G. Estimating On-farm Food Loss at the Field Level: A Methodology and Applied Case Study on a North Carolina Farm. *Resour. Conserv. Recycl.* **2018,** *137*, 243–250.
Johnston, M.; Davison, I. *Study: $750 MA Year Wasted on Food*; NZ Herald, 2011.
Jones, T. Addressing Food Wastage in the US. Interview: The Science Show, 2006; p 8.
Jörissen, J.; Priefer, C.; Bräutigam, K.-R. Food Waste Generation at Household Level: Results of a Survey among Employees of Two European Research Centers in Italy and Germany. *Sustainability* **2015,** *7* (3), 2695–2715.
Kantor, L. S.; Lipton, K.; Manchester, A.; Oliveira, V. Estimating and Addressing America's Food Losses. *Food Rev./Natl Food Rev.* **1997,** *20* (1482-2016-121447), 2–12.
Khalid, S.; Naseer, A.; Shahid, M.; Shah, G. M.; Ullah, M. I.; Waqar, A. *et al.;* Assessment of Nutritional Loss with Food Waste and Factors Governing This Waste at Household Level in Pakistan. *J. Cleaner Prod.* **2019,** *206*, 1015–1024.
Kitinoja, L.; AlHassan, H.; Saran, S.; Roy, S. Identification of Appropriate Postharvest Technologies for Improving Market Access and Incomes for Small Horticultural Farmers in Sub-Saharan Africa and South Asia. Part, 2010.
Koivupuro, H. K.; Hartikainen, H.; Silvennoinen, K.; Katajajuuri, J. M.; Heikintalo, N.; Reinikainen, A. et al. Influence of Socio-Demographical, Behavioural and Attitudinal Factors on the Amount of Avoidable Food Waste Generated in Finnish Households. *Int. J. Consumer Stud.* **2012,** *36* (2), 183–191.
Kummu, M.; De Moel, H.; Porkka, M.; Siebert, S.; Varis, O.; Ward, P. J. Lost Food, Wasted Resources: Global Food Supply Chain Losses and Their Impacts on Freshwater, Cropland, and Fertiliser Use. *Sci. Total Environ.* **2012,** *438*, 477–489.
Left-Out, M. C. *An Investigation of the Causes & Quantities of Crop Shrink*; Natural Resources Defense Council, 2012.
Lipinski, B.; Hanson, C.; Lomax, J.; Kitinoja, L.; Waite, R.; Searchinger, T. Reducing Food Loss and Waste, Installment 2 of Creating a Sustainable Food Future. *World Research Institute Working Paper*, 2013; p 40.
Liu, C.; Mao, C.; Bunditsakulchai, P.; Sasaki, S.; Hotta, Y. Food Waste in Bangkok: Current Situation, Trends and Key Challenges. *Resour. Conserv. Recycl.* **2020,** *157*, 104779.

Liu, J.; Lundqvist, J.; Weinberg, J.; Gustafsson, J. Food Losses and Waste in China and Their Implication for Water and Land. *Environ. Sci. Technol.* **2013,** *47* (18), 10137–10144.

Lundqvist, J.; de Fraiture, C.; Molden, D. *Saving Water: From Field to Fork: Curbing Losses and Wastage in the Food Chain*; Stockholm International Water Institute: Stockholm, 2008.

Marangon, F.; Tempesta, T.; Troiano, S.; Vecchiato, D. Food Waste, Consumer Attitudes and Behaviour. A Study in the North-Eastern Part of Italy. *Ital. Rev. Agric. Econ.* **2014,** *69* (2–3), 201–209.

MINZ, W. New Zealand Food Waste Audits: Waste Not Consulting, 2015.

Mondiale, B. Missing Food: The Case of Postharvest Grain Losses in Sub-Saharan Africa. Report Number, 2011.

Muth, M. K.; Birney, C.; Cuéllar, A.; Finn, S. M.; Freeman, M.; Galloway, J. N. et al. A Systems Approach to Assessing Environmental and Economic Effects of Food Loss and Waste Interventions in the United States. *Sci. Total Environ.* **2019,** *685*, 1240–1254.

Nahman, A.; de Lange, W. Costs of Food Waste along the Value Chain: Evidence from South Africa. *Waste Manage.* **2013,** *33* (11), 2493–2500.

Nahman, A.; De Lange, W.; Oelofse, S.; Godfrey, L. The Costs of Household Food Waste in South Africa. *Waste Manage.* **2012,** *32* (11), 2147–2153.

Oelofse, S. H.; Nahman, A. Estimating the Magnitude of Food Waste Generated in South Africa. *Waste Manage. Res.* **2013,** *31* (1), 80–86.

Östergren, K.; Gustavsson, J.; Bos-Brouwers, H.; Timmermans, T.; Hansen, O.-J.; Møller, H. et al. FUSIONS Definitional Framework for Food Waste. *Projekt FUSIONS (Food Use for Social Innovation by Optimising Waste Prevention Strategies)*, Europäische Union, 2014.

Papargyropoulou, E.; Steinberger, J. K.; Wright, N.; Lozano, R.; Padfield, R.; Ujang, Z. Patterns and Causes of Food Waste in the Hospitality and Food Service Sector: Food Waste Prevention Insights from Malaysia. *Sustainability* **2019,** *11* (21), 6016.

Parfitt, J.; Barthel, M.; Macnaughton, S. Food Waste within Food Supply Chains: Quantification and Potential for Change to 2050. *Phil. Trans. R. Soc. B* **2010,** *365* (1554), 3065–3081.

Pekcan, G.; Köksal, E.; Küçükerdönmez, Ö.; Özel, H. *Household Food Wastage in Turkey*; FAO: Rome, Italy, 2006.

Pimentel, D. Amounts of Pesticides Reaching Target Pests: Environmental Impacts and Ethics. *J. Agric. Environ. Ethics* **1995,** *8* (1), 17–29.

Pimentel, D.; Houser, J.; Preiss, E.; White, O.; Fang, H.; Mesnick, L. et al. Water Resources: Agriculture, the Environment, and Society. *BioScience* **1997,** *47* (2), 97–106.

Porpino, G. Household Food Waste Behavior: Avenues for Future Research. *J. Assoc. Consumer Res.* **2016,** *1* (1), 41–51.

Pradhan, P.; Reusser, D. E.; Kropp, J. P. Embodied Greenhouse Gas Emissions in Diets. *PloS One* **2013,** *8* (5), e62228.

Principato, L. *Food Waste at Consumer Level: A Comprehensive Literature Review*; Springer, 2018.

Reynolds, C. J.; Mirosa, M.; Clothier, B. New Zealand's Food Waste: Estimating the Tonnes, Value, Calories and Resources Wasted. *Agriculture* **2016,** *6* (1), 9.

Richter, B.; Bokelmann, W. Explorative Study about the Analysis of Storing, Purchasing and Wasting Food by Using Household Diaries. *Resour. Conserv. Recycl.* **2017,** *125*, 181–187.

Ridoutt, B.; Juliano, P.; Sanguansri, P.; Sellahewa, J. The Water Footprint of Food Waste: Case Study of Fresh Mango in Australia. *J. Cleaner Prod.* **2010,** *18* (16–17), 1714–1721.

Roka, K. Environmental and Social Impacts of Food Waste, 2019.

Sage, C. The Interconnected Challenges for Food Security from a Food Regimes Perspective: Energy, Climate and Malconsumption. *J. Rural Stud.* **2013,** *29*, 71–80.
Salihoglu, G.; Salihoglu, N. K.; Ucaroglu, S.; Banar, M. Food Loss and Waste Management in Turkey. *Bioresour. Technol.* **2018,** *248*, 88–99.
Schindler, D. W.; Carpenter, S. R.; Chapra, S. C.; Hecky, R. E.; Orihel, D. M. *Reducing Phosphorus to Curb Lake Eutrophication Is a Success*; ACS Publications, 2016.
Schneider, F. Review of Food Waste Prevention on an International Level; ICE Publishing, 2013; pp 187–203.
Schott, A. B. S.; Andersson, T. Food Waste Minimization from a Life-Cycle Perspective. *J. Environ. Manage.* **2015,** *147*, 219–226.
Secondi, L.; Principato, L.; Laureti, T. Household Food Waste Behaviour in EU-27 Countries: A Multilevel Analysis. *Food Policy* **2015,** *56*, 25–40.
Serafini, M.; Lee, W.; Toti, E.; Bucatariu, C.; Fonseca, J.; Van Otterdijk, R. et al. Global Variations in Micro-Nutrient Losses in the Fruit and Vegetables Supply Chain, 2015.
Silvennoinen, K.; Korhonen, O. *Food Waste Volume and Composition in Helsinki Region Households*; SE, 2013.
Silvennoinen, K.; Katajajuuri, J.-M.; Hartikainen, H.; Heikkilä, L.; Reinikainen, A. Food Waste Volume and Composition in Finnish Households. *Br. Food J.* 2014.
Song, G.; Li, M.; Semakula, H. M.; Zhang, S. Food Consumption and Waste and the Embedded Carbon, Water and Ecological Footprints of Households in China. *Sci. Total Environ.* **2015,** *529*, 191–197.
Stuart, T. *Waste: Uncovering the Global Food Scandal*; WW Norton & Company, 2009.
Tenneson, M. Phosphorus Fields. *Discover*. Dec 2009.
Thönissen, R. Food Waste: The Netherlands, 2009.
Thyberg, K. L.; Tonjes, D. J.; Gurevitch, J. Quantification of Food Waste Disposal in the United States: A Meta-Analysis. *Environ. Sci. Technol.* **2015,** *49* (24), 13946–13953.
Timmermans, A. FUSIONS Food Waste Data Set for EU-28: New Estimates and Environmental Impact, 2015.
Tonini, D.; Albizzati, P. F.; Astrup, T. F. Environmental Impacts of Food Waste: Learnings and Challenges from a Case Study on UK. *Waste Manage.* **2018,** *76*, 744–766.
Trang, P. T. T.; Dong, H. Q.; Toan, D. Q.; Hanh, N. T. X.; Thu, N. T. The Effects of Socio-Economic Factors on Household Solid Waste Generation and Composition: A Case Study in Thu Dau Mot, Vietnam. *Energy Procedia* **2017,** *107*, 253–258.
Tucker, C.; Farrelly, T. Household Food Waste: The Implications of Consumer Choice in Food from Purchase to Disposal. *Local Environ.* **2016,** *21* (6), 682–706.
van Geffen, E.; van Herpen, H.; van Trijp, J. Quantified Consumer Insights on Food Waste: Pan-European Research for Quantified Consumer Food Waste Understanding: D1. 4. REFRESH, 2017.
Vanham, D.; Bouraoui, F.; Leip, A.; Grizzetti, B.; Bidoglio, G. Lost Water and Nitrogen Resources Due to EU Consumer Food Waste. *Environ. Res. Lett.* **2015,** *10* (8), 084008.
Venkat, K. The Climate Change and Economic Impacts of Food Waste in the United States. *Int. J. Food Syst. Dyn.* **2011,** *2* (4), 431–446.
Vinikoor-Imler, L. C.; Davis, J. A.; Luben, T. J. An Ecologic Analysis of County-Level PM2. 5 Concentrations and Lung Cancer Incidence and Mortality. *Int. J. Environ. Res. Public Health* **2011,** *8* (6), 1865–1871.
Visschers, V. H.; Wickli, N.; Siegrist, M. Sorting Out Food Waste Behaviour: A Survey on the Motivators and Barriers of Self-Reported Amounts of Food Waste in Households. *J. Environ. Psychol.* **2016,** *45*, 66–78.

Williams, H.; Lindström, A.; Trischler, J.; Wikström, F.; Rowe, Z. Avoiding Food Becoming Waste in Households–The Role of Packaging in Consumers' Practices across Different Food Categories. *J. Cleaner Prod.* **2020,** 121775.

Wohner, B.; Pauer, E.; Heinrich, V.; Tacker, M. Packaging-Related Food Losses and Waste: An Overview of Drivers and Issues. *Sustainability* **2019,** *11* (1), 264.

Woods, J.; Williams, A.; Hughes, J. K.; Black, M.; Murphy, R. Energy and the Food System. *Phil. Trans. R. Soc. B* **2010,** *365* (1554), 2991–3006.

WRAP, W. The Food We Waste. Report No. 1-84405-383-0, 2008.

WRAP, W. *Household Food and Drink Waste in the UK*; Report Prepared by WRAP; WRAP: Banbury, UK, 2009.

Wrap. *New Estimates for Household Food and Drink Waste in the UK*; Wrap Banbury, 2011.

WRAP. *Strategies to Achieve Economic and Environmental Gains by Reducing Food Waste*; Waste and Resources Action Programme Banbury, 2015.

Figures:

Images in Figures downloaded from google images.

a href="https://www.freepik.com/vectors/business">Business vector created by macrovector_official-www.freepik.com</a

https://www.123rf.com/clipart-vector/crop_sprayer.html?sti=mmj2sf3hamj55zu9p3|&mediapopup=114799313

https://www.pinterest.com/pin/716987203157402987/?nic_v2=1a4fUT9Br

https://www.dreamstime.com/illustration/roastbeef.html

https://www.nicepng.com/ourpic/u2q8q8e6q8u2r5u2_fruit-and-vegetables-clipart-fruit-clipart-png-fruit/

https://www.shutterstock.com/image-vector/industrial-cold-store-scene-vector-illustration-319667207

https://www.shutterstock.com/search/grocery+store+icon

https://www.kindpng.com/imgv/Tmhhxwm_tribal-fish-bones-clipart-png-black-and-white/

https://www.publicdomainpictures.net/en/view-image.php?image=30745&picture=tea-bag-clip-art

http://clipart-library.com/banana-peel-cliparts.html

https://www.google.com/imgres?imgurl=https%3A%2F%2Fillustoon.com%2Fphoto%2F414.png&imgrefurl=https%3A%2F%2Fillustoon.com%2F%3Fid%3D414&tbnid=PM_jfx7-8B4wUM&vet=12ahUKEwiZrIzHq4TuAhVD0RoKHWPjCAIQMyg0egQIARBe..i&docid=GuNXJes1EjW1xM&w=960&h=960&q=fish%20clipart&hl=en&ved=2ahUKEwiZrIzHq4TuAhVD0RoKHWPjCAIQMyg0egQIARBe

http://clipart-library.com/clipart/1557503.htm

CHAPTER 11

Thermal Waste Management Techniques

MUHAMMAD SAJID[1*], MUHAMMAD IRFAN AHAMAD[2], ADNAN UL REHMAN[2], MUHAMMAD SAIF UR REHMAN[3], and MUHAMMAD MOHSIN AZIM[4]

[1]*Department of Materials and Chemical Engineering, Yibin University, Yibin 644000, Sichuan, China*

[2]*Shaanxi Key Laboratory of Earth Surface System and Environmental Carrying Capacity, College of Urban and Environmental Sciences, Northwest University, Xi'an 710127, China;*

[3]*Department of Chemical Engineering, Khwaja Fareed University of Engineering and Information Technology, Abu Dhabi Road, Rahim Yar Khan, Pakistan*

[4]*Department of Chemistry, Norwegian University of Science and Technology, Trondheim, Norway*

[*]*Corresponding author. E-mail: engr.sajid80@gmail.com; drsajid@yibinu.edu.cn*

ABSTRACT

Waste management (WM) is now the most critical issue in modern societies concerning community health and safety. This issue is increasing with socioeconomic development and with the implication of more sophisticated measures. Different techniques have been applied for a long to overcome this sociotechnical problem. Among these, thermal management has the potential to shrink the waste volume and render it innocuous by mitigating pathogens and other toxic moieties. Incineration, pyrolysis, gasification, and plasma are the mostly applied thermal methodologies. Within the thermal

Waste Problems and Management in Developing Countries. Umair Riaz, Shazia Iqbal, & Moazzam Jamil (Eds.)

WM hierarchy, incineration with energy recovery has been considered and employed desirably at an industrial level in different countries. Pyrolysis, being lower in temperature than gasification results in truncated volatilization of carbon (C), heavy metals, and dioxin precursors. Gasification is an energy efficient and eco-friendly technique with a broader development perspective. Plasma is the most expensive and typical process behaving at extremely high temperatures. Therefore, plasma gasification is mostly used for special purposes such as the conversion of organic matters to syngas. Mostly the process economics of thermal management is directed by the energy recovery process. Therefore, the selection of thermal technique is governed by careful energy balance and efficient economic analysis. This chapter deals with the management of wastes with the application of thermal technologies, comparative analysis, and application feasibilities.

11.1 INTRODUCTION

Waste generation is inevitably related to life, which is increasing with modernization, socialization, and urbanization (Pfaltzgraff et al., 2013). It has been reported that accumulated global waste generation was 267.8 million tons in 2017 of which only 12.7% was thermally treated for energy recovery. The landfill was observed as the major waste disposal techniques having >52% proportion. According to the statically reported trends, waste generation is increasing exponentially and will reach up to 1.7 times of 2017 in 2050. These reports are alarming with increasing life. Global waste management (WM) will become gradually convoluted and exorbitant with the cumulative Global Waste Index (GWI) and compelling to search for sustainable WM techniques concerning the living environment (Sajid et al.,2020a, 2020b; Yan et al., 2016).

WM continues to be a high priority for local, national, and international authorities while establishing future planning. Considering the socioeconomic impact of WM, the WM framework has been defined as reduce, reuse, recycle, recovery, and disposal as shown in Figure 11.1. Among these, the major focus was to reduce, reuse, and recycle to reduce the waste magnitude before entering the mainstream.

Landfills remain the major waste disposal technique for since long because of process simplicity, low cost, and land availability (Elwan et al., 2014). Making an allowance for GWI, the area demand for the establishment of new landfill sites is up to 2.8 million m^2/year, which is extremely high (Ahamad et al., 2020a, 2020b; Ouda et al., 2016). The abandoned release of methane to the environment, leachate formation triggering subversive water pollution, release of odorous compounds, and production of pathogenic microorganisms along with land availability are serious concerns associated with landfill WM (Matsakas et al., 2017; Yan et al., 2016). Emphasizing the recycle and recovery objective; biochemical and biological such as compositing was adopted as alternative WM techniques. However, WM employing thermal treatment techniques has the potential to shrink the size of the waste and to condense it harmless, by the devastation of pathogens and poisonous substances (Yan et al., 2016). Thermal waste management (TWM) can be used widely with varying composition of solid wastes. If the construction industry contributed heavily to solid waste, thermal processing is not a viable option (Edo et al., 2016). However, if the major contribution is from agricultural and residual waste such as municipal solid waste (MSW), thermal processing is the best option with integrated energy recovery.

FIGURE 11.1 Conceptual design of waste management (European Environment Agency, 2013; Matsakas et al., 2017).

In this chapter, we have summarized the principal TWM techniques, their application principles, comparative and associated merits, and demerits. Processes are described with the help of process diagrams considering the latest developments concerning the current energy and environment-sensitive scenario.

11.2 SALIENT FEATURES OF TWM

TWM works on the principle of recovery and disposal. Therefore, this methodology has many advantages over other WM processes (Luz et al., 2015; Pfaltzgraff et al., 2013; Sharholy et al., 2008); such as

a) Integrated energy recovery.
b) Volume reduction.
c) Detoxification, mostly toxic moieties are reformulated.
d) Can be coupled for integrated electricity generation.
e) Mitigate organic contamination simultaneously.

f. Equally applicable for diverse feed range.
g. Lowland requirement.
h. Capable to cope with toxic gases onsite.

TWM is a comparatively applicable process especially for developing countries because it has many associated benefits. However, these techniques also have some drawback such as (Fabry et al., 2013; Sharholy et al., 2008):

a. High capital and operating cost.
b. Waste segregation and classification are required.
c. Skilled labor is required to operate the system.
d. Ash pollution needs careful onsite management and disposal system.
e. Toxic air pollutants (incomplete combustion and products of/or heavy metals).

11.3 TWM TECHNIQUES

Considering the advantages associated with TWM techniques, these are vital gears of assimilated WM (Gehrmann et al., 2017). The wide applicability of these techniques has dragged ample research and development in this field and many of them are now operative efficaciously as confirmed by numerous reports (Vallero, 2011; Yan et al., 2016). However, the sorting of solid waste before thermal treatment is required for efficient operation. Thermal processing of solid wastes operated at high temperatures with high conversion rates than biochemical and physiochemical processing. Typically hydrocarbons are converted to burning products with energy recovery as shown in eqs 11.1 and 11.2. It has been estimated that 1 ton of MSW can produce 500 kWh of electricity. Hence, its energy potential is equivalent to 200 kg of oil and 2.5 tons of steam at 40 bar and 400°C (Kumar, 2016).

$$\text{Hydrocarbons} + O_2 \rightarrow CO_2 + H_2O + \text{Heat Energy} \tag{11.1}$$

$$\begin{aligned} &C_{20}H_{32}O_{10} + x_1O_2 + x_2\ H_2O_\Delta \rightarrow y_1C + y_2CO_2 + y_3 \\ &CO + y_4H_2 + y_5CH_4 + y_6H_2O + y_7C_nH_m \end{aligned} \tag{11.2}$$

Selection of WM technique is crucial as there is no single technique that is equally applicable for managing all waste streams in all circumstances. Therefore, technique selection is an art considering the waste composition and hierarchy (Figure 11.1). Seeing the WM hierarchy, most considered TWM techniques are outlined in Figure 11.2 and have been described comprehensively in this chapter.

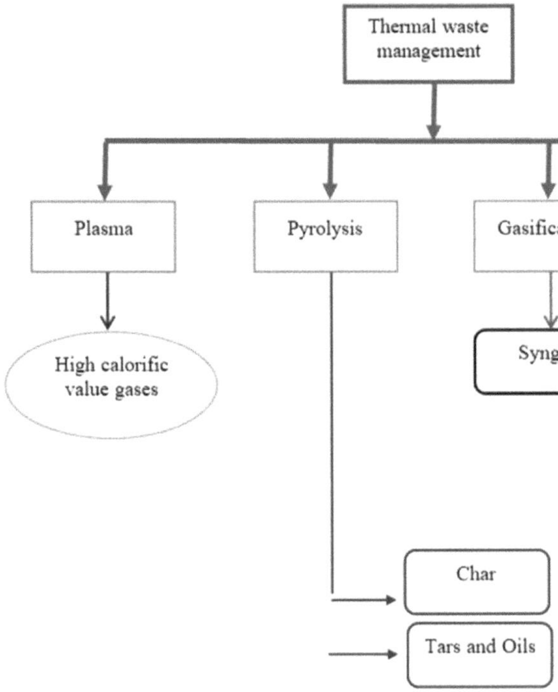

FIGURE 11.2 Thermal waste management techniques.

11.4 GASIFICATION

Thermal decomposition of hydrocarbons to syngas (H_2 and CO) employing low oxygen and high temperature is defined as gasification.

The history of the gasification technique dated back to the early 1800s to produce tar. Since then, this technology has been evaluated and practiced worldwide. In recent decades, gasification technology gained extra attention at the industrial scale considering the volume reduction of waste with simultaneous production of cleaner energy as syngas over the conventional incineration/combustion (Basu, 2020). Different gasification technologies have been developed considering the available materials (MSW, industrial waste, and agricultural wastes) during the last few decades in order to minimize the waste collected as well as to establish the natural carbon cycle.

Gasification is an alternative to more mature methods of waste streams into electricity and other useful products. Under certain applications and conditions, there are various benefits of gasification, especially the use of waste to generate clean electricity (Figure 11.3) (Rutberg et al., 2011). Different gasifier design has been developed and employed considering the waste composition and quality (Figure 11.4). The application of particular technology involves many technical (feasibility, reliability, sustainability, process efficiency, raw material, and product flexibility), economic (cost and availability), environmental (carbon neutral and other sources of secondary pollution), and political factors (national energy security, public relations, and public policy) (Galeno et al., 2011).

Technically gasification is quite similar to the incineration process but with a controlled supply of oxygen to achieve partial oxidation. Whereas conventional combustion processes need excess oxygen for complete oxidation; however, this process needs pretreatment of wastes. Therefore, principally gasification can be considered in-between traditional combustion and pyrolysis with continuous development (Fichtner Consulting Engineers Ltd., 2004). Gasification is preferably used for the thermal decomposition of agricultural waste, sludge, and MSW. During the gasification process, carbon species (carbon-based feedstock) can be converted into syngas, which has protracted industrial utilization. Syngas as the name implied is a source of synthetic natural gas and is colorless, odorless, and flammable, containing carbon monoxide (CO) and hydrogen (H_2). It can be used directly to generate electric power, in fertilizers, for the production of chemicals, and directly as a fuel. Syngas being rich in H_2, can be used directly as a gaseous fuel as well as for the synthesis of gasoline and diesel. Gasification technology can produce an unprecedented number of products at production plants (Shahabuddin et al., 2020). The generalized schematic illustration of the gasification process is given in Figure 11.3 and different designs of gasifiers are available in Figure 11.4.

11.4.1 FUNDAMENTALS OF GASIFICATION PROCESS

The general combustion process works on the principle of excess oxygen and high oxygen/air input is required; whereas, the gasification process is an efficient process considering the oxygen-to-fuel ratio to achieve sinusoidal conversion (Nzihou, 2010). The amount of oxygen used for gasification is less than half of the oxygen vital for combustion. The detailed process and the reactions that occurred at each stage are outlined in Figure 11.5.

During the gasification process, the following processes occur at different stages;

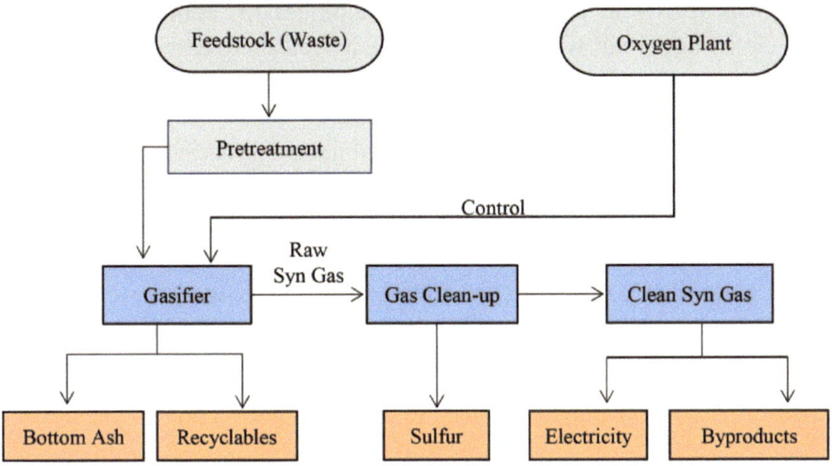

FIGURE 11.3 A general illustration of the gasification process (Moustakas and Loizidou, 2010; Shahabuddin et al., 2020).

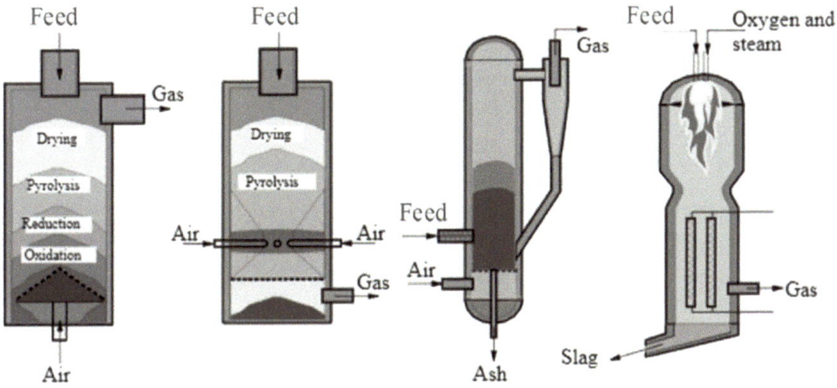

FIGURE 11.4 Different gasifier designs used for the thermal conversion of wastes.

Source: Illustration by Kopiersperre, 2018. https://creativecommons.org/licenses/by-sa/3.0/

- **Drying and Dehydration:** All the dispersed and associated water in the feed (raw waste stream) evaporates leaving the dry matter for subsequent processing in the later stages. Absorbed and adsorbed water evaporates completely along with some chemical water.
- **Pyrolysis:** Pyrolysis occurs when raw materials are exposed to high temperatures in the gasifier. Bonds breaking and formation of new products occur at this stage releasing volatile gases such as methane and H_2 with tar production.

- **Combustion:** Along with partial oxidation, complete combustion may occur but with reduced intensity releasing carbon dioxide (CO_2) and water (H_2O). This reaction in the chorus delivers the heat energy necessary for the gasification reaction.
- **Distribution:** Some of the residual material reacts to extract CO and H_2 with CO_2 and steam.
- **Water and methane exchange:** This is an independent and reversible gas-phase reaction that occurs instantaneously under gasification conditions. These are secondary reactions with little effect on the gasifiers; therefore, considering the product requirements, the water under the carburetor can be methanized (Zhao et al., 2010).

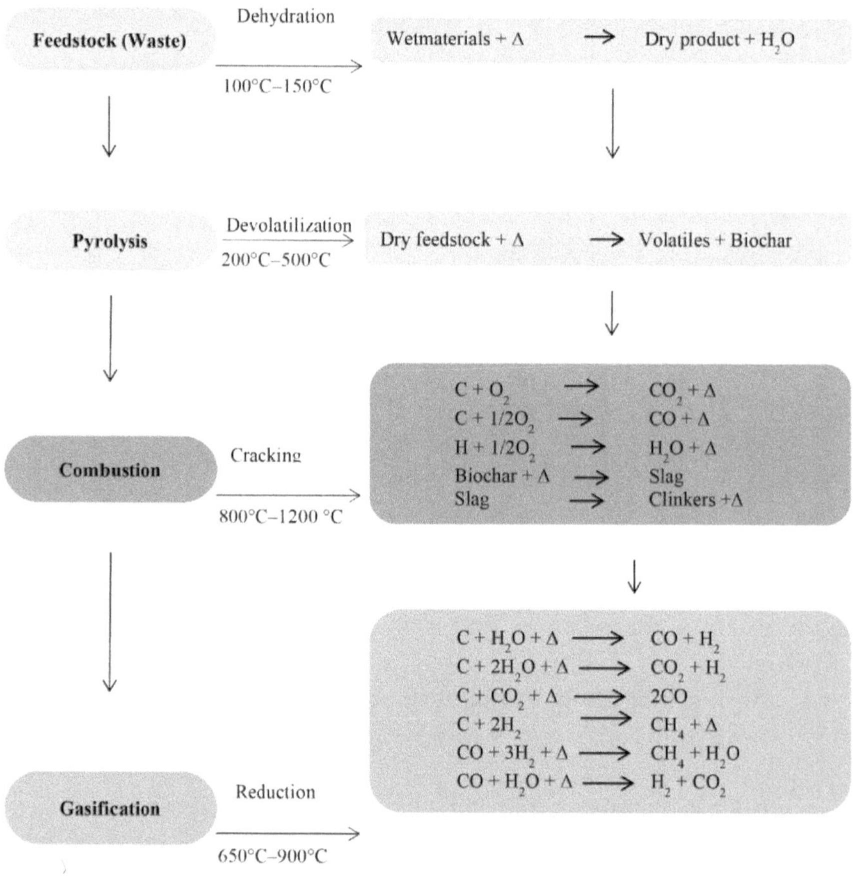

FIGURE 11.5 The processes and chemical reactions involved during the waste gasification (Begum et al., 2014).

Based on the temperature differences and waste types, the gasification process can be divided into two main categories, (1) *low-temperature gasification (LTG),* which is further subdivided into three categories as direct, indirect, and thermoset gasification accordingly but the first two gasification technologies have a vast application and are commonly used, (2) *high-temperature gasification (HTG)* (Yan et al., 2016). Different gasifier designs have been employed considering the design applicability, feasibility, process efficiency, and process smoothness as shown in Figure 11.4. All these gasification technologies have their own operational merits and demerits. The detail of each gasification process is given below;

11.4.2 LOW-TEMPERATURE GASIFIER (LTG)

In LTG, the gas and feed both flow down cocurrently. Generally, low flow devices are easy to operate at low temperatures (788°C–1200°C/1450–2200°F) and create low tar and particulate sins (Rutberg et al., 2013). In this process, low moisture contents are essential as the air degasser has a small capacity and cannot withstand more than 20% moisture. LTG loses elasticity of the feed materials and is considered less efficient as compared with HTG. These gasifiers are mostly required sorting of feedstock; therefore, have limited scope. The following are the merits and demerits of this type of gasifier.

Advantages	Disadvantages
Small size.	The feedstock's need to be sorted.
These are portable.	There is a production of hazardous ash/char that needs to be removed.
These are modular.	It has size limitations.
	It has a low moisture compatibility.

11.4.3 HIGH-TEMPERATURE GASIFIER (HTG)

HTGs are more efficient with complete conversion and operational easiness. Previously, plasma gasification technology was only used as HTG technology actively (Galeno et al., 2011). In recent years, this technology offers a variety of HTG process along with plasma furnaces.

11.4.4 FLUIDIZED BED GASIFICATION

These gasifiers are typically employed with the solution of granular moieties such as sand, limestone, dolomite, or aluminum. The temperature of these devices is maintained between 540°C–980°C (1000°F–1800°F). The waste, air, and particles are mixed and react in a gasification field, then the remaining gas and layers are separated in a cyclone separator as shown in Figure 11.6 (Yassin et al., 2009). This method provides a certain elasticity of raw materials and is particularly effective for agricultural and forest residues (Nishikawa et al., 2006). These gasifiers are economical at a large scale and are used for commercial applications as compared with downdraft gasification (Saghir et al., 2018).

Advantages	Disadvantages
It is an ideal choice for biomass.	Stocking feedback is essential.
It offers some feedback flexibility.	Narrow variety of process feedstock.
It produces uniform syngas with low tar content at uniform temperature distribution.	Ash is produced as a byproduct, hence disposal problem.
The efficiency of tar removal can be enhanced by the use of catalyst.	Feedstock needs drying.
	Low moisture compatibility.
	It has a limited efficiency.

11.4.5 FASTOX GASIFICATION

FastOx gasification was designed to overcome the limitations of high and low temperatures gasifiers. FastOx is a fixed-field gasification system capable of rapidly hitting high concentrations of oxygen and vapor flow. These gasifiers have the potential to degrade waste feedstock at the molecular level. Thus, the name FastOx came from the combination of oxygen and steam as injections in the gasifier. The FastOx gasifier works equally for wide variety of waste materials except hazardous materials such as explosives and radioactive materials.

The feedstock (waste) is pretreated and fed through an airlock at the top of the gasifier and injectants (oxygen and steam) at the bottom. This injection of oxygen and steam produces very high temperatures (~2200°C/4000°F) inside the chamber resulting in the oscillation of partial oxidation of materials present in the chamber. Different by-products are extracted at different stages during the descending flow of materials. The highest temperature exists at the bottom of the gasifier. Here, the left carbonaceous material reduces to char; inorganic material and metals are also reduced/melted because of high temperature. FastOx gasification converts 90% of the waste (organic part) into pure and high-quality synthetic gas directly having 60% CO and 30% H_2(Rutberg et al., 2011). The high temperature of the FastOx gasifier allows the remaining 10% of the waste (inorganic parts and metal compounds) to be dissolved into recycled metals and inert stones (slag) that can be used as building aggregates and in concretes.

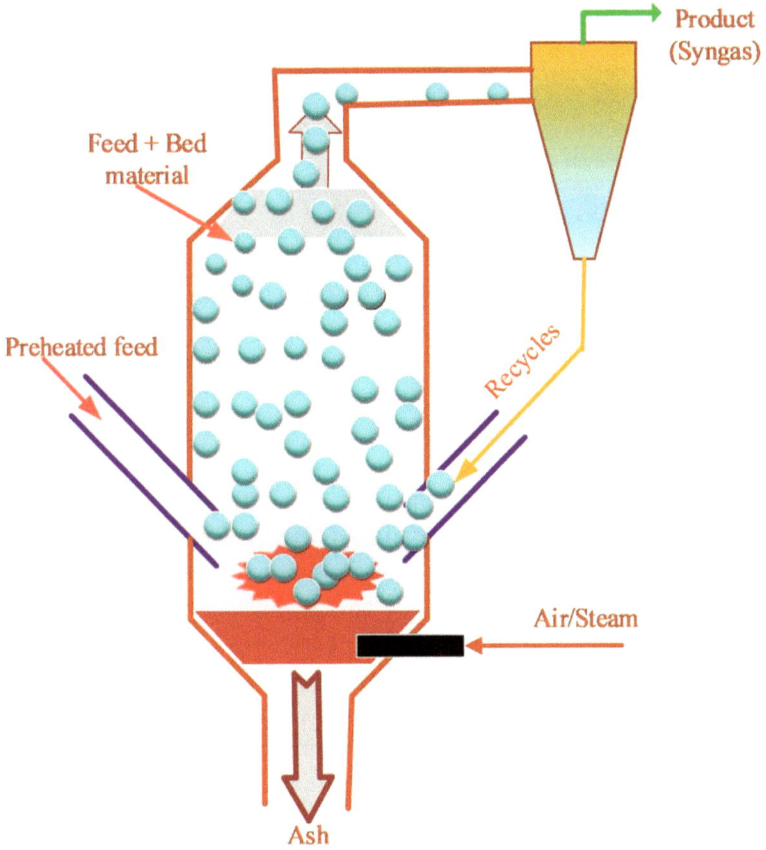

FIGURE 11.6 The schematic diagram for fluidized bed gasifier.

The FastOx gasification process is mainly divided into three stages: preparation of the waste, syngas production, and cleaning of gases. The schematic flow diagram is shown in Figure 11.7.

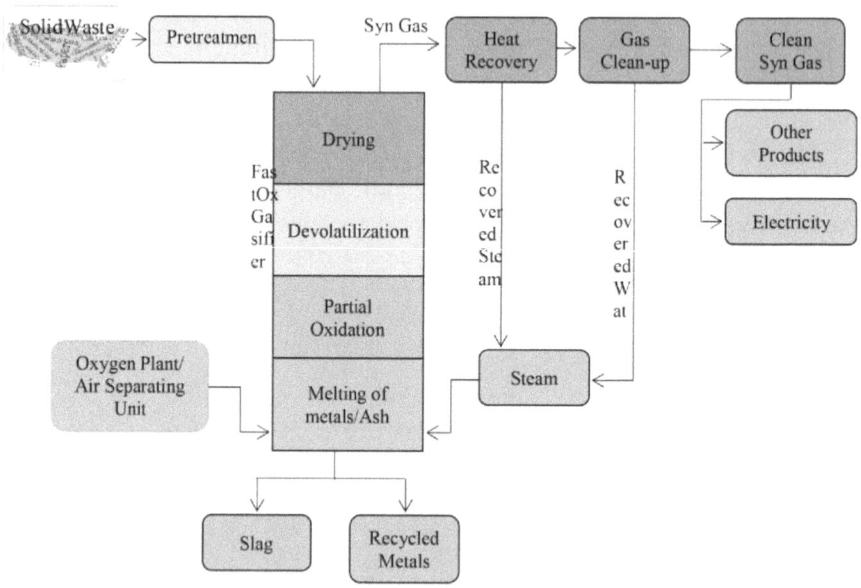

FIGURE 11.7 A schematic flow diagram of the FastOx gasification process.
Source: Adapted from Shahabuddin et al., (2020) and Sierra Energy (2020).

11.5 PLASMA GASIFICATION

Plasma was identified as a semineutral gas with a combined behavior of charged particles and neutral particles. There sis evidence that the thermal plasma process is not harmful to the environment, but only produces inert slag and fully complies with green energy protocols. During the thermal plasma process/gasification, a high temperature (1400°C) may be used to melt the ash and maximize the production/amount of CO and H_2(Prado et al., 2020a, 2020b). MSW is mostly composed of organic and inorganic species. During this process, the organics part having hydrocarbons is converted to syngas while the inorganic portion is reduced to molten slag (Prado et al., 2020a, 2020b).

The plasma process was originally developed for aerospace and military applications but is now used in metallurgy and waste heat treatment processes (Ahamad et al., 2020a, 2020b; Sajid et al., 2020a, 2020b). Thermal waste has previously been shown to be compatible with plasma and can eliminate hazardous wastes such as asbestos or chemicals (Prado et al., 2020a, 2020b). The vitrification of radioactive waste is another mature application.

The plasma process can be regarded as a new process because it is still developing considering the process and energy efficiency. Principally, plasma gasifiers operate by passing a pressurized gas through a high voltage arc. This arc generates high temperatures up to 4425°C (8000°F) in the chamber. This high temperature rupture the species present at the molecular level. The crawler carburetor produces synthetic gas and slag, similar to the FastOx carburetor in the comfort range of raw and recyclable materials (Nzihou, 2010). The electrode at the bottom of the gasifier triggers an electric arc and the required temperature is maintained by this spark requiring a great deal of electrical input. Therefore, this technology is compelled by high capital and operating costs. This energy security problem has been solved partially with the concept of "classical" pyrolysis reactions. In this process, a separate gasification step is performed at medium temperature with low flow. This plasma phase is then used for gas treatment and residual materials vitrification. This process has its own merits and demerits, which are summarized below.

Advantages	**Disadvantages**
A small number of oxidants are needed to form syngas.	High capital investment.
It results in low emissions.	High operating costs.
It is well proven.	It has a complicated design and process.
It reduces the sorting needs.	It contains large footprints.
It results in very little or no landfill disposal.	Multiple unite as required for large footprints.
It generates very useful materials.	
High energy density, suitable temperature, and reaction time.	
Shorten start and shutdown times.	

11.5.1 PLASMA GASIFIERS

Plasma gasification is principally a type of gasification technique but differs in operational parameters and gasifier design. An important method of treating solid waste is gasification and vitrification associated with energy reclamation and reduction of volume. Plasma is produced by the electronic collision that can crumble residues excellently. Different thermal and nonthermal plasma gasifier designs have been used such as pulsed corona, dielectric barrier discharges, microwave plasma, RF plasma, DC corona discharges, and RF-DC hybrid system (Shahabuddin et al., 2020; Tang et al., 2013). Similarly, different reactors' designs have been investigated considering the plasma efficiency and heat flux. Commonly reactor designs having fixed bed and moving bed were used initially but with the development of plasma technology, other reactor types such as entrained-flow reactors, spout bed reactors, and spout-fluid bed reactors have also been practiced considering reactors advantages such as heat transfer efficiencies in conventional gasification processes (Tang et al., 2013).

11.5.2 PLASMA GASIFICATION PROCESS

Figure 11.8 illustrates the composition of other processes used for the schematic evaporation of thermal plasma. Thermal plasma is a process of conversion of carbonaceous compounds to gaseous products and solid residue in limited oxygen and high-temperature environment (Nishikawa et al., 2006). Waste materials are injected into a reactive high-temperature zone, come in contact with a cloud of excited molecules, ions, and electrons. The gas temperature in the plasma for heating the furnace is very high (> 1200 °C) (Morrin et al., 2012). Pretreated MSW is injected into the plasma furnace using a thermal plasma torch. Liquefied petroleum gas burners are sometimes installed to increase the initial heat and maintain the required temperature. Coal or any other fuel can also be used with flexible adaptability. Feed temperature increased rapidly due to plasma, which results in cracking of material fed. Gases (H_2, CO_2, CO, and H_2O) and low molecular hydrocarbons (CH_4, C_2H_2, etc.) with little biochar/ash are produced instantaneously (Nishikawa et al., 2006). Air (oxygen) and/or steam is provided to maintain an oxidizing environment. Product quality (gas composition) can be standardized given the downstream consumption. The gaseous stream utilization is flexible because of sophisticated quality control and can be used in power generation (gas turbine or gas engine) or as a feed for the synthesis of chemicals or liquid fuels (Cheng et al., 2011). The plasma process is a multifarious route involving many systematic junctures; however, for easiness, the entire process can be pigeonholed as

i. Rapid heating of feed with plasma.
ii. Violent release of volatiles from the feed material.
iii. Rapid gasification of homogeneous phase due to swift mass/heat transfer.
iv. Biochar gasification due to numerous gasiform constituents.

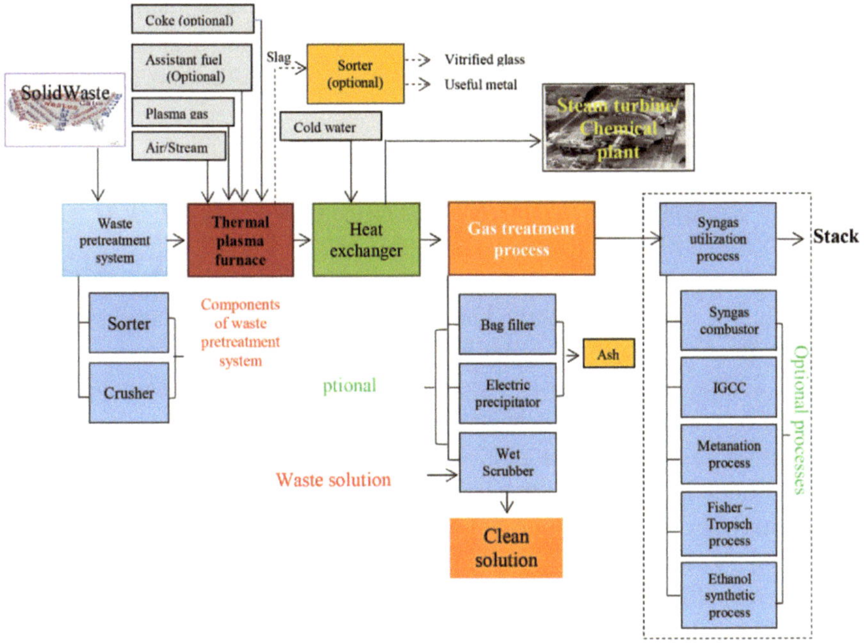

FIGURE 11.8 Schematic diagram of plasma gasification.

Source: Adapted with permission from Li et al. (2016) © 2020 Elsevier.

Along with downstream utilization flexibility, plasma gasification has many operational advantages. Heat efficiency can be increased by placing a heat exchanger to heat the plasma employing heat recovery from the product gases. This recovered heat can also be used as a secondary energy source for steam turbines. The refrigerant gas passing through the heat exchanger must be cleaned to remove clean breasts. Wet scrubbers are commonly applied to eradicate fly ash and acid gases (Gomez et al., 2009; Heberlein and Murphy, 2008; Li et al., 2016). In addition, the plasma process does not emit NO_x hence equipped with a preferred advantage over thermal gasification. Therefore, the refined synthesis gases can be regarded as an environmentally benign energy source. Firstly, this can be achieved with electric and gas turbines. To generate electricity from a steam turbine, steam is only produced by burning abysses injected into the steam turbine (Jan, 2010). If other equipment is needed, such as gas purifiers and synchronous compressors, gas turbines can be used to generate synchronous power. Secondly, by combining chemical processes, it can produce valuable chemicals in parallel. As the Fisher–Tropsch process that produced chemical waxes. Hydrogen produced is of high purity and has extended utilization in the electrochemical process and fuel cells. With continuous development, the thermal plasma gasification process is continuously expanding the scope of application (Prado et al., 2020a, 2020b).

11.5.3 PLASMA TECHNOLOGY FOR NUCLEAR WASTE TREATMENT

Revolutionary development was witnessed in the nuclear industry since the beginning of the 20th century. This development extended nuclear applications in the field of energy, medical, and other industrial processes (Prado et al., 2020a, 2020b). This extended utilization boosted nuclear waste demanding intelligent nuclear waste disposal techniques. Development and application of special protocols were necessary for the efficient management of nuclear wastes for the safekeeping of life and the atmosphere. Due to extensive protecting equipment, 80% of radioactive waste is compactable (e.g., safety aprons, goggles, protecting sheets, gloves, and other safety auxiliaries). Therefore, the plasma technique is a favorable route to achieve the volume decrease of nuclear wastes (Jan, 2010; Prado et al., 2017). A study conducted by Parado et al. elucidated that employing plasma at 1400°C can achieve a very high volume reduction factor (1:99) in only 0.5 h (Prado et al., 2020a, 2020b). Hence, a very high volume reduction of nuclear waste eases the waste handling and disposal of nonactive species. The plasma process cremates the ignitable fragments of nuclear waste and vitrifies the noncombustible fractions concurrently producing glossy slags (Ghiloufi, 2009).

Plasma technology has many operational advantages in radioactive WM such as volumetric shrinking, destruction, and radionuclide holding (Heberlein and Murphy, 2008). In addition, plasma technology offers a commercial incentive in terms of nuclear-waste packing and rapid processing over conventional handling procedures. Cascading of plasma torch with a traditional burner can be applied considering the environmental safety standards. However, besides these worth mentioning advantages, plasma processing of radioactive wastes have some limitation. High electricity cost and proper management of volatile radionuclides from exhaust gases are process constraints (Heberlein and Murphy, 2008). Therefore, small-scale processing becomes uneconomical due to high capital and operational costs. Hence a reliable balance between plant capacity and expenses is necessary to achieve a breakeven point.

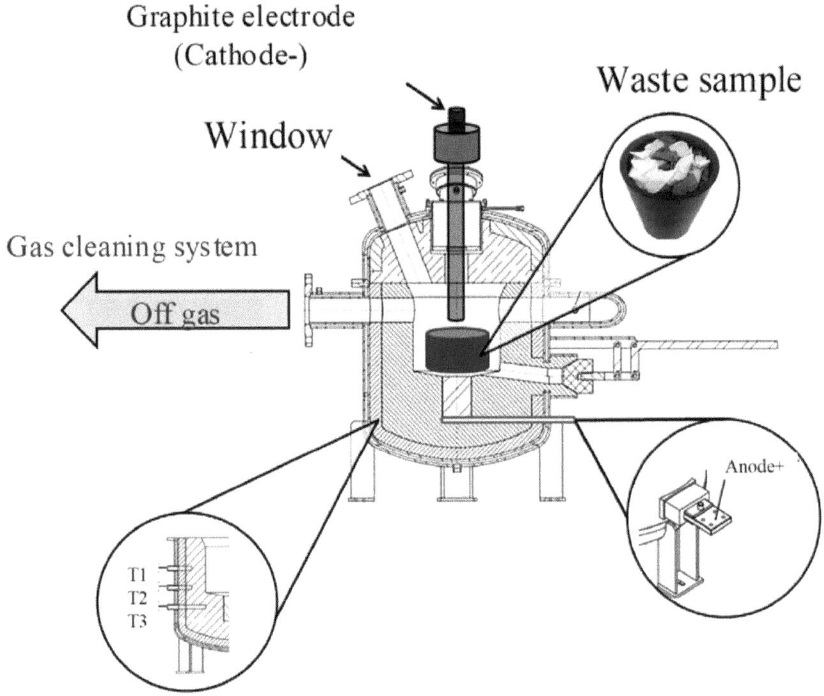

FIGURE 11.9 Plasma gasifier for radioactive waste gasification.

Source: Reprinted with permission from Prado et al. (2020a, 2020b) © 2020 Elsevier.

11.5.4 PLASMA TECHNOLOGY FOR MEDICAL WASTE TREATMENT

All wastes produced in any medical facility is designated as medical waste and needs careful treatment considering environmental pollution and life sustainability (Organization, 1999). Medical waste generation varied from 0.2 to 0.5 kg/day/medical facility from low- to high-income countries (Cai and Du, 2020). Generally, hazardous medical waste contributes nearly 15% of overall waste spawned by a community that enormously increased abruptly during any disease outbreak. Furthermore, the composition of medical waste usually may vary; causing an additional problem for WM. As witnessed with the COVID-19 outbreak, an abrupt increase in inpatient density resulted in a high volume of medical waste with added hazardous pathogens. In Wuhan city China, medical waste increased from 40 tons per day to ~247 tons per day. Hence, proper treatment and disposal of these hazardous wastes are continuous challenge with continuous technology development necessities.

Due to increased risk, medical wastes can be treated effectively with the provision of onsite treatment facilities. Previously, incinerators were used to meet the disposal standards. But with the implementation of stern regulation of risk management and environmental pollution; medical WM is a leading task. This problem is partially solved with the application of plasma; similar to radioactive waste. The plasma technique advantageous is having a compact design and high temperature. Hence, it is an encouraging route for onsite and offsite handling of medical waste.

Plasma process reported by Rutberg et al. (2002) having a capacity of 150–200 kg/hwas used for the disinfection of medical waste. Two AC torches were placed in a rotary kiln to maintain a temperature between 1000°C and 1200°C. Secondary air and hot water/steam were used to control the temperature and exit air quality considering the depression of NO_x and generation of CO and H_2. Mostly, medical treatment facilities are equipped with afterburners. The purpose of this afterburner is to convert the hazardous gases produced during the combustion of medical waste to their respective stable complexes. Exhaust gases from the rotary kiln are passed through a secondary burner maintaining high temperature (1200°C–1400°C) and a short time (~2 s). These produced gases are then passed through scrubbers at reduced temperatures (≤950°C) avoiding free chlorine formation. Hazardous gases are scrubbed using the dilute urea solution. Heat is recovered by heating inflow air and cool exhaust gases are passed through demister and filters to remove fly ash and particulate matters. Cleaned gases are then passed to stack after diluting with fresh air conforming to the emission standards.

Similarly, many other studies have reported the efficient treatment of biomedical waste employing plasma gasification (Messerle et al., 2018). The reported results have shown this process promising for the effective treatment of potentially pathogenic structures such as microbes, viruses, prisons, and strains. Resultant gases are clean and nontoxic for the living environment and inorganic potions are converted to neutral slag. Hence, the plasma process can be used effectively for the treatment of biomedical wastes conforming the emission protocols.

11.6 INCINERATION

Incineration is a widely employed traditional method of thermal waste disposal. It involves the combustion of organics available in waste material. It works on the conversion of wastes into heat and flue gases without discharging any ash to the atmosphere without the engrossment of any additional processing.

In this process, the byproduct is mainly fuel and oxygen. The type of waste produced can be even greater and use effectively through appropriate treatment reactions analysis (Tang et al., 2013). Commonly, there are traces of fuel, nitrogen, carbon dioxide, and sulfur dioxide in incineration products. When incineration is used, the best conditions for the production of bio-oil are high heat rate, low shelf life, and temperature ~500 °C (Jeong et al., 2016). Relative humidity and inert materials tend to increase the heat loss; increasing temperature, size, and shape of the substance (Wareing et al., 2017). The energy can be used for direct heating, electricity generation employing a steam turbine generator.

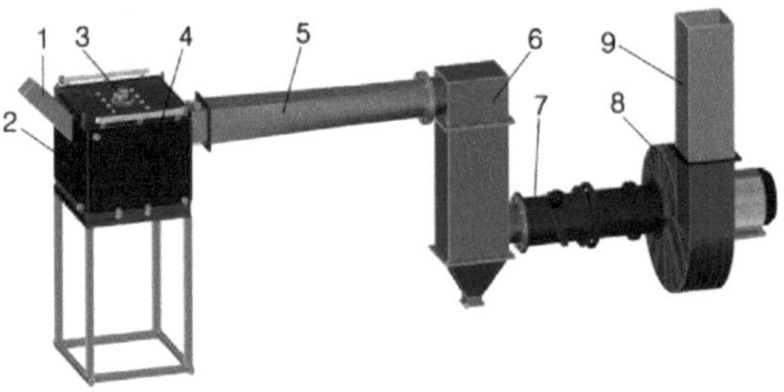

1 – Feed inlet
2 – Plasma reactor
3 – Electric-arc direct current plasmatron
4 – Gasification zone
5 – Heat exchanger
6 – Gas-cleaning unit with a bag filter
7 – Exhaust gas tube
8 – Exhaust fan
9 – Stack

FIGURE 11.10 Plasma gasification system for the treatment of biomedical wastes.

Source: Reprinted with permission from Messerle et al. (2018) © 2020 Elsevier.

Figure 11.11 describes the heat and energy recovery process for the incineration of wastes. The furnace is designed to completely mix the combustion air with the gas and steam from the combustion waste. However, incomplete combustion may occur near the walls of the furnace causing fly ash or other micro particles mixing in the flue gases due to the nonuniform mixing of the steams (Danthurebandara et al., 2015).

Incinerators use a variety of conventional methods to achieve accurate combustion and waste reduction (Lombardi et al., 2015). To achieve efficient combustion elusive control of supply ratio and the temperature is essential. Inefficient combustion leads to increased fly ash and other micro and macroparticles in exhaust gases. Optimal furnace design and operation require attention to combustion temperatures and turbulence (Lee and Ojovan, 2013). To achieve efficient combustion, part of the overall airflow must reach an adequate temperature for an extended duration and there must be sufficient fuel and oxygen to mix in proper proportion.

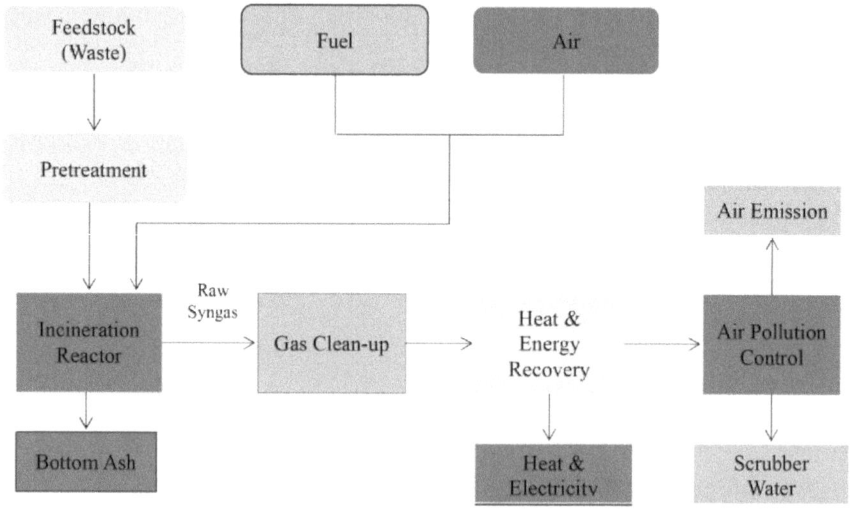

FIGURE 11.11 Process diagram of incineration-based waste treatment and integrated energy recovery.

The temperature obtained is the result of the heat generated by the oxidation process. Temperatures should be high to achieve complete combustion, but not high enough to damage the heater or release NO_x. The temperature is usually synchronized by regulating the feed rate; to ensure that the temperature is at the anticipated level, and then the conditions are created by changing the amount of excess air (Galeno et al., 2011).

The commotion is obligatory to ensure an adequate link between macromersions and the molecular level of gas and oxygen combustion. The availability of sufficient oxygen in the oven and thorough mixing of incoming steams is the guarantee of appropriate processing. Cold stains are sometimes observed near the oven wall due to inefficient operation. First, remove the heat from the combustion process. In a water-cooled fireplace, such cold stains on the wall are more than fire-resistant stoves (Prado et al., 2020a, 2020b).

Many new design features and treatments have been adopted that increase heat transformation, prolong life, and increase excitement when burning waste with increased combustion efficiency and some other benefits (e.g., improving ash quality) (Changming et al., 2018). In addition to the functions and methods discussed, they include efficient combustion systems, waste pretreatment methods (such as decomposition and mixing), and oxygen enrichment. Much attention has also been given to measuring and monitoring key process conditions to better manage the entire combustion process in order to further optimize the process (Prado et al., 2020a, 2020b).

11.6.1 TYPES OF INCINERATORS

A wide range of incinerators is available that can be used for the processing of wastes. These include very complex high-temperature processing plants to the most basic low-temperature combustion devices. These units should be carefully selected based on available sources, local conditions, and risk factors to achieve the health benefits of treatment and the increased risk of air or groundwater contamination (Zhao et al., 2010). Especially considering the treatment of medical waste, the following technologies are of paramount importance:

i. Two-chamber furnace for incineration of infectious health waste (Vorona et al., 2015).
ii. Single chamber stoves with static doors can only be used if they cannot withstand combustion.
iii. Rotary kilns operating at high temperatures can cause the decomposition of toxic genetic material and heat-resistant chemicals.
iv. Specially designed medical waste treatment furnaces range between 900°C and 1200°C (Vorona et al., 2015).

Many researchers are focusing on the efficient treatment of hazardous wastes especially, biomedical wastes. De Montfort University (UK) (Jeong et al., 2016) is developing a simple and economical high-temperature cooker, especially for low-income countries. Similarly, a portable oven for the treatment of hospital wastes in Brazil has been tested and approved at the pilot scale (Lombardi et al., 2015). The test results of the device with a capacity of 30 kg/h are satisfactory in terms of function, performance, and air pollution. This facility eliminates the hazards associated with the transportation of infectious medical wastes.

In many countries/regions, burning high-temperature chemical cement or pharmaceutical waste in industrial kilns on cement or steel is a valuable option that does not require additional investment to obtain a free supply of industrial fuel. Specific waste parameters should be assessed at the planning stage to determine the most appropriate type and amount of incineration:

The current state of production and medical waste:

a) anticipate future waste production;
b) eliminate the daily burning of waste.

All physical parameters determine whether the waste is suitable for combustion, such as low heat and low humidity.

i. Pyrolytic Incinerators

The safest and most commonly used treatment for medical waste is pyrolysis combustion, also known as controlled air combustion or double chamber combustion. Pyrolysis system having a processing capacity of 1–8 tons per day are mainly designed for continuous operation. They can also be adopted as fully automated, including moving waste in-house, removing ash, and burning waste (Czajczyńska et al., 2017).

As is often the case in hospitals, a limited number of properly maintained functional pyrolysis burners do not require exhaust gas treatment plants. Bottom ash can be disposed of in the landfills. However, it is recommended not to mix chlorinated plastic bags or other chlorinated compounds to prevent dioxin formation during the process (Surisetty et al., 2012).

Large fuel incinerators should ideally be located in industrial areas intended for hazard factories. Such areas have good road access, electricity, and water supply and are usually a long way from home. In any event, the burner must be at least 500 m away from any living area (Yazicioğlu and YaÇar Katircioğlu, 2017).

ii. Rotary Kilns

The rotary kiln can work continuously and is suitable for various loading devices. Ideally, waste prepared for the disposal of toxic waste should be segregated to special landfills and disposed of in industrial buildings or "parks" (Materazzi et al., 2016).

11.6.2 MUNICIPAL INCINERATORS

Infectious social medical waste is economically attractive, if the community has an acceptable distance from the sanitary kiln. As the caloric value of medical waste is many times higher than household waste, using a relatively small amount of medical waste will not affect the operation of municipal incinerators. City furnaces generally accept two-chamber designs (Tang et al., 2013).

There are several rules and recommendations for the disposal of medical waste in municipal establishments. Care should be taken in order to abide by the laws of the local government. Some general guidelines are listed below:

i. When transporting medical waste to the incinerator, care should be taken in its packaging, loading/unloading, and during the traffic flow.
ii. Healthy waste should not be disposed of in a cylindrical container as it may turn over the grill.
iii. The incinerator must be able to clean and disinfect vehicles, including vehicle management systems and personals.
iv. It is not recommended to store waste in ordinary containers. The waste bag may damage when the tap moves the waste container into the oven. Therefore, medical waste should be sent directly to the stove with careful shortest arrangements.
v. Using automatic devices for loading medical bags and containers instead of manual loading/unloading may ensure workers' safety.
vi. The storage time of medical waste in the incinerator should be no more than 24 h, and a longer storage period requires refrigeration equipment to avoid the development of harmful microbes.
vii. When burning medical waste, it must reach at least 97% (Wareing et al., 2017).
viii. Medical waste should only be included in the furnace under normal incineration conditions—never at the beginning or end of the combustion process.
ix. The process should be designed to prevent the pollution of medical waste from ash or sewage.
x. Extra care is highly recommended during any pandemic such as COVID-19.

The waste that does not need to be incinerated is equal to the waste listed in the pyrolysis burner

11.6.3 CHARACTERISTICS

On the one hand, some of the benefits of incinerators in the past have helped to manage MSW. In areas with limited land and high transportation costs, gas stations may be the best MSW Treatment and Energy Conversion Options.

Incineration may (1) reduce energy waste, (2) save lands wasted by landfills, (3) recover energy from municipal waste, (4) use solid waste, and (5) remove potential contaminants from waste.

Renewable energy from the incineration, thereby using renewable thermal energy to generate electricity and/or heat for industrial, commercial, or residential heating applications as hot water or steam. Conventional combustion technologies include large-scale combustion and incineration (Jeong et al., 2016). Incineration of waste can also help municipalities reduce the number of landfills. In turn, this helps reduce waste gas and groundwater pollution. In addition, the incineration of solid waste does not need any scientific pretreatment; hence has the energy and economical advantages (Yassin et al., 2009) (Table 11.1).

Along with many advantages, incineration possesses some disadvantages (Rutberg et al., 2011). The ash produced by the incinerator contains a large amount of organic matter and smoke polygon. Because the efficiency of steam production depends on the quality of the fuel, so less efficient. The combustion process can also cause some health problems.

11.7 PYROLYSIS

Pyrolysis is a process of rupture of molecules at moderate temperatures in the absence of air or oxygen. Compared with combustion, pyrolysis is less efficient at lower temperatures and emits fewer air pollutants (e.g., polybrominated diphenyl ethers). Besides, the range of pyrolysis units is more flexible than fuel. It can also use high-value chemicals to convert low-energy materials into high-density biofuels.

Pyrolysis can be defined as an intermediate technique between torrefaction and gasification (Matsakas et al., 2017). Mostly, pyrolysis yields liquids due to a high rate of heat transfer and relatively moderate process temperature. Mostly pyrolysis of waste materials is conducted from 300 °C to 650 °C that produces gases (condensable) and biochar. The composition of the product stream (also known as bio-oil) depends upon the composition of the feed stream and operating parameters. Bio-oil can be used directly as fuel after upgradation or as an intermediary for the production of downstream moieties. Similarly, biochar has extended industrial utilization including soil adjustment, carbon seizure, and energy production.

Considering the current energy and environment-sensitive scenario, much attention has been paid to improve the design and operation of pyrolysis especially focusing the solid WM. The pyrolysis process is classifieds as

TABLE 11.1 Comparative Characteristics of Thermal Waste Management Techniques (Kumar, 2016; Yan et al., 2016).

Components	Incineration		Gasification	Pyrolysis	Plasma
Objective	High-temperature flue gases		High calorific value gases	putrefaction of feed material	Gases with high energy density and inert solid-phase slag
Process parameters	800°C–1450°C and 1 atm.		500°C–1800°C and 1 –45 atm.	250°C–900°C and 1 atm.	1200°C–2000°C and 1 atm.
Process environment	Oxidizing air		Oxidizing O_2 and/or H_2O	Inert	Oxidizing O_2, H_2O Plasma gas, O_2, N_2, Ar
Stoichiometric ratio	>1		<1	NA	<1
Process outputs	**Gaseous**	CO_2, H_2O, N_2, and O_2	H_2, CO, CO_2, N_2, CH_4, and H_2O	Hydrocarbons, H_2, N_2, CO, and H_2O	H_2, CO, CO_2, N_2, CH_4, and H_2O
	Solids	Ash and slag	Ash and slag	Ash and coke	Ash and slag
	Liquids	No product	No product	Pyrolysis oil and water	No product

(i) fast pyrolysis and (ii) slow pyrolysis. Fast pyrolysis produces gas and bio-oil observing high heat rate and very low residence time (few seconds). Whereas, slow pyrolysis produces biochar observing high contact time and low heat rate (Basu, 2020). Ample attention to process efficiency leads to develop integrated conceptual designs. Flash pyrolysis with an exceptionally high heating rate has been applied effectively for the manufacture of syngas in a MSW treatment facility. microwave integrated vacuum conditioning pyrolysis process can produce a gaseous stream with high heating value with the added advantages of short residence time and uniform heat distribution (Yin, 2012).

The major advantage of this method is that it can use many types of raw materials, including industrial and household waste (Vorona et al., 2015). Nature of feedstock such as physical characteristics, conductivity, dielectric properties, and operating parameters (temperature, heating uniformity, and residence time) demonstrates the efficiency of pyrolysis. Much conventional and advanced design of pyrolysis reactors are being used considering the process requirements. Some of them are discussed below.

11.7.1 FIXED BED REACTOR

A fixed bed reactor is the most convenient and traditional reactor used from bench scale to an industrial level. In this type of reactor, the material is placed in the reactor and heated from the outside source. N_2 is mostly used to establish an inert environment. The electric oven is a highly used solution. Anyway, stationary reactors are a good way to provide experimental information about pyrolysis parameters and products (Jeong et al., 2016). However, these reactors are now being replaced with other types of reactors having higher efficiencies and operational flexibility.

11.7.2 FLUIDIZED BED REACTORS

Fluidized bed technology is mostly used for the thermal disposal of wastes and this technology has replaced fixed bed technology in almost every field. Fluidized beds are good with uniform heat distribution. The rate of heat and mass transfer is high in fluidized bed reactors as compared with others. However, considering the waste potential and volume frequency, this technique has some limitations. Firstly, the feed to the reactor must be small enough to float in the liquid; hence, shredding and drying are required. The required feed essential is in the size range of biomass ≤1 cm, plastic, and paper ≤5 mm. Therefore, the preparation of feed is an associated problem. Figure 11.12 demonstrates an array of steps intricate in the pyrolysis of waste encompassing a fluidized bed reactor and the structure is available in Figure 11.4C.

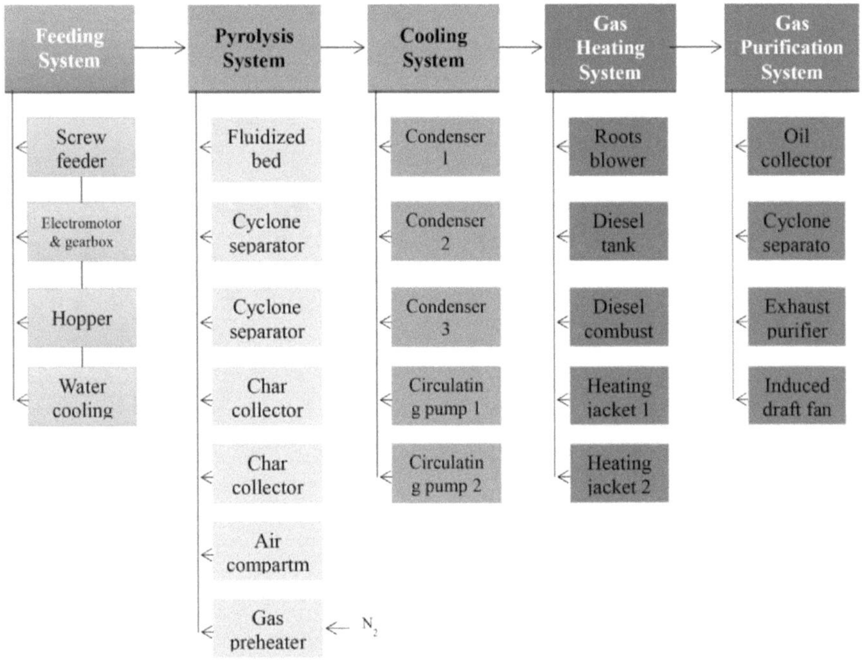

FIGURE 11.12 The layout scheme of the waste pyrolysis process.

11.7.3 DISCHARGE BED REACTOR

The discharge reactor bed has many operational advantages. It can be used for the processing of a robust feed size ranging from irregular shapes to fine particles, adhesive solid particles, and miscellaneous particles (Czajczyńska et al., 2017). Moreover, the system has many airflow functions and can operate in a short time. In addition, solids flow in the reactor causes a high rate of heat transfer between stages, making the reactor suitable for smoke pyrolysis. This reactor is equally suitable for continuous operation, which is important for large pyrolysis processes having a high feed rate. Due to its operational robustness, discharge bed reactors have been successfully used for the pyrolysis of polymeric wastes (polystyrene, polyethylene, polypropylene, and polyethylene terephthalate) and biomass.

With this type of reactor, the waste plastic dissolves when it enters the reactor. Due to the periodic movement, the waste forms a uniform covering around the sand. High heat transfer between phases and low flow of adhesive solids made of plastic can also be carried out. The shape of the breast layer and the effect of a nozzle reduce the formation of agglomerates. A cone reactor design with cone nozzles for plastic pyrolysis is shown in Figure 11.13.

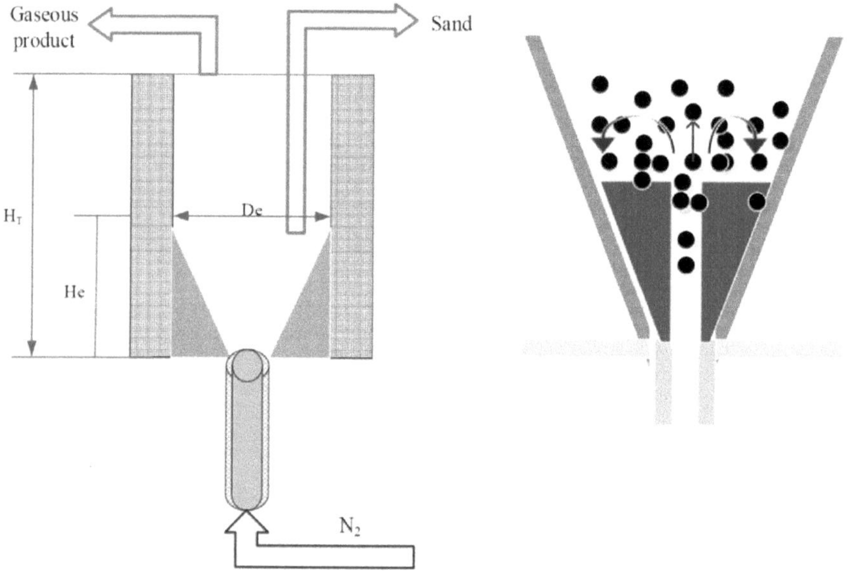

FIGURE 11.13 Conical spouted bed reactor for pyrolysis of polymeric materials.
Source: Reprinted with permission from Czajczyńska et al. (2017) © 2020 Elsevier.

11.7.4 ROTARY KILN REACTOR

The rotatory kilns have long been used for the handling of wastes, especially MSW. These are very significant as they can handle wastes of different size, shape, and caloric value with continuous loading. It provides improved thermal conductivity of raw materials than fixed bed reactors and heats up more easily than fluidized bed reactors. In the pyrolysis process, the feed length of the chamber is an important factor because it determines the energy at a given temperature. The handling capacity and exhaust gas composition can be controlled with feed rate and speed of rotation. The rotary kiln has various comparative advantages as compared to a fluidized bed and fixed bed. For instance, the gentle movement of the sloping chamber may mix the waste materials well to obtain a uniform pyrolysis product. Similarly, flexible shelf-life adjustment can induce the pyrolysis reaction under optimal conditions.

A schematic diagram has been provided in Figure 11.14 for the pyrolysis of waste employing rotary kiln. The collected waste is fed to the rotary kiln through the feeding system. Exit gases are passed to gas turbines after passing through a battery of cyclone separators. Cyclone separators are incorporated to remove any ash or other solid particles entrained. The off-gases from the turbine can be used to preheat feeding material for better energy efficiency. Part of these gases is forwarded to stack conforming the environmental standards.

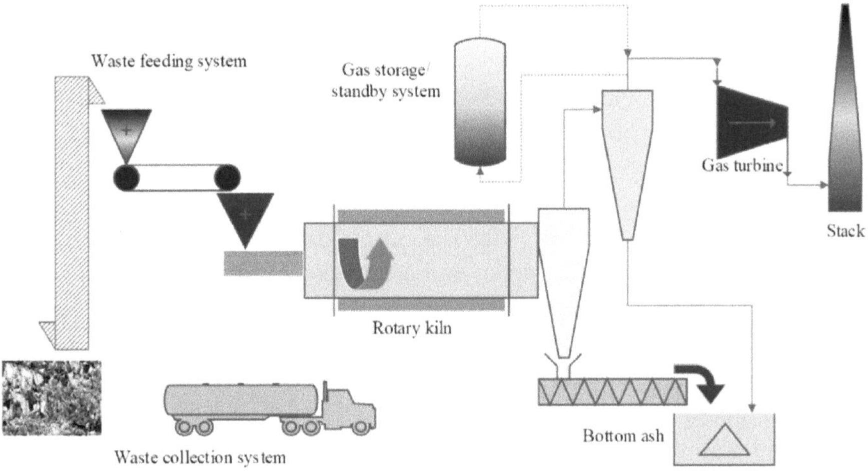

FIGURE 11.14 A rotary kiln reactor system is used for the pyrolysis of wastes (Czajczyńska et al., 2017).

Rotary kilns are used for the pyrolysis of industrial waste, as they adequately transmit temperature and reduce energy consumption. It is promising as it provides a mixture of modern and traditional applications of fixed bed.

11.7.5 MICROWAVE AUXILIARY REACTOR

The integration of microwave with pyrolysis is used widely due to associated operational benefits. Elimination of hot spots and uniform temperature distribution is the most important advantage in a short time. Similarly, handling of large particles, flexible feed composition, rapid internal heat, rapid startup, immediate reaction stop, high energy efficiency, and no requirement for mixing device are some additional merits. It is proposed that this solution ensures efficient pyrolysis of paper, biomass, as well as different types of plastics. However, the associated cost of microwave heating is a point of great concern for large-scale applications. Therefore, this technology has not been fully developed concerning commercial applications.

Pyrolysis helps to the use of all organic and inorganic CO_2 materials, unlike traditional biological waste treatment methods. Rapid pyrolysis is more complex, but it can be used. The product quality can be improved by adding inorganic waste materials (e.g., rubber, industrial textile, etc.). This can enhance the quality of the product as the gases contain less CO_2, lighter hydrocarbons, less water, and less oil. Both, organic and inorganic waste is produced in municipal wastes, so pyrolysis is a suitable technique to treat MSW. The contact time of MSW can be varied from a few seconds to 2 h and the heating rate from 10°C to 600°C per min. It has been proved that longer reaction time and high temperature lead to improve syngas production and a decrease in char production (Matsakas et al., 2017). Operating temperature also affects the biochar characterization altering the downstream utilization. If pyrolysis waste contains polyvinyl chloride, polyurethane, rubber, or electrical and electronic equipment, its waste should be carefully monitored, as the product may contain chlorides, H_2 sulfide, bromides, and some nitrogenous compounds.

Besides all these characteristics, it can be concluded that pyrolysis is still a promising alternative for MSW treatment. The emission of heavy metals, organics, and inorganics can be controlled with the provision of a gas scrubbing system.

11.8 SUMMARY

Population growth and improvement in living standards result in increasing waste generation exponentially. Therefore, waste disposal is a permanent challenge with increasing intensity. The research and development community is continuously struggling to handle this issue managerially as well as scientifically. Environmental concern and energy security compelled to draft a waste disposal solution having waste to energy potential. Considering these circumstances, many TWM techniques have been practiced with simultaneous energy recovery. Old waste disposal techniques such as combustion and landfilling are absolutes due to their environmental hazards. Pyrolysis, gasification, incineration, and plasma gasification have been used considering the waste composition, volume, and hazardous nature. For biomedical and radioactive waste, plasma gasification is the preferred solution. However, plasma gasification is an energy extensive process. Whereas, for municipal and industrial waste, gasification and pyrolysis are preferred. All these processes except combustion have the advantages that these can be coupled with an integrated electricity production system or with the chemical synthesis unit for the production of useful chemicals and fuels.

ACKNOWLEDGMENTS

We pay our sincere gratitude to the Department of Materials and Chemical Engineering, Yibin University, Yibin Sichuan, China for supporting us to write this manuscript.

KEYWORDS

- **gasification**
- **incineration**
- **plasma**
- **pyrolysis**
- **thermal waste management**

REFERENCES

Ahamad, M. I.; Zafar, Z.; Arsalan, M.; Rehman, A.; Sajid, M.; Zulqarnain, R. M.; Mehmood, M. S.; Abdal, S.; Aslam, M. Effects of Temperature and Pressure on Reservoir Fluids and Seismic Properties of Reservoir Rocks. *Int. J. Pharm. Sci. Rev. Res.* **2020a,** *63*, 36–43.

Ahamad, M. I.; Song, J.; Sun, H.; Wang, X.; Mehmood, M. S.; Sajid, M.; Si, P.; A. J. K. Contamination Level, Ecological Risk, and Source Identification of Heavy Metals in the Hyporheic Zone of the Weihe River, China. *Int. J. Environ. Res. Public Health* **2020b,** *17*, 1070.

Basu, P. *Biomass Gasification, Pyrolysis and Torrefaction: Practical Design and Theory*, 2nd ed.; Elsevier Inc.; London, 2020.

Begum, S.; Rasul, M. G.; Akbar, D.; Cork, D. An Experimental and Numerical Investigation of Fluidized Bed Gasification of Solid Waste. *Energies* **2014,** *7*, 43–61.

Cai, X.; Du, C. Thermal Plasma Treatment of Medical Waste. *Plasma Chem. Plasma Process* 2020.

Changming, D.; Chao, S.; Gong, X.; Ting, W.; Xiange, W. Plasma Methods for Metals Recovery from Metal–Containing Waste. *Waste Manage.* **2018,** *77*, 373–387.

Cheng, T. W.; Tu, C. C.; Ko, M. S.; Ueng, T. H. Production of Glass-Ceramics from Incinerator Ash Using Lab-Scale and Pilot-Scale Thermal Plasma Systems. *Ceram. Int.* **2011,** *37*, 2437–2444.

Czajczyńska, D.; Anguilano, L.; Ghazal, H.; Krzyżyńska, R.; Reynolds, A. J.; Spencer, N.; Jouhara, H. Potential of Pyrolysis Processes in the Waste Management Sector. *Therm. Sci. Eng. Prog.* **2017,** *3*, 171–197.

Danthurebandara, M.; Van Passel, S.; Vanderreydt, I.; Van Acker, K. Environmental and Economic Performance of Plasma Gasification in Enhanced Landfill Mining. *Waste Manage.* **2015,** *45*, 458–467.

Edo, M.; Budarin, V.; Aracil, I.; Persson, P.-E.; Jansson, S. The Combined Effect of Plastics and Food Waste Accelerates the Thermal Decomposition of Refuse-Derived Fuels and Fuel Blends. *Fuel* **2016,** *180*, 424–432.

Elwan, A.; Arief, Y. Z.; Muhamad, N. A.; Bashir, N.; Adzis, Z. Solid Waste as a Renewable Feedstock: A Review. *ARPN J. Eng. Appl. Sci.* **2014,** *9*, 1297–1310.

European Environment Agency, Managing Municipal Solid Waste—A Review of Achievements in 32 European Countries—Managing, **2013**.

Fabry, F.; Rehmet, C.; Rohani, V.; Fulcheri, L. Waste Gasification by Thermal Plasma: A review. *Waste Biomass Valorization* **2013,** *4*, 421–439.

Fichtner Consulting Engineers Ltd. The Viability of Advanced Thermal Treatment of MSW in the UK, ESTET; London, 2004.

Galeno, G.; Minutillo, M.; Perna, A. From Waste to Electricity through Integrated Plasma Gasification/Fuel Cell (IPGFC) System. *Int. J. Hydr. Energy* **2011**, *36*, 1692–1701.

Gehrmann, H. J.; Hiebel, M.; Simon, F. G. Methods for the Evaluation of Waste Treatment Processes. *Hindawi J. Eng.* **2017,** 3567865.

Ghiloufi, I. Simulation of Radioelement Volatility during the Vitrification of Radioactive Wastes by Arc Plasma. *J. Hazard. Mater.* **2009,** *163*, 136–142.

Gomez, E.; Rani, D. A.; Cheeseman, C. R.; Deegan, D.; Wise, M.; Boccaccini, A. R. Thermal Plasma Technology for the Treatment Of wastes: A Critical Review. *J. Hazard. Mater.* **2009,** *161*, 614–626.

Heberlein, J.; Murphy, A. B. Thermal Plasma Waste Treatment. *J. Phys. D. Appl. Phys.* **2008,** *41*, 053001.

Jan, D. Treatment of Low-Level Radioactive Waste by Plasma: A Proven Technology? *Proc. ASME 13th Int. Conf. Radioact. Waste Manage. Environ. Remediat.* **2010**.

Jeong, J.; Baik, M. H.; Kang, M. J.; Ahn, H. J.; Hwang, D. S.; Hong, D. S.; Jeong, Y. H.; Kim, K. Radiological Safety Assessment of Transporting Radioactive Wastes to the Gyeongju Disposal Facility in Korea. *Nucl. Eng. Technol.* **2016,** *48*, 1368–1375.

Kopiersperre - Own work, C. B.-S. Gasification, 2018. https://commons. wikimedia. org/w/ index. php?curid=37292688 (accessed Oct 10, 2020).

Kumar, S. *Municipal Solid Waste Management in Developing Countries, Municipal Solid Waste Management in Developing Countries*; CRC Press, Taylor & Francis, 2016; pp 35–55.
Lee, W. E.; Ojovan, M. I. Fundamentals of Radioactive Waste (RAW): Science, Sources, Classification and Management Strategies; Chapter 1. In *Radioactive Waste Management and Contaminated Site Clean-Up: Processes, Technologies and International Experience*; Elsevier Inc, 2013; pp 3–49.
Li, J.; Liu, K.; Yan, S.; Li, Y.; Han, D. Application of Thermal Plasma Technology for the Treatment of Solid Wastes in China: An Overview. *Waste Manage.* **2016,** *58*, 260–269.
Lombardi, L.; Carnevale, E.; Corti, A. A Review of Technologies and Performances of Thermal Treatment Systems for Energy Recovery from Waste. *Waste Manage.* **2015,** *37*, 26–44.
Luz, F. C.; Rocha, M. H.; Lora, E. E. S.; Venturini, O. J.; Andrade, R. V.; Leme, M. M. V.; Del Olmo, O. A. Techno-Economic Analysis of Municipal Solid Waste Gasification for Electricity Generation in Brazil. *Energy Convers. Manag.* **2015,** *103*, 321–337.
Materazzi, M.; Lettieri, P.; Taylor, R.; Chapman, C. Performance Analysis of RDF Gasification in a Two Stage Fluidized Bed-Plasma Process. *Waste Manage.* **2016,** *47*, 256–266.
Matsakas, L.; Gao, Q.; Jansson, S.; Rova, U.; Christakopoulos, P. Green Conversion of Municipal Solid Wastes Into Fuels and Chemicals. *Electron. J. Biotechnol.* **2017,** *26*, 69–83.
Messerle, V. E.; Mosse, A. L.; Ustimenko, A. B. Processing of Biomedical Waste in Plasma Gasifier. *Waste Manage.* **2018,** *79*, 791–799.
Morrin, S.; Lettieri, P.; Chapman, C.; Mazzei, L. Two Stage Fluid Bed-Plasma Gasification Process for Solid Waste Valorisation: Technical Review and Preliminary Thermodynamic Modelling of Sulphur Emissions. *Waste Manage.* **2012,** *32*, 676–684.
Moustakas, K.; Loizidou, M. Solid Waste Management through the Application of Thermal Methods. In *Waste Management*; InTech, 2010; pp 90–123.
Sajid, M.; Ahmad, M. I.; Shafqat, S. S.; Pasha, M. K. De-Colorization of Azo Dyes: C. B-10X Blue and C. B. M-10X Blue by Ozonation for Waste Water Treatment. *Univ. Wah J. Sci. Technol.* **2020a,** *4*, 1–7.
Sajid, M.; Ahmed, M. I.; Shafqat, S. S.; Shahi Mulk, M. K. P. Study of Phosphorous Oxide (P2O5) and Iron Oxide (Fe_2O_3) in Phosphorite of Hazara Basin, Pakistan. *Int. J. Agrochem.* **2020b,** *6*, 46–51.
Nishikawa, H.; Ibe, M.; Tanaka, M.; Takemoto, T.; Ushio, M. Effect of DC Steam Plasma on Gasifying Carbonized Waste. *Vacuum* **2006,** *80*, 1311–1315.
Nzihou, A. Toward the Valorization of Waste and Biomass. *Waste Biomass Valorization* **2010,** *1*, 3–7.
Organization, W. H. *Safe Management of Wastes from Health-Care Activities*; World Health Organization: Geneva, 1999; pp 20–22.
Ouda, O. K. M.; Raza, S. A.; Nizami, A. S.; Rehan, M.; Al-Waked, R.; Korres, N. E. Waste to Energy Potential: A Case Study of Saudi Arabia. *Renew. Sustain. Energy Rev.* **2016,** *61*, 328–340.
Pfaltzgraff, L. A.; De Bruyn, M.; Cooper, E. C.; Budarin, V.; Clark, J. H.; Green Chemistry Food Waste Biomass: A Resource for High-Value Chemicals. *Green Chem.* **2013,** *15*, 307–314.
Prado, E. S.; Dellamano, J.; Carneiro, A. L. G.; Santos, R. C.; Petraconi, G.; Potiens Jr, A. J. Technical Feasibility Study on Volumetric Reduction of Radioactive Wastes Using Plasma Technology. In *2017 International Nuclear Atlantic Conference—Associação Brasileira De Energia Nuclear—Aben*, Belo Horizonte, MG Brazil, 2017.

Prado, Eduardo S. P.; Miranda, F. S.; de Araujo, L. G.; Petraconi, G.; Baldan, M. R. Thermal Plasma Technology for Radioactive Waste Treatment: A Review. *J. Radioanal. Nucl. Chem.* **2020,** *325* (2), 331–342.

Prado, E. S. P.; Miranda, F. S.; Petraconi, G.; Potiens, A. J. Use of Plasma Reactor to Viabilise the Volumetric Reduction of Radioactive Wastes. *Radiat. Phys. Chem.* **2020,** *168*, 108625.

Rutberg, P. G.; Bratsev, A. N.; Kuznetsov, V. A.; Popov, V. E.; Ufimtsev, A. A.; Shtengel', S. V. On Efficiency of Plasma Gasification of Wood Residues. *Biomass Bioenergy* **2011,** *35*, 495–504.

Rutberg, P. G.; Bratsev, A. N.; Safronov, A. A.; Surov, A. V.; Schegolev, V. V. The Technology and Execution of Plasmachemical Disinfection of Hazardous Medical Waste. *IEEE Trans. Plasma Sci.* **2002,** *30*, 1445–1448.

Rutberg, P. G.; Kuznetsov, V. A.; Serba, E. O.; Popov, S. D.; Surov, A. V.; Nakonechny, G. V.; Nikonov, A. V. Novel Three-Phase Steam-Air Plasma Torch for Gasification of High-Caloric Waste. *Appl. Energy* **2013,** *108*, 505–514.

Saghir, M.; Rehan, M.; Nizami, A.-S. Recent Trends in Gasification Based Waste-to-Energy; Chapter 6. In *Gasification for Low-Grade Feedstock*; Yun, Y.; Ed.; InTech., 2018; pp 97–113.

Shahabuddin, M.; Alam, M. T.; Krishna, B. B.; Bhaskar, T.; Perkins, G. A Review on the Production of Renewable Aviation Fuels from the Gasification of Biomass and Residual Wastes. *Bioresour. Technol.* **2020,** *312*, 123596.

Sharholy, M.; Ahmad, K.; Mahmood, G.; Trivedi, R. C. Municipal Solid Waste Management in Indian Cities—A Review. *Waste Manage.* **2008,** *28*, 459–467.

Sierra Energy—What Is FastOx Gasification? | All Sustainable Solutions, 2020. https://allsustainablesolutions. com/fastox-gasification-by-sierra-energy (accessed Oct 24, 2020).

Surisetty, V. R.; Kozinski, J.; Dalai, A. K. Biomass, Availability in Canada, and Gasification: An Overview. *Biomass Convers. Biorefinery* **2012,** *2*, 73–85.

Tang, L.; Huang, H.; Hao, H.; Zhao, K. Development of Plasma Pyrolysis/Gasification Systems for Energy Efficient and Environmentally Sound Waste Disposal. *J. Electrostat.* **2013,** *71* (5), 839–847.

Vallero, D. A. *Thermal Waste Treatment, Waste*; Elsevier Inc., 2011; pp 219–231.

Vorona, N. A.; Gavrikov, A. V.; Samokhin, A. A.; Smirnov, V. P.; Khomyakov, Y. S. On the Possibility of Reprocessing Spent Nuclear Fuel and Radioactive Waste by Plasma Methods. *Phys. At. Nucl.* **2015,** *78*, 1624–1630.

Wareing, A.; Abrahamsen-Mills, L.; Fowler, L.; Grave, M.; Jarvis, R.; Metcalfe, M.; Norris, S.; Banford, A. W. Development of Integrated Waste Management Options for Irradiated Graphite. *Nucl. Eng. Technol.* **2017,** *49*, 1010–1018.

Yan, J.; Huang, Q.; Lu, S.; Li, X.; Chi, Y. Thermal Treatment Techniques: Incineration, Gasification, and Pyrolysis. *Sustain. Solid Waste Manage. ASCE* **2016,** 149–186.

Yassin, L.; Lettieri, P.; Simons, S. J. R.; Germanà, A. Techno-Economic Performance of Energy-from-Waste Fluidized Bed Combustion and Gasification Processes in the UK Context. *Chem. Eng. J.* **2009,** *146*, 315–327.

Yazicioğlu, Ö.; YaÇar Katircioğlu, T. Application of Plasma Technology in Energy Sector. *Kirklareli Univ. J. Eng. Sci.* **2017,** *3*, 18–44.

Yin, C. Microwave-Assisted Pyrolysis of Biomass for Liquid Biofuels Production. *Bioresour. Technol.* **2012,** *120*, 273–284.

Zhao, Y.; Sun, S.; Zhou, H.; Sun, R.; Tian, H.; Luan, J.; Qian, J. Experimental Study on Sawdust Air Gasification in an Entrained-Flow Reactor. *Fuel Process. Technol.* **2010,** *91*, 910–914.

PART III

Modern Techniques for Waste Management

CHAPTER 12

Vermicomposting: A Sustainable and Environment-Friendly Approach for Organic Waste Management

ZUBAIR ASLAM[1*], SAFDAR BASHIR[2], KORKMAZ BELLITURK[3], SAMI-UR-REHMAN[4], LIXIN ZHANG[4], QAMAR UZ ZAMAN[5], and ALI AHMAD[1]

[1]*Department of Agronomy, University of Agriculture, Agriculture University Road, Faisalabad, Pakistan*

[2]*Institute of Soil and Environmental Sciences, University of Agriculture, Agriculture University Road, Faisalabad, Pakistan*

[3]*Department of Soil Science and Plant Nutrition, Namık Kemal Üniversitesi, Kampüs Cad No:1, 59030 Süleymanpaşa/Tekirdağ, Turkey*

[4]*College of Life Science, Northwest A&F University, 3 Taicheng Rd, Yangling District, Xianyang, Shaanxi, China*

[5]*Department of Environmental Sciences, University of Lahore, 1- Km Raiwind Rd, Sultan Town, Lahore, Punjab, Pakistan*

[*]*Corresponding author. E-mail: zauaf@hotmail.com*

ABSTRACT

Variations in intensive agricultural management practices, lifestyle, urbanization, and population increase have recently enhanced the synthesis of organic wastes in ecosystem. These organic wastes have a detrimentally negative impact on the ecosystem. The essential nutrients attained from organic wastes, for instance, agriculture, kitchen, and municipal wastes

Waste Problems and Management in Developing Countries. Umair Riaz, Shazia Iqbal, & Moazzam Jamil (Eds.)

are pivotal part of their management and for reducing declination of the ecosystem. Innovative vermicomposting technology is an encouraging tool for recycling of organic wastes and is an environmentally sustainable technology, ethically acknowledgeable, and financially accessible around the world. Earthworms' species like *Eisenia fetida* are associated with the synthesis of vermicompost and organic waste recycling technologies. Vermicompost is a procedure to which earthworms are used to transform different plants and animals waste into organic fertilizer for agricultural use. Vermicompost is nutritionally rich with essential amount of both minor and major nutrients such as Zn, Mn, Mg, Ca, B, N, K, P, etc., beneficiary soil-dwelling microbes including (phosphate-solubilizing and nitrogen-fixing bacteria) and growth-regulating hormones, namely, cytokinins, gibberellins and accordingly can be utilized as organic fertilizers for appropriate plant growth. The utilization of vermicompost as an organic fertilizer has completely abandoned the recommendation of chemical fertilizers. Therefore, vermicompost technology is considered as an innovative tool like two birds can be killed by one stone as suitable management of organic waste and at the same time creation of organic fertilizer. The management of waste is undertaken to recycle the organic wastes to diminish the ill effects of wastes on health, environment, and aesthetics. Vermicomposting is verified viable technique aimed at renovating diverse biological waste to valuable bio-fertilizer. The sustainability of agriculture significantly depends on the application of vermicompost.

12.1 INTRODUCTION

An increased population has increased the production of solid wastes especially organic wastes which require extra labor and financial resources to reduce their detrimental (Yadav and Garg, 2011). The sustainability of solid waste management practices is mandatory to regulate ecosystem clean and secure (Singh et al., 2011).

The condition of solid waste generation has deteriorated throughout the world. It was conclusively noticed that 1,800,000 tons of solid waste will predictively be produced till 2025 annually (Visvanathan and Trankler, 2003). During the annual of 2009, approximately 243 tons of urban solid waste generation unexpectedly recorded in the USA. Regarding various research analysis, 23 under developing countries generate averagely 770 g/person/day solid waste (Troschinetz and Mihelcic, 2009). It is recently estimated to enhance 3 billion tons of global solid waste generation up to

2025 annually (Charles et al., 2009). Large quantities of solid wastes have not been completely exploited due to economic and technical problems. The developing countries have no accessible technology for financial recycling whereas numerous innovative strategies have been executed for the management of solid waste recycling and generation, consisting of source depletion, material recovery, waste-to-energy transformation, curbside recycling, incineration, composting, and landfill dumping (Chang and Davila, 2008). Some of these disposal methods may create consequential climatic problems. Regarding numerous information of (Mor et al., 2006; Tsai et al., 2007; Cadena et al., 2009), it was observed that solid waste placed in exposing dumps or landfill-induced contamination of groundwater owing to leachate process of inorganic and organic matter in solid waste. Dumping of the landfill has also an adverse effect on greenhouse (Lee et al., 2009). Furthermore, treatment relating to incineration of solid waste is limited due to its less-caloric value and the price of fuel advancements (Tsai et al., 2007). The sewage sludge is directly dumped on agricultural land on account of its elevated phosphorous and nitrogen concentration utilized as organic fertilizer; nevertheless, it may induce toxicity for plants and have oppressive impacts on the soil-microbes metabolism (Tsai et al., 2007). In these circumstances, vermicompost technology is economically and ecologically sound. It also becomes eco-friendly for processing organic wastes in organic-enriched farmyard manure (FYM). Vermicompost technology is associated with the degradation of organic detritions of waste material in an environmentally sound way that could be easily reserved, handled, and incorporated on agricultural lands except for detrimental effects (Singh et al., 2011; Khwairakpam and Bhargava, 2009; Aira et al., 2002).

The mutual functionality of earthworm and microbes in eco-friendly conditions generates vermicompost, therefore, to attain nonthermophilic decaying and stability of solid waste organic fraction (Deka et al., 2011; Edwards and Lofty, 1977; Pramanik, 2010; Pramanik et al., 2009; Ronald and Donald, 1977). Generally, earthworm converts organic waste into nutrient-enriched vermicompost through the biotechnological process (Benıtez et al., 2008). The presence of microbes and earthworms in the soil are associated with biochemical breakdown of organic wastes, alternation of biological activity in soil, and conditioning of substrate are regulated by earthworms. This is an inexpensive technology for the processing and conversion into useful vermicompost generation by using earthworms (Hand et al., 1988). Distinct kinds of earthworm dwell in soil with distinct kinds of nature and nutrition mode owing to which the proportion of decomposition is influenced (Khwairakpam and Bhargava, 2009).

Vermicompost process will be majorly recommended in the situation of municipal solid waste (MSW) on account of its hygienic and secure decomposition of major quantity of organic matter in solid waste (Singh et al., 2011). Vermicompost is an important slow-release organic fertilizer for the protection of soils. In contrast, vermiculture focuses on the worms themselves and promotes the production of worm biomass may be for sale as feed for animals or sale to new vermicomposting operations. Vermi-technology is a promising technique for organic waste recycling.

The major intention of this knowledge is to deliver a review of the potentiality of vermicomposting in dealing with soil waste management in an eco-friendly manner. This chapter explores the potentiality of vermicomposting and integrated composting approaches and optimization of vermicomposting processing through co-metabolism of organic substrates. Moreover, these associated factors have been mentioned here that adversely affect the vermicomposting process. The potentiality of vermicompost technology offers the chances of managing the organic excreta on first hand and generation of energy to the second hand.

Vermi-fertilizer offers numerous benefits like boosting the water holding capacity and soil fertility; thus increasing the soil biodiversity, agricultural productivity, decreasing ecological, and environmental risks.

12.2 WASTE PROBLEMS AND THEIR MANAGEMENT IN DEVELOPING AND DEVELOPED COUNTRIES

The solid waste creation is an inevitable result of anthropological activity that affects the surrounding environment and human health. MSW is all the semisolid or solid materials disposed of by the domestic and business community apart from harmful waste material. Currently, world generation of municipal solid waste is around 1,300,000 tons per year and is predictable to escalate to the 2,200,000 tons per year in 2025 (Hoornweg and Bhada-Tata, 2012). The production rate of MSW depends on population size, industrialization, per capita income, climate, and public health. The developing countries are facing the challenges of increased solid waste generation due to immigration of people from countryside to urban areas and increasing urban population rates (Bongaarts, 2009). The MSW includes paper, vegetable waste, textile, rubber, and plastic wastes. Solid wastes have serious environmental impacts as aerobic decomposition of organic wastes results in methane generation and its impacts on climate change along with affecting the water quality. Open dumping of organic wastes pollutes the adjacent water bodies with organic pollutants. This results in a disease outbreak by people exposed to harmful products. The total amount and generation rate of the municipal solid waste in the advanced and evolving areas are expressed below (Table 12.1).

TABLE 12.1 Total Quantity and Production Rate of the Municipal Solid Waste in Developing and Developed Countries of the World.

Country	Waste generation rate (kg/capita/day)	Total quantity of waste (million tons/year)	References
		Developing country	
Bangladesh	0.57	5.20	Shams et al. (2017)
China	0.72	215.21	NBS (2019)
India	0.45	62.00	Sharma and Jain (2019)
Indonesia	0.34	68.39	Gunaruwan and Gunasekara (2016), Raharjo et al. (2018)
Malaysia	1.12	1.21	Yiing Chiee and Latifah (2017), Hassan (2002)
Nepal	0.35	0.52	Yiing Chiee and Latifah (2017), Khajuria et al. (2010)
Pakistan	0.60	32.30	Korai et al. (2017)
Sri Lanka	0.59	4.50	Yiing Chiee and Latifah (2017), Asian Development Bank (2013)
		Developed country	
Australia	2.23	67.00	Yildiz et al. (2013), Palanivel and Sulaiman (2014)
Germany	2.11	47.70	Palanivel and Sulaiman (2014), Shekdar (2009)
Japan	0.96	43.00	Helmy et al. (2006), Staley and Barlaz (2009)
South Korea	1.00	17.80	Mühle et al. (2010), Defra (2010)
USA	2.01	267.80	Helmy et al. (2006), Hai and Ali (2005)

12.3 FORMATION OF MUNICIPAL SOLID WASTES

The formation of MSW is important to evaluate the possible environmental effect on society and nature. The production and the formation of solid wastes is vital information for the decision-maker of any specific region.

The physical composition of waste is affected by the economic growth of the country, population, food habit, cultural habit, and the effect of globalization. The characterization of waste is helpful to determine the equipment requirements, management system, and plans. The waste disposal and energy recovery system depend on chemical characteristics of waste (Siami et al., 2019). The percentage composition of MSW of different countries is shown in Table 12.2.

TABLE 12.2 Percentage Formation of MSW of Developed and Developing Countries.

Country	Municipal solid waste formation						
	Organic wastes	**Paper and card waste**	**Textile waste**	**Plastic wastes**	**Metals wastes**	**Glass wastes**	**References**
			Developing country				
Bangladesh	76.0	4.11	1.9	4.6	1.1	0.8	Hoornweg and Bhada-Tata (2012)
India	40.0	10.0	3.6	2.0	1.0	0.2	Bourtsalas et al. (2019)
Indonesia	35.5	15.2	2.6	17.8	0.8	1.5	Ryu and Shin (2013)
Malaysia	40.0	15.0	23.0	15.0	3.0	4.0	Niyati (2015)
Oman	33.6	25.2	5.2	11.8	2.4	8.0	USEPA (2018)
Pakistan	56.3	2.1	9.2	0.6	0.1	0.6	Govet. of Australia (2018)
			Developed country				
Germany	30.0	24.0	22.0	13.0	1.0	10.0	Azam et al. (2019)
Turkey	53.7	16.7	2.2	2.1	0.7	2.9	Yan et al. (2020)
United Kingdom	41.0	18.0	19.0	7.0	8.0	7.0	Halder et al. (2014)
USA	17.7	23.8	4.2	15.8	1.4	4.9	Dasgupta et al. (2013)

12.4 MANAGEMENT OF ORGANIC WASTES

Different methods of waste management are used across the globe. In Australia, the waste disposal method is landfill, incineration, and landfill in the USA whereas recycling and incineration methods of waste management are used in Japan (Belliturk et al., 2020). Open landfill sites threaten human health by attracting disease vectors. Incineration of waste material releases different types of pollutants such as furan and dioxins (Tchobanoglous and

Kreith, 2002; White et al., 1999). Vermicomposting is a viable choice to tackle solid waste in an environmentally benign method. It prevents methane emissions and reduces environmental pollution. Vermicomposting is a biotechnology process of stabilizing agriculture, industrial, domestic, and solid wastes by the interaction of earthworms and microbes and transforms to manure (Figure 12.1). It is an advanced era of earthworm biotechnology, propagation, breeding, and the use of earthworms casting has become a best way of waste recycling. The final product of vermicomposting of municipal solid wastes is odorless, economic, and contains high nutrient contents, and free from pathogenic microbes with low greenhouse gas emission (Singh et al., 2011).

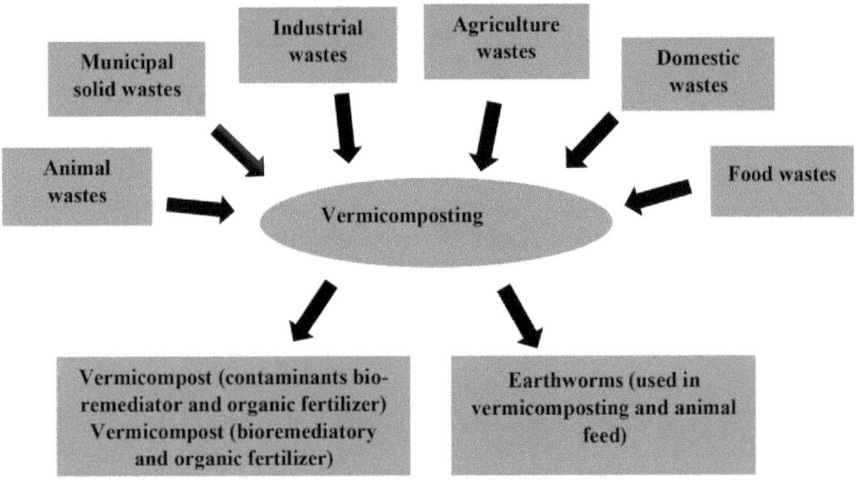

FIGURE 12.1 Vermicomposting of different wastes and their uses.

12.5 EARTHWORMS

Earthworms are ecological engineers which improve the soil properties contaminated with residues of pesticides and heavy metals (Sˇrut et al., 2019; Sanchez-Hernandez, 2019). They contribute to the renovation of contaminated soil to ameliorate health of soil and act as an indicator of soil. Vermiculture (farming of earthworms) is another bioprocess for organic waste transformation to organic manure (Anand and Sinha, 2020). Almost 3320 species are found across the globe. However, all species of earthworm cannot be used in vermicomposting as different earthworms feed on different types of wastes. Only 8–10 earthworms' species are considered suitable for vermicomposting. Earthworms are generally categorized into three main groups, that is, anecic, endogeic, and epigeic based on their behavior (Sherman, 2003). Anecic earthworms are brown and live in permanent vertical burrows. They cast on the surface of their burrows. Their life cycle is long; therefore, it requires long time to produce vermicompost. *Lumbricus friend* and *Lumbricus terrestris* are anecic earthworms. Endogeic earthworms are pink gray and live in horizontal burrows and feed there. So, it is also called deep feeders. *Octolasion rosea* and *Allolobophora chlorotica* are endogeic earthworms. Epigeic earthworms are known as surface feeders that feed on surface leaf litter and are red. Epigeic earthworms such as *Eudrilus euginiae, Perionyxhawayana,* and *Eisenia fetida* are helpful in stabilization of wastes. Usually, *E. euginiae* is used in tropical and subtropical areas whereas *E. fetida* is extensively used around the globe (Kumar, 2005).

12.5.1 LIFE CYCLE OF EARTHWORMS (E. FETIDA)

E. fetida is a dark brown earthworm with yellow color on the tip of the tail. The adult earthworms are 0.3–0.5 and 5–7 cm long in diameter along with 0.5–0.6 g biomass. Earthworms at adulthood progresses ring-shaped clitellum that secret a dense mucus to form a cocoon, these are hermaphrodites. The cocoon hatching rate is 0.35/worm/day. The newly formed cocoon is light in color which changed to light brown before the hatching. The cocoons of *E. fetida* range between 3.2 and 4.0 mm in length and about 2.0–2.7 mm in diameter and biomass of 12.65 mg [93]. The number of hatching per cocoon range between 3 and 5 and the hatching success rate of *E. fetida* ranges from 70 to 86%. During the first 3–4 weeks of hatching, earthworms grow slowly and then their growth rate increased speedily. In *E. fetida,* the clitellum developed after third weeks of hatching (Ali and Kashem, 2018). The average standard time for *E. fetida* to reach sexual maturity is about 40–60 days and cocoons production starts after 5–7 days of mating. Each cocoon produced three to five hatchlings after the incubation period of 14–21 days (Figure 12.2).

12.5.2 REPRODUCTION AND GROWTH OF EARTHWORMS IN ORGANIC WASTES

The earthworms boost the decomposition rate during the vermistabilization of waste materials. They can ingest various kinds of waste material such as cow dung, sewage sludge, crop residues, oil palm waste, and other agro-industrial wastes. The type and quality of waste material is a basic condition that influences the earthworm's reproduction and growth. The accessibility of food material at early stages enhanced the biomass of the worms. The type and biochemical quality of food material affect earthworm's reproduction and biomass. In vermicomposting process, microbial population and organic waste decomposition are important parameters to determine the earthworm's growth and reproduction. The optimum stocking density of 27–53 earthworms/kg food material is effective for the vermicomposting process (Niyati, 2015). The cocoon production is more in cattle dung and is lowest in horse dung due to its low nutritional status (Ryu and Shin, 2013). Earthworm's reproduction and biomass boost with reducing the C:N ratio in food material. The waste material with high nutrients content enhances the earthworm's growth and reproduction.

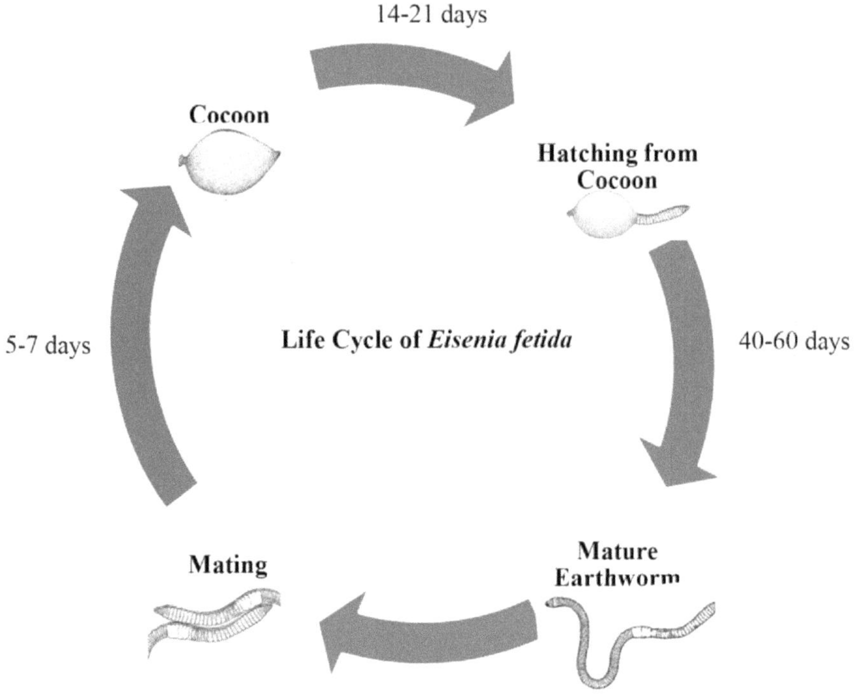

FIGURE 12.2 Life cycle of earthworm (*E. fetida*).

The different feed materials such as cow dung with press mud caused a considerable enhance in weight of body, production of cocoon, and creation of hatchlings. Higher feed mixtures reduced the degradation rate and affected the worms by declining and delaying the cocoon production. The feed material that provides the worms with an adequate quantity of nonassimilated carbohydrates and organic matter boosts the earthworm's reproduction and growth. In addition to the nature of waste material, breakdown and microbial biomass during the process of vermicompost are significant in determination of cocoon and biomass production. The sperms of earthworms are damaged due to heavy metals present in the substrates and lower the fertility and viability of cocoons. Earthworms eat on plant and animal material and use a little portion for their growth and evacuate big amounts of decomposed material in the form of vermicast. This procedure includes biochemical and physical actions. The grinding and mixing contain physical activity, while the microbial decomposition in the earthworm's intestine contains biochemical process. In fast-growing vermi-beds, feeding is essential at every 3–5 days with an optimum rate of feeding, 0.75 kg feed/kg worms/day. Overfeeding is prevented as it leads to higher fermentation in bed and causes the earthworms to death. It also attracts more mites to the vermicomposting system which competes with earthworms. Aerobically digested organic wastes are crucial for growth of earthworms. So, the predigestion of cow dung can be important for growth as it can contain toxic components.

12.5.3 EARTHWORM'S ROLE IN VERMICOMPOSTING

The earthworms play a pivotal part in vermicomposting of discarded material by consumption, digestion, and assimilation of wastes. Earthworms are ravenous feeders that consume diverse organic material in bulk amounts. The population of human with 1000 kg weight can eat food 20 kg/day while the earthworm can eat 500 kg of eating material with the same biomass (Bhawalkar, 1993). Earthworm acts as a grinder, chemical degrader, and biostimulator of organic wastes (Singleton, et al., 2003). It hosts millions of microbes (Nagavallemma et al., 2004), hormones, and hydrolytic enzymes that enhance the decomposition of organic wastes into vermicompost within 30–60 days (Johnson et al., 2002) in contrast with conventional composting that is completed in about 5 months (Munnoli et al., 2010). The mechanism of vermicomposting is completed in the following phases; the organic wastes ingested by earthworm are softened by its saliva. More softening and breaking down of ingested food are done by gizzard resulted in the reduction of food particle size less than 2 μ which increases the surface area for microbial priming. Earthworm's gut transit time plays a key function in the fate of ingested food and it depends on species of earthworm, food quality, and temperature (Hendriksen, 1991). Gut transit time for *E. fetida* and *Allolobophora rosea* is 1–3 h, whereas for *Eudrilus eugeniae* is 6 h and for *L. rubellus* is more than 8 h (Ba, 1989). Number, growth, and activation of bacteria increased from interior to posterior region during gut transit (Scho ̈nholzer et al., 1999).

The digestion in earthworm is a complex process including the mutualistic relationship between earthworms and microorganisms (Curry and Schmidt, 2007). The earthworm activates the microbes in biomass by providing optimum conditions, that is, water concentration, mucus, and aeration. Generally, the microorganisms are in resting state and get activated when ingested in the gut by priming effect (Brown et al., 2004). Furthermore, microbes linked with intestine assist the conversion of complex biomolecules into simple ones. The different enzymes in earthworm gut also significantly alter the vermicomposting. The activity of enzymes in earthworm is influenced by ingested substrate, age, physiological function, and microorganism activity. The finally grounded material mixes with many enzymes, for example, chitinase, lipase, cellulase, protease, and amylase secreted in the small intestine and lumen (Dominguez and Edwards, 2010). The consumed material is absorbed only 5–10% in the form of earthworm tissue for growth and the remainder is removed as vermicompost.

12.6 SUBSTRATES USED IN VERMICOMPOSTING

Vermicomposting is fascinating and ecofriendly technology of managing wastes even in poor socioeconomic conditions. Therefore, several kinds of organic wastes have been managed through vermicomposting (Table 12.3) and nutrient-rich biofertilizer is also produced. The quality of obtained biofertilizer largely depends on the composition of substrate mixture. So, the use of the components that are comprised of toxic pollutants should be reduced in vermicomposting. For example, sewage sludge is highly contaminated with different chemicals such as polychlorinated biphenyl, hydrocarbons, pharmaceutical, personal care products, and other heavy metals and pathogens (Fijalkowski et al., 2017). In the same instance, sewage and sludge is a potential source of nutrients and organic matter, making it best growth medium for the earthworms. The appropriate composition of waste material assures the dilution of toxic components, which is very important for the health of earthworms (Rorat et al., 2013). This permits the mixing of physicochemical factors to attain quality products in a brief period. The organic component of MSW, sawdust, green wastes, and cattle manure are different types of co-substrate used in vermicomposting. Some industrial sludges such as paper and pulp industry, food industry, tannery, sugar industry, and dairy industry are principal ingredients for vermicomposting. The worms were effectively employed to renovate different organic wastes, that is, kitchen wastes, paper wastes, sugarcane residues, medicinal herbal residues, coal fly ash, and several kinds of manures.

TABLE 12.3 Vermicomposting of the Mixture of Various Organic Wastes with Different Substrates Using *E. fetida* Earthworms.

Type of wastes	Substrate	Time (days)	Result	References
Paper industry sludge	Cow dung	48	Increased the available essential plant nutrients	Mohan (2017)
Tomato stems	Cow dung	50	Enhanced the degradation of organic matter	Yang et al. (2017)
Medicinal plant residues	Cattle manure	49	Effective stabilization of medicinal plant residues with cattle manure, extraordinary influence on the microbial composition.	Chen et al. (2018)
MSW		45	Earthworms transferred MSW into dark brown vermicompost (organic fertilizer)	Manyuchi et al. (2018)
Sewage sludge	Cow dung	80	Earthworms increased organic matter degradation, suppressed pathogens and promoted growth of *Planctomycetes* and *Acidbacteria*.	Lv et al. (2018)
Sawdust	Cattle blood powder	120	Increased stabilization and maturity of sawdust, high NPK contents.	Najjari and Ghasemi (2018)

Recently, various investigations have revealed the benefits of spending diverse essences related to plants and animals substrates. For example, addition of biochar to the crop residues increases the earthworm's growth, reproduction, and enzymatic activities along with enhancing decomposition of organic wastes. The application of biochar prepared from bamboo biomass and green wastes completely affected the earthworms augmented the lignin decomposition and dissolved the organic carbon. This reduced the time to vermicompost maturity and made it as effective fertilizer (Gong et al., 2018). The mixture of ornamental rock dust with waste material is good for plant growth as it contains nitrogen, phosphorus, potassium, magnesium, calcium, and sulfur along with other minor-nutrients (Rodrigues et al., 2017). The process of vermicompost formation through a combination of cattle's blood powder and sawdust had positive effects on nutritional status of the final product (Najjari and Ghasemi, 2018). The incorporation of combination of strains of bacteria such as *Lactobacillus pantheries, Bacillus anthracis,* and *Bacillus thuringiensis* for the vermicomposting of wastes of paper resulted in superior quality vermicompost in a smaller period (Arumugam et al., 2018).

12.6.1 FORBIDDEN MATERIAL FOR VERMICOMPOSTING

Earthworms are voracious feeders, but they are unable to eat some materials such as dairy products, animal byproducts, greasy or oily food, bones, fat, fish, tobacco, donkey and cat feces and processed meat. The presence of such material in the vermicomposting system invites rodents, flies, and other pests which are dangerous for the earthworms.

12.7 METHODOLOGY OF VERMICOMPOSTING

Vermicomposting is an aerobic and bio-oxidative method which comprises of the joint work of earthworms and microbes that ingest and decompose the organic material into humus like nutrient-rich material called vermicast. It is humus like porous material having high water retention capacity, ridiculous in plant nutrients, growth regulators of plant, and microbes that boost plant growth. The process of vermicomposting begins with the ingestion of organic wastes and the breakdown of these complex molecules into plant-available form. Most used vermicomposting methods are described below:

12.7.1 BIN COMPOSTING

This is most prevalent form of vermicomposting. Different materials such as recycled, wooden and plastic containers can be used to build bin. It may be of different shapes and sizes but average 45 × 30 × 45 cm along with small holes in bottom and sides for drainage and aeration (Figure 12.3). This is for small scale and women can do easily by using kitchen waste and this vermicompost can be used for kitchen gardening.

FIGURE 12.3 Earthworms bin for vermicomposting.

12.7.2 WINDROW COMPOSTING

The compost creation by support of organic wastes in long rows (windrow) called windrow composting. It comprises of spreading a layer of bedding material and earthworms on the ground or floor to start. The height of composting material reaches to 2–2.5 ft and width 4–8 ft (Figure 12.4). About 0.5 kg of *E. fetida* earthworms is spread on 1 ft^2 of surface area. Firstly, bedding material is spread in a layer, and earthworms are settled on the material and then apply waste material layer of 1–1.5 inches. The waste material has been ingested by earthworms before placing more organic material for composting.

12.7.3 PIT COMPOSTING

Pit composting has been used for many years. This is a simple structure for vermicomposting as no special construction is needed. Generally, the dimension of pit for vermicomposting is 1.5 m width, 1 m depth, whereas length of pit varies as required. Like other vermicomposting structures, moist bedding material is laid down in a pit. Then earthworms are added to this bedding material and a thin layer of feedstock is positioned on the top. Another benefit of using pits is that it provides insulation against cold and hot weather. Pits can be saturated with rainwater. However, rains can creat problem.

FIGURE 12.4 Windrow method of vermicomposting.

12.7.4 HEAP COMPOSTING

In this method of vermicomposting, organic wastes are spread on a polyethylene sheet on floor and then enclosed with FYM. The dimension of heap should be 2.0 × 0.6 ×0.5 m (length × width × height) as high heaps resulted in lower number of cocoons and earthworms. Then earthworms @ 1 kg per 50 kg of wastes are spread on the heap and then watering is done by sprinkler method.

12.7.5 CEMENT RING COMPOSTING

Vermicompost can be prepared by making cement rings above the ground. The ring of cement should be 90 cm in diameter and 30 cm in depth. The bottom of ring is covered with a polythene sheet and waste substance is spread on the sheet. The cement ring is filled with cow dung and waste material up to three-fourth of the cement ring (Rathore and Srinivasulu, 2018). After 15–20 days of precomposting, earthworms are released for vermicomposting. Gunny bags are used to cover the material to prevent the worms from birds. Water is sprinkled after three days to maintain temperature and moisture (Nagavallemma et al., 2004).

12.7.6 AUTOMATED FLOW-THROUGH REACTOR

This system runs by the following procedure. Earthworms live in a rectangular raised box having 1 m width, 10 m length, and depth not more than 1 m. In an automated flow-through reactor, preliminary bedding material of 10 cm thickness is used. The reactor then feeds with worms and feeding stock of 20 cm thickness. Organic material is continuously added to the top once a week. The earthworm's population in vermi-reactor reach about 2–9 kg/m^2 (Bellitürk, 2018a, 2018b). Such vermi-reactor (Figure 12.5) can entirely process all the materials they contain for 3–4 months. There is no need for agitation of material as earthworm moves on organic material resulting in the mixing in vermi-reactor. The term "flow-through" described that the earthworms are not distressed in the container and feed stuff travel to the uppermost and final product is collected from the bottom through a hydraulically driven breaker bar. No worms left at the bottom of reactor with vermicasts as they move to the top due to the attraction of organic wastes.

12.8 STEPS INVOLVED IN VERMICOMPOSTING

Before starting the vermicomposting, the predigestion of organic wastes is done to help the pathogen reduction, mass reduction, and waste stabilization (Nair et al., 2006; Frederickson et al., 2007). This also results in bioremediation of toxic compounds present in organic wastes. The thermophilic composting before the vermicomposting avoids the mortality of earthworms. After the predigestion, material gets cool and transferred to bed or container for vermicomposting. The following steps are involved in the vermicomposting of organic wastes (Figure 12.6).

FIGURE 12.5 Automated flow-through reactor.

12.8.1 SELECTION OF SITE

Vermicompost can be produced at any place having shade, high humidity, and a cool environment. For example, artificial shading, abandoned poultry shed, cattle shed, and unused buildings.

12.8.2 COLLECTION OF ORGANIC WASTES

Carbon and nitrogen-rich organic wastes such as animal dung, poultry manure, agriculture residues, textile mill wastes, dairy industry wastes, beverage and sugar industry wastes, paper wastes, municipal and domestic wastes can be used for vermicomposting. The nature of the initial wastes determines the nutrient status of the final product.

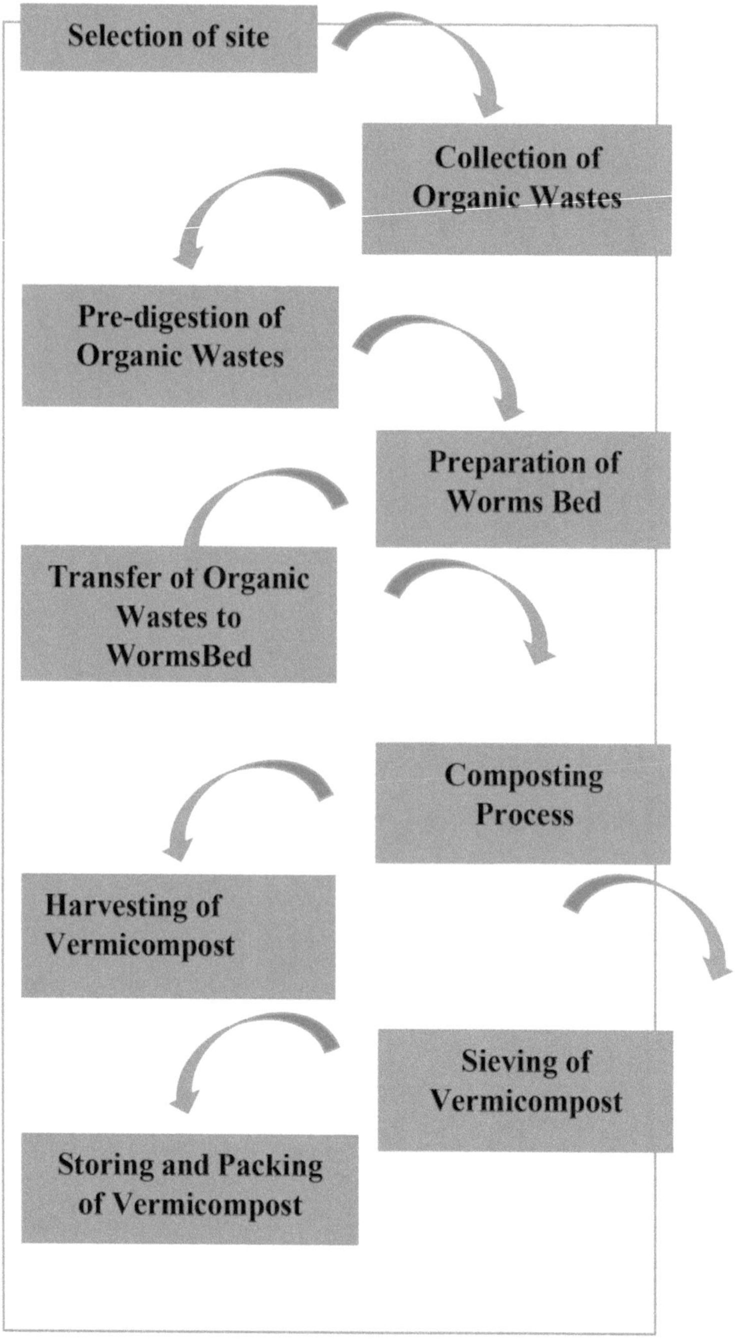

FIGURE 12.6 Steps involved in vermicomposting process.

12.8.3 ORGANIC WASTES PROCESSING

The collected organic material should be processed to remove metal, ceramics, glass, bones, and other materials that affect the activity of earthworms. Earthworms do not like acidic and citrus wastes; therefore, these materials should be removed from the vermicomposting system before starting to produce vermicompost.

12.8.4 MIXING SUBSTRATE WITH ORGANIC WASTES

The combination of substrate and waste material alters the physicochemical nature of wastes, which ultimately affects the vermicomposting. These organic substrates create such the nature of wastes that enhanced the earthworm's efficiency in vermicomposting system. It improves the vermicompost maturity, reproduction, and growth of worms during vermicomposting. The ability of earthworms to live in wastes of industry decreased critically and consequently, there is necessity for nutritious organic sources such as farmyard manure, slurry, and poultry manure to blend through wastes of industry to hasten the process of vermicompost preparation by supplying an adequate nutrients and microbial inoculum (Garg et al., 2005). When sawdust and cow dung are mixed with waste of guar as a biological supplement, they offer a favorable environment for the worms and successfully increased the rate of wastes degradation (Suthar, 2007). High amount of distillery sludge disturbs the decomposition proficiency of earthworms. The cattle manure about 40% as a preliminary substrate in wastes of vegetable could yield best feature vermicompost. Wastes of Tannery are toxic for the worms; therefore, cow dung is added to augment the contents of nutrient for the superior endurance of worms in the system. The resulted vermicompost is enriched with nutrients having low EC and C:N ratio.

12.8.5 PRECOMPOSTING OF WASTES MATERIAL

Precomposting of waste material for 20–25 days should be run before mixing organic wastes with substrates, for example, cow manure. Making the material fit for worm's consumption, regular watering is necessary. Moisture level required for the effective predigestion ranges from 55 to 65%. Heat produced during predigestion reduces the pathogen level and moisture contents of wastes. This is beneficial as high moisture resulted in putrefaction of organic wastes (Kristiana et al., 2004). The C:N ratio of organic wastes should be 30:1 or 35:1 as it gives final product of high stability, lowest pollution potential, and high nutrient status. When a material has high C:N ratio (>40:1), which is suggested to incorporate nitrogenous increments for efficient decomposition. The changes in the C:N ratio during predigestion are the results of carbon loss as CO_2 (Tripathi and Bhardwaj, 2004).

12.8.6 PREPARATION OF WORMS BED

Worms bed is a large version of worm bin, an ecosystem for raising and feeding the worms within a stable habitat. The bedding material should be of high absorbency (to uphold water for earthworm), bulking material with high porosity for earthworm's respiration and good C:N ratio. A newspaper is considered best bedding material having high absorbance capacity. Vermiculture beds can be prepared by spreading the newspaper and soil layer at the bottom of bed. However, ink of newspaper might be toxic to earthworms. The floor of bed should be flattened to avoid the movement of earthworms to outside of the bed.

12.8.7 TRANSFER OF ORGANIC WASTES TO BED

When a layer of precomposted waste material is placed, then next a substrate narrow layer such as manure of cow is added on the layer of discarded substance as earthworm start feed. Earthworm (*E. fetida*) is spread on the layer of cow dung and this reduces the waste volume up to 40–60%. Earthworms can consume all the organic wastes and ingest the wastes per day equivalent to their weight of body. The casting is rich in the plant required nutrients such as nitrogen, phosphorus, potassium, calcium, and magnesium (Misra et al., 2003).

12.8.8 COMPOSTING PROCESS

The thermophilic process ends after 15 days of the transfer of earthworms to the bed. So, the material should be turned over to 2–3 times after 72 h to maintain temperature for optimum activity of earthworm. Earthworms are sensitive to temperature and moisture. The optimum temperature for growth, reproduction, activity, and cocoons production is 25°C–37°C. Earthworms can tolerate 5°C–29°C temperature (Sinha et al., 2002). The earthworm rate of growth is directly dependent on the moisture content. Optimum moisture for the effective vermicomposting is 50–80% (Neuhauser et al., 1979). So continuously sprinkling of water should be done. The gunny bags should be used to cover vermicomposting bed that protect the material from water loss and predators such as birds. No smell is produced from composting site as pure vermicompost gives odor of wet soil. The fully prepared vermicompost could be mediated by visual observation of the granular creation of vermicasts on the surface of material.

12.8.9 HARVESTING OF VERMICOMPOST

Harvesting of vermicompost should be started by observation of vermicompost granules on waste material. Different method of vermicomposting harvesting is used.

12.8.9.1 VERMICOMPOST HARVESTING BY LIGHT METHOD

Make a heap of vermicompost containing worms on the polyethylene sheet and place it in sunlight for 1–2 h. Worms are sensitive to light, so they will move away from light and gather at the bottom of heap. Vermicompost can be collected from the top of the heap and earthworms are collected at the bottom.

12.8.9.2 VERMICOMPOST HARVESTING BY FOOD MIGRATION METHOD

Another method of vermicompost harvesting is the migration of worms to a fresh food source. The habit of earthworms is that it leaves the heap deficient of diet and translocating toward fresh food sources. So, the exhausted pile is harvested for vermicompost.

12.8.9.3 VERMICOMPOST SIEVING

The vermicompost sieving is prepared by hand with a sieve. A small portion of vermicompost is shifted into the filter and it is vibraten. Small black granules of earthworm's vermicompost fall on the floor and uncomposted material and earthworms remain on the sieve. Thus, earthworms are separated manually from the vermicompost pile. At this stage, it is necessary to pay attention that there are no earthworms and/or cocoons in the sieved vermicompost (Figure 12.7).

FIGURE 12.7 Manual sieving of vermicompost.

12.8.9.4 MECHANICAL SIEVE

A mechanical sieve is used to isolate the worms from vermicast in a large-scale production system. A slanted cylindrical motor-driven harvesting machine is used for commercial vermicompost harvesting (Figure 12.8). A motor turns the cylinder which is made of large screen. The composted material from the vermicomposting system is fed to the raised end. As the material moves down the cylinder, the screen allows the vermicompost to move down the floor and earthworms left on the sieve within the cylinder. The eggs and earthworms are returned to the fresh bedding.

FIGURE 12.8 Mechanical sieving of vermicompost.

12.8.10 PACKING AND STORING OF VERMICOMPOST

The vermicompost after sieving should be stocked in cool, dry, and dark places as light of sun results in nutrient and moisture loss. It should be stored in an aerated place as it contains beneficial aerobic microbes that require oxygen for respiration. Packing should be done at selling time and in plastic-coated packets. If vermicompost is stored in an open area, then a periodic sprinkling of water should be done to sustain beneficial microbes' population and moisture levels. It can be stored for a longer time by maintaining its moisture contents between 30% and 40%.

12.9 ABIOTIC FACTORS AFFECTING THE VERMICOMPOSTING

The abiotic issues that disturb the efficiency of vermicomposting are moisture, C:N ratio, temperature, pH, and stocking density.

12.9.1 TEMPERATURE

Although, earthworms show different responses to change in temperature yet the growth, reproduction, respiration, and metabolism of earthworms is affected by temperature (Edwards and Bohlen, 1996). However, the temperature should not increase to 25°C–30°C for the development of a higher population of earthworms. *E. fetida* can survive very low temperatures but could not reproduce. Vermicomposting is greatly influenced through change in temperature in contrast to conventional composting. The hot weather in vermicomposting is responsible to nitrogen loss as ammonia due to volatilization, whereas the low temperature fails to kill the pathogenic organisms.

12.9.2 MOISTURE

The degree of growth of earthworms is critically affected by the wetness of vermicomposting system. Ideal moisture content for effective vermicomposting is considered as 70–80%. Earthworm's sexual development is delayed by low moisture contents. Optimum moisture level for *E. fetida* is 75%.

12.9.3 pH

Neutral pH is considered as the best for the functioning of earthworms, but it can survive the pH range of 5.0–9.0. The chemical and physical properties of waste and earthworm's sensitivity affect the pH. The alterations in physicochemical properties of plants and animals wastes are owing to microbial action during the decomposition and mineralization of nutrients. At the initial stage, pH of vermicomposting system is alkaline which shifts to acidic or neutral at the end of vermicompost. This slight decrease in pH is due to CO_2 evolution and accumulation of organic acids.

12.9.4 EARTHWORM'S STOCKING DENSITY

Quantity and quality of substrate, moisture, and temperature of the composting system influences the earthworm's stocking density. The earthworm copulation frequency is low at high stocking density, whereas it is high at low density. The vermicompost preparation, stocking density of 1.60 kg worms/m^2, is considered ideal.

12.9.5 C:N RATIO

The substrate C:N ratio influences the growth and metabolism of earthworms. The waste stabilization is indicated by the C:N ratio and is mostly adopted as a maturation index of compost. The C:N ratio of less than 20 reflects good quality of compost. During vermicomposting process, the decline in the C:N ratio is due to quick organic decomposition and mineralization. The microbial respiration lost carbon while earthworm's mucus and nitrogenous excretory material release nitrogen which leads to an overall decrease in the C:N ratio.

12.10 IMPORTANCE OF VERMICOMPOST AS GREAT AGRONOMIC VALUES

12.10.1 HIGHER LEVEL OF PLANT AVAILABLE NUTRIENTS

The most essential nutrients present in vermicompost that are in available forms for plants are potassium, phosphorus (P), nitrogen (N), and exchangeable calcium (Ca), magnesium (Mg), iron (Fe), and sulfur (S). The microsites for strong retention of nutrients and microbial activities are done by vermicompost, which is due to large particulate surface area. Vermicomposting of cow dung preferred nitrification, resulted in the fast alteration of nitrogen to nitrate-nitrogen by ammonification. The microbes in the gut of earthworms hasten the phosphatase activity which makes the phosphorus available to plants due to mobilization (Garg et al., 2006).

Solid vermicompost results of these investigations are elaborated in Table 12.4.

TABLE 12.4 Different Solid Vermicompost Chemical Compositions Which Was Manufactured in Turkey.

Contents	Raw material of vermicompost (main ingredient fed to earthworms)			
	Pruned olive tree waste	**Mix food waste**	**Sheep manure**	**Cow dung**
Organic material (%)	48.00	41.00	42.00	51.00
Nitrogen (%)	1.80	1.95	1.60	1.75
Phosphorus (%)	0.23	2.75	0.45	1.29
Potassium (%)	1.74	1.10	0.32	1.12
Zinc (mg kg^{-1})	545.00	354.00	405.00	427.00

12.10.2 PATHOGEN AND TOXIC CHEMICALS FREE ORGANIC FERTILIZER

Vermicompost is free from pathogens as vermicomposting reduced the injurious microorganisms by demolishing them and by the release of coelomic fluid that acts as antibiotic. Earthworms effectively biodegrade and bioaccumulate several chemicals such as heavy metals like polycyclic aromatic hydrocarbon (PAHs) from vermicomposting system.

12.10.3 RICH WITH BENEFICIAL SOIL MICROBES

The soil and crops beneficial microbes such as fungi, bacteria, and actinomycetes present excess amount in vermicompost. Fresh vermicompost contains 32 million per gram a bacterial count as compared to soil bacterial count of 6–9 million per g. This increase in the number of microbes is due to an increase in respiration rate (90%) in vermicompost as compared to soil. Earthworms ingest beneficial soil microbes, namely, *Azotobacter, Rhizobium,* and *Azosprillium,* as well as rhizosphere soil and these microbes become stimulated due to supreme microenvironment in earthworms' gut.

The organic matter is an important source for soil fertility and crop productivity that increase the activity of beneficial microbes in soil. The soil enzymatic activities enhanced by organic amendments like vermicompost can foster beneficial microorganisms.

12.10.4 IMPORTANCE OF VERMICOMPOST BACTERIA IN SOLID WASTE MANAGEMENT

Vermicomposting of wastes goes under the biological conversion and organic matter stabilization and significantly reduces the pathogens. Vermicomposting is not a thermophilic process; thus can be effectively used for the management of solid wastes. The earthworm's activities reduced the odor of putrefaction and hasten the sludge stabilization. Earthworms encourage beneficial decomposers fungi and bacteria in the effluent and work as crushers, aerators, chemical degraders, biostimulators, and grinders (Sinha et al., 2002).

12.10.5 IMPORTANCE OF VERMICOMPOST IN SOIL FERTILITY

The soil fertility and structure are improved by the addition of vermicompost along with suppressing the soil-borne pathogens thus reduces the diseases. Vermicompost is used as a supernumerary for sealing media due to its economical, physicochemical characteristics, and excellent nutrients contents. The amendment of vermicompost affected the soil enzyme activities, soil pH, microbial population, and reduced the environmental contamination by reducing the number of water-soluble chemicals. Vermicompost enhances the association of soil air–water by improving the macro-pore space from 50 to 500 μm and enhanced the plant growth (Marinari et al., 2000).

Vermicompost acts as a "conditioner for soil" significantly, and its constant use in years show entire enhancement in the agricultural soil quality, even the tarnished Thrace Region of soils in Turkey. The backbone of Turkey economy is reliant on farming, creating it the major consumer of water. Thus, the extreme force is on the farming zone proficient use of water resources for the irrigation. It is suggested by agricultural scientist to apply vermicompost in drought-prone regions to overcome this problem in such drought periods because vermicompost is highly porous and spongy in nature and store water in it when plant require water then it release.

12.10.6 VERMICOMPOST AS A PLANT GROWTH PROMOTER

Vermicompost increased the growth of seedling, flowering, fruiting, and productivity and used as a supplement media, of extensive range of crop varieties. The growth promoters of plant such as humic acids, gibberellins, auxins, and cytokinin are existing in adequate amount in vermicompost. Microorganisms such as *Arthrobacter* and *Bacillus* spp. produce gibberellins and *Azospirillum brasilense* produces auxins that affect the growth of Poaceae family. The biological yield of oat and cotton is enhanced by humic substances. The fulvic and humic acid in vermicompost solubilize the nutrients in partially decomposed material and make them easily accessible for the crops. The application of vermicompost increased the leaf area (23.1%), plant spread (10.7%), dry matter contents (20.7%), and enhanced the fruit yield of strawberry up to 32.7% (Singh et al., 2008). The vermicompost application in many crops enhances early germination. The pot experiment results showed that Wavy seeds of lettuce germinated earlier in vermicompost treatment than the other organic compost. The higher nutrients accessibility in vermicompost may be the cause of early germination in plants.

12.10.7 VERMICOMPOST AS BIOPESTICIDE

Organic matter deficient and low microbial population soils are sensitive to roots diseases of crops and application of biological substances efficiently reduced the plant diseases. The major reasons for suppression of disease might be due to microbial antagonism as biological improvements increase the diversity and population of microorganisms. Earthworm's ingestion of organic material cuts the plant pathogen survival such as *Fusarium* spp. and densities of filamentous actinomycetes increased manifold. It also reduced the attack of insects that are sucking type under conditions of field and inhibited the destruction of two-spotted spider mite, mealybug and aphids under greenhouse condition. The mechanism behind the pest control by vermicompost application is that the vermicompost provides nutrients at slow rate; thus plants grown with organic amendments have low nitrogen levels and high phenol contents that make the plant resistance to pest attack.

Application of vermicompost considerably reduced the occurrence of "color rot," "powdery mildew," plus "Yellow Vein Mosaic" in the Okra (Agarwal et al., 2010). The vermicompost mixing with soil released large amounts of plant-available nutrients which consequently boosted the natural defense of plants against diseases. Moreover, vermicompost enhances the biological resistance of plants by stimulating the beneficial microbes in soil and protect the plants from pathogens, the existence of antifungal and antibiotic compounds in vermicompost, and initiation of antifungal compound biosynthesis (Sahni et al., 2008). The plant extracted from antifungal compounds have aromatic rings structure, that is why vermicompost with an excess amount of aromaticity plus humification is therefore imperative for origination of substances production that are antifungal and most of them are primary and secondary metabolites (Arif et al., 2011). The combined use of antagonistic bacteria strains and vermicompost decreased the occurrence of collar rot of chickpea. The green house experiment of chickpea showed that the instant application of *Pseudomonas syringae* and vermicompost initiated a bargain mortality in seedling (Sahni et al., 2008). The soil-applied vermicompost increased the manganese, iron, and phosphorus contents in soil and this augmentation seemed to encourage the growth of *Pseudomonas syringae*. The increase in the growth of *Pseudomonas mosselii* using vermicompost promoted the effective plant defense against pathogens, and improved the growth (Singhai et al., 2011). Aqueous extract from vermicompost is proficient in diminishing the attack of farming diseases on tomatoes and cucumbers. The actinomycetes obtained from crop residues vermicompost tested positively against the *Fusarium oxysporum*, a toadstool that creates numerous plants diseases.

12.10.8 VERMICOMPOSTING SET UP IN DIFFERENT COUNTRIES

Vermicomposting process is known all over the world, although in limited areas. As a viable technology for managing organic residues, it characterizes an alternative method in management of wastes. It is an important process as the unwanted material is neither burned nor landfilled but is reflected as a recyclable store. The methods of vermicompost preparation services have now entered in national and international markets in advanced nations like the USA, Italy, Canada, Japan, and Turkey. The process for vermicompost preparation was initiated in 1970 in Canada and is now managing almost 75 tons of wastes in a week. The AEC (American Earthworm Company) initiated a vermicompost farm in 1978–1979 by 500 tons processing capability per 30 days. The Japan (Aoka Sangyo Co. Ltd.) has started to process the wastes from food industries and pulp of about 3000 tons per month. In addition to these, around 3000 further units of vermicomposting with 5–50 tones wastes processing capacity per month in Japan. That is also initiated in the Philippines plus Italy for waste management (Aaalok et al, 2008).

After 2011 commercial production of vermicompost began in Turkey, the cumulative production of vermicompost in 2017 in Turkey was about 20,000 tons which were manufactured by 15 vermicomposting units with certified production permits through the Turkey Ministry of Agriculture and Forestry. By 2023, Turkey intention is to be among the top three producers of organic amendments. Besides, there are around 4200 unlicensed units of vermicompost in Turkey. However, the interest and creation of vermicompost increased during the last decades (Bellitürk, 2018a, 2018b). In Pakistan, very few studies were conducted on vermicomposting. Up till now, no vermicomposting commercial unit is working in the country. Currently, a "Vermicompost Center" has been developed at the University of Agriculture Faisalabad, Pakistan with the cooperation of Higher Education Commission of Pakistan and researchers from Turkey (Rehman et al., 2020; Aslam et al., 2020a, 2020b; Aslam and Ahmad, 2020).

12.11 SUMMARY

Solid organic wastes management is a serious problem that affects the surrounding environment and human health. Approximately, 1,300,000 tons per year global municipal solid waste is produced and is predictable to rise to 2,200,000 tons per year in 2025. Solid wastes have serious environmental impacts as aerobic decomposition of organic wastes results in methane generation which has serious impacts on climate change. Open dumping of organic wastes pollutes the adjacent water bodies with organic pollutants; thus affecting the groundwater quality. This results in a disease outbreak by people exposed to harmful products. Different methods of waste management such as recycling, landfill, incineration are used across the globe. The incineration of waste material releases different types of pollutants such as furan and dioxins. Vermicomposting is a viable choice to manage solid waste which is now environment responsive method. It prevents methane emissions and reduces environmental pollution. The final product of vermicomposting of municipal solid wastes is odorless, economic, and contains high nutrient contents, and free from pathogenic microbes with low greenhouse gas emission. The earth worm's casting contains the maximum essential nutrients for plants available forms as nitrogen (N), phosphorus (P), potassium (K), zinc (Zn), calcium (Ca), magnesium (Mg), iron (Fe) and sulfur (S), etc. The worm's castings also contain beneficial bacteria, fungi, actinomycetes, growth promoters and regulators, and certain beneficial enzymes for plants. Vermicomposting offers good prospective to turn unwanted material into a treasured soil improvement. The key objective of the chapter is to show and share proper methodology and benefits and practical causes for augmented curiosity in earthworms and vermicompost.

ACKNOWLEDGMENTS

We are thankful to Higher Education Commission of Pakistan who financially supported the vermicompost project (Project no. 7527/Punjab/NRPU/R&D/HEC/2017) entitled with "Vermicomposting: A Resourceful Organic Fertilizer to Improve Agriculture Production and Soil Health."

KEYWORDS

- ***Eisenia fetida***
- **municipal solid organic waste**
- **nutrients**
- **organic fertilizer**
- **vermicomposting**

REFERENCES

Aalok, A.; Tripathi, A. K.; Soni, P. Vermicomposting: A Better Option for Organic Solid Waste Management. *J. Hum. Ecol.* **2008,** *24* (1), 59–64.

Agarwal, S.; Sinha, R. K.; Sharma, J. Vermiculture for Sustainable Horticulture Agronomic Impact Studies of Earthworms, Cow Dung Compost and Vermicompost vis-a-vis Chemical Fertilizers on Growth and Yield of Lady's Finger (*Abelmoschus esculentus*). *Int. J. Global Environ. Iss.* **2010,** *10* (3–4), 366–377.

Aira, M.; Monroy, F.; Domınguez, J.; Mato, S. How Earthworm Density Affects Microbial Biomass and Activity in Pig Manure. *Eur. J. Soil Biol.* **2002,** *38*, 7–10.

Ali, S.; Kashem, M. A. Life Cycle of Vermicomposting Earthworms *Eisenia fetida* and *Eudrilus eugeniae* under Laboratory Controlled Condition. *Biomed. J. Sci. Tech. Res.* **2018,** *10* (5), 8110–8113.

Anand, K.; Sinha, P. B. Vermitechnology: A Solution for Agricultural Waste. In *Innovative Waste Management Technologies for Sustainable Development*; IGI Global, 2020; pp 273–290.

Arif, T.; Mandal, T. K.; Dabur, R. Natural Products: Antifungal Agents Derived from Plants, in Opportunity, Challenge and Scope of Natural Products in Medicinal Chemistry, Research Signpost, Kerala, India, 2011.

Arumugam, K.; Renganathan, S.; Babalola, O. O.; Muthunarayanan, V. Investigation on Paper Cup Waste Degradation by Bacterial Consortium and *Eudrillus eugeinea* through Vermicomposting. *Waste Manage.* **2018,** *74*, 185–193.

Asian Development Bank. *Solid Waste Management in Nepal: Current Status and Policy Recommendations*; Asian Development Bank, 2013.

Aslam, Z.; Ahmad, A. Effects of Vermicompost, Vermi-Tea and Chemical Fertilizer on Morpho-Physiological Characteristics of Maize (*Zea mays* L.) in Suleymanpasa District, Tekirdag of Turkey. *J. Innov. Sci.* **2020,** *6* (1), 41–46.

Aslam, Z.; Ahmad, A.; Bellitürk, K.; Iqbal, N.; Idrees, M.; Rehman, W. U.; Akbar, G.; Tariq, M.; Raza, M.; Riasat, S.; Rehman, S. U. Effects of Vermicompost, Vermi-Tea and Chemical Fertilizer on Morpho-Physiological Characteristics of Tomato (*Solanum lycopersicum*) in Suleymanpasa District, Tekirdag of Turkey. *Pure Appl. Biol.* **2020a,** *9* (3), 1920–1931.

Aslam, Z.; Ahmad, A.; Idrees, M.; Iqbal, N.; Akbar, G.; Ali, U.; Ibrahim, M. U.; Bellitürk, K.; Naeem, S.; Nawaz, M.; Nadeem, M.; Waqas, M.; Rehman, W. U.; Sajjad, M.; Samiullah.; Akram, Y. Comparative Analysis of Nutritional Sources on the Morpho-Physiological Characteristics of Mung Bean (*Vigna radiata*). *Int. J. Food Sci. Agric.* **2020b,** *4* (3), 314–322.

Ayuso, M.; Pascual, J. A.; Garcıa, C.; Hernandez, T. Evaluation of Urban Wastes for Agricultural Use. *Soil Sci. Plant Nutr.* **1996,** *42*, 105–111.

Azam, M.; Jahromy, S. S.; Raza, W.; Jordan, C.; Harasek, M.; Winter, F. Comparison of the Combustion Characteristics and Kinetic Study of Coal, Municipal Solid Waste, and Refuse–Derived Fuel: Model-Fitting Methods. *Energy Sci. Eng.* **2019,** *7* (6), 2646–2657.

Ba, C. C. Biomass and Vermicompost Production by the Earthworm *Eudrilus eugeniae* (Kinberg). *Rev. Biol. Trop.* **1989,** *37* (1), 11–13.

Bellitürk, K. Some Evaluations about Use of Vermicompost in Agricultural Activity of Thrace Region, Turkey: A Review. *J. Rice Res.* **2018a,** *6* (2), 100–193.

Bellitürk, K. Vermicomposting in Turkey: Challenges and Opportunities in Future. *Eurasian J. Forest Sci.* **2018b,** *6* (4), 32–41.

Belliturk, K.; Aslam, Z.; Ahmad, A.; Rehman, S. Alteration of Physical and Chemical Properties of Livestock Manures by Eisenia fetida (Savigny, 1926) and Developing Valuable Organic Fertilizer. *J. Innov. Sci.***2020,** **6** (1), 47–53.

Benıtez, E.; Nogales, R.; Masciandaro, G.; Ceccanti, B. Isolation by Isoelectric Focusing of Humic–Urease Complexes from Earthworm (*Eisenia fetida*) Processed Sewage Sludges. *Biol. Fertil. Soils*, **2000,** *31*, 489–493.

Bhawalkar, U. *Turning Garbage into Gold: An Introduction to Vermiculture Biotechnology*; Bhawalkar Earthworm Research Institute, 1993.

Bongaarts, J. Human Population Growth and the Demographic Transition. *Phil. Trans. R. Soc. B* **2009,** *364* (1532), 2985–2990.

Bourtsalas, A. T.; Seo, Y.; Alam, M. T.; Seo, Y. C. The Status of Waste Management and Waste to Energy for District Heating in South Korea. *Waste Manage.* **2019,** *85*, 304–316.

Brown, G. G.; Doube, B. M.; Edwards, C. A. Functional Interactions between Earthworms, Microorganisms, Organic Matter, and Plants. *Earthworm Ecol.* **2004,** 213–239.

Cadena, E.; Colon, J.; Sanchez, A.; Font, X.; Artola, A. A Methodology to Determine Gaseous Emissions in a Composting Plant. *Waste Manage.* **2009,** *29*, 2799–2807.

Chang, N. B.; Davila, E. Municipal Solid Waste Characterizations and Management Strategies for the Lower Rio Grande Valley, Texas. *Waste Manage.* **2008,** *28*, 776–794.

Charles, W.; Walker, L.; Cord-Ruwisch, R. Effect of Pre-aeration and Inoculum on the Start-up of Batch Thermophilic Anaerobic Digestion of Municipal Solid Waste. *Bioresour. Technol.* **2009,** *100*, 2329–2335.

Chen, Y.; Chang, S. K. C.; Chen, J.; Zhang, Q.; Yu, H. Characterization of Microbial Community Succession during Vermicomposting of Medicinal Herbal Residues. *Bioresour. Technol.* **2018,** *249*, 542–549.

Curry, J. P.; Schmidt, O. The Feeding Ecology of Earthworm Sea Review. *Pedobiologia* **2007,** *50* (6), 463–477.

Dasgupta, B.; Yadav, V. L.; Mondal, M. K. Seasonal Characterization and Present Status of Municipal Solid Waste (MSW) Management in Varanasi, India. *Adv. Environ. Res. (AER), Int. J.* **2013,** *2* (1), 51–60.

Defra, U. K. *Food Security Assessment: Detailed Analysis*; Department for Environment Food, and Rural Affairs, 2010.

Deka, H., Deka, S.; Baruah, C. K.; Das, J.; Hoque, S.; Sarma, N. S. Vermicomposting of Distillation Waste of Citronella Plant (*Cymbopogon winterianus* Jowitt.) Employing *Eudrilus eugeniae*. *Bioresour. Technol.* **2011,** *102*, 6944–6950.

Dominguez, J.; Edwards, C. A. Relationships between Composting and Vermicomposting: Relative Values of the Products. In *Vermiculture Technology: Earthworms, Organic Waste and Environmental Management*; Edwards, C. A., Arancon, N. Q., Sherman, R. L., Eds.; Boca Raton, FL, 2010; pp 1–14.

Edwards, C. A.; Bohlen, P. J. *Biology and Ecology of Earthworms*, Vol. 3; Springer Science & Business Media, 1996; p 115.

Edwards, C.; Lofty, J. R.; The Influence of Invertebrates on Root Growth of Crops with Minimal or Zero Cultivation. *Ecol. Bull.* **1977,** 348–356.

Fijalkowski, K.; Rorat, A.; Grobelak, A.; Kacprzak, M. J.; The Presence of Contaminations in Sewage Sludge—The Current Situation. *J. Environ. Manage.* **2017,** *203*, 1126–1136.

Frederickson, J.; Howell, G.; Hobson, A. M.; Effect of Pre-composting and Vermicomposting on Compost Characteristics. *Eur. J. Soil Biol.* **2007,** *43*, 320–326.

Garg, P.; Gupta, A.; Satya. S.; Vermicomposting of Different Types of Waste Using *Eisenia foetida*: A Comparative Study. *Bioresour. Technol.* **2006,** *97*, 391–395.

Garg, V.; Chand, S.; Chhillar, A.; Yadav, A.; Growth and Reproduction of *Eisenia foetida* in Various Animal Wastes during Vermicomposting. *Appl. Ecol. Environ. Res.* **2005,** *3*, 51–59.

Gong, X.; Cai, L.; Li, S.; Chang, S. X.; Sun, X.; An, Z.; Bamboo Biochar Amendment Improves the Growth and Reproduction of *Eisenia fetida* and the Quality of Green Waste Vermicompost. *Ecotox. Environ. Safe.* **2018,** *156*, 197–204.

Government of Australia. National Waste Report, **2018,** 8–9.

Gunaruwan, T. L.; Gunasekara, W. N.; Management of Municipal Solid Waste in Sri Lanka: A Comparative Appraisal of the Economics of Composting. *NSBM J. Manage.* **2016,** *2* (1), 27–45.

Hai, F. I.; Ali, M. A.; A study on Solid Waste Management System of Dhaka City Corporation: Effect of Composting and Landfill Location. *UAP J. Civil Environ. Eng.* **2005,** *1* (1), 18–26.

Halder, P. K.; Paul, N.; Hoque, M. E.; Hoque, A. M.; Parvez, M. S.; Rahman, M. H.; Ali, M. Municipal Solid Waste and Its Management in Rajshahi City, Bangladesh: A Source of Energy. *Int. J. Renew. Energy Res.* **2014,** *4* (1), 168–175.

Hand, P.; Hayes, W.; Frankland, J.; Satchell, J. Vermicomposting of Cow Slurry. *Pedobiologia* **1988,** *31*, 199–209.

Hassan, M. N.; Solid Waste Management in Malaysia: Can We Charter Future Strategies. In *Proceedings of International Conference Environmental Management: Ten Years after Rio*; 2002.

Helmy, M.; Laksono, T. B.; Gardera, D. 1026 3R Implementation in Indonesia. Paper Presented at the Senior Official Meeting on the 3R Initiative, J1CA, Tokyo, Japan, 2006.

Hendriksen, N. B. Gut Load and Food-Retention Time in the Earthworms *Lumbricus festivusand* L. Castaneus: A Field Study. *Biol. Fertil. Soils*, **1991,** *11* (3), 170–173.

Hoornweg, D.; Bhada-Tata, P. What a Waste: A Global Review of Solid Waste Management. *The World Bank*, **2012,** 1–116.

Johnson, L. M.; Cao, X.; Jacobsen, S. E. Interplay between Two Epigenetic Marks: DNA Methylation and Histone H3 Lysine 9 Methylation. *Curr. Biol.* **2002,** *12* (16), 1360–1367.

Khajuria, A.; Yamamoto, Y.; Morioka, T. Estimation of Municipal Solid Waste Generation and Landfill Area in Asian Developing Countries. *J. Environ. Biol.* **2010,** *31* (5), 649–654.

Khwairakpam, M.; Bhargava, R. Vermitechnology for Sewage Sludge Recycling. *J. Hazard. Mater.* **2009,** *161*, 948–954.

Korai, M. S.; Mahar, R. B.; Uqaili, M. A. The Feasibility of Municipal Solid Waste for Energy Generation and Its Existing Management Practices in Pakistan. *Renew. Sustain. Energy* **2017,** *72*, 338–353.

Kristiana, R.; Nair, J.; Anda, M.; Mathew, K. Monitoring of the Process of Composting of Kitchen Waste in an Institutional Scale Worm Farm. *Water Sci. Technol.* **2004,** *51* (10), 171–177.

Kumar, A. *Verms & Vermitechnology*; APH Publishing: New Delhi, India, 2005.

Lee, D. H.; Behera, S. K.; Kim, J. W.; Park, H. S. Methane Production Potential of Leachate Generated from Korean Food Waste Recycling Facilities: A Lab-Scale Study. *Waste Manage.* **2009,** *29*, 876–882.

Lv, B.; Xing, M.; Yang, J. Exploring the Effects of Earthworms on Bacterial Profiles during Vermicomposting Process of Sewage Sludge and Cattle Dung with High-Throughput Sequencing. *Environ. Sci. Pollu.* **2018,** *25*, 12528–12537.

Manyuchi, M. M.; Mbohwa, C.; Muzenda, E. Resource Recovery from Municipal Waste and Bio Solids (Digestate) through Vermicomposting: A Waste Management Initiative. In *IEEE International Conference on Industrial Engineering and Engineering Management* **2018,** 2054–2057.

Marinari, S.; Masciandaro, G.; Ceccanti, B.; Grego, S. Influence of organic and mineral fertilisers on soil biological and physical properties. *Bioresour. Technol.***2000,** *72 (1),* 9–17.

Misra, R. V.; Roy, R. N.; Hiraoka, H. *On-Farm Composting Methods*; UN-FAO: Rome, Italy, 2003.

Mohan, S. M. Vermicomposting of Paper Mill Sludge with *Eisenia fetida* for Its Conversion to Nutrient Using Different Seed Materials. *J. Instn. Engr.* **2017,** *98*, 545–553.

Mor, S.; Ravindra, K.; Dahiya, R.; Chandra, A. Leachate Characterization and Assessment of Groundwater Pollution Near Municipal Solid Waste Landfill Site. *Environ. Monit. Assess.* **2006,** *118*, 435–456.

Mühle, S.; Balsam, I.; Cheeseman, C. R. Comparison of Carbon Emissions Associated with Municipal Solid Waste Management in Germany and the UK. *Resour. Conserv. Recycl.* **2010,** *54* (11), 793–801.

Munnoli, P. M.; Da Silva, J. A. T.; Saroj, B. Dynamics of the Soil-Earthworm-Plant Relationship: A Review. *Dyn. Soil Dyn. Plant* **2010,** *4* (1), 1–21.

Nagavallemma, K. P.; Wani, S. P.; Lacroix, S.; Padmaja, V. V.; Vineela, C.; Rao, M. B.; Sahrawat, K. L. Vermicomposting: Recycling Wastes into Valuable Organic Fertilizer. *Global Theme Agroecosyst. Rep.* **2004,** 8.

Nair, J.; Sekiozoic, V.; Anda, M. Effect of Pre-composting on Vermicomposting of Kitchen Waste. *Bioresour. Technol.* **2006,** *97*, 2091–2095.

Najjari, F.; Ghasemi, S. Changes in Chemical Properties of Sawdust and Blood Powder Mixture during Vermicomposting and the Effects on the Growth and Chemical Composition of Cucumber. *Sci. Hortic.* **2018,** *232*, 250–255.

NBS, N. B. O. S. *NBS China Statistical Year*; Book China Statistics Press: Beijing, 2019.

Neuhauser, E. F.; EF, N.; DL, K. Life History of the Earthworm. *Eudrilus eugeniae* **1979,** *16* (4), 525–534.

Niyati, M. A Comparative Study of Municipal Solid Waste Management in India and Japan. *J. Soc. Stud.* **2015,** *25*, 48–61.

P, K. Singhai.; B, K. Sarma.; J, S. Srivastava. Biological Management of Common Scab of Potato through *Pseudomonas* Species and Vermicompost. *Biol. Control*, **2011,** *57,* 150–157.

Palanivel, T. M.; Sulaiman, H. Generation and Composition of Municipal Solid Waste (MSW) in Muscat, Sultanate of Oman. *APCBEE Procedia* **2014,** *10*96–1102.

Pramanik, P. Changes in Microbial Properties and Nutrient Dynamics in Bagasse and Coir during Vermicomposting: Quantification of Fungal Biomass through Ergosterol Estimation in Vermicompost. *Waste Manage.* **2010,** *30*, 787–791.

Pramanik, P.; Ghosh, G.; Banik, P. Effect of Microbial Inoculation during Vermicomposting of Different Organic Substrates on Microbial Status and Quantification and Documentation of Acid Phosphatase. *Waste Manage.* **2009,** *29*, 574–578.

Raharjo, S.; Ruslinda, Y.; Bachtiar, V. S.; Regia, R. A.; Fadhil, M.; Rachman, I.; Matsumoto, T. S. Investigation on Municipal Solid Waste Characteristics from Commercial Sources and Their Recycling Potential in Padang City, Indonesia. *Mater. Sci. Eng.* **2018,** *288*, 012134.

Rathore, M. S.; Srinivasulu, Y. Vermicomposting Bed Types for Recycling of Sericultural Waste. *Int. Res. J. Eng. Technol.* **2018,** *5* (8), 1484–1488.

Rehman, S.; Aslam, Z.; BELLİTÜRK, K.; Ahmad, A.; Nadeem, M.; Waqas, M. Vermicomposting in Pakistan: Current Scenario and Future Prospectives. *Mod. Cep. Devel. Agron.* **2020,** *6* (1), 629.

Rodrigues, A. S. D. L.; Mesak, C.; Silva, M. L. G.; Silva, G. S.; Leandro, W. M.; Malafaia, G. Organic Waste Vermicomposting through the Addition of Rock Dust Inoculated with Domestic Sewage Wastewater. *J. Environ. Manage.* **2017,** *196*, 651–658.

Ronald, E.; Donald, E. *Earthworms for Ecology and Profit: Scientific Earthworm Farming*; Bookworm Publishing Company: Ontario, 1977; p 1.

Rorat, A.; Kacprzak, M.; Vandenbulcke, F.; Płytycz, B. Soil Amendment with Municipal Sewage Sludge Affects the Immune System of Earthworms *Dendrobaena veneta*. *Appl. Soil Ecol.* **2013,** *64*, 237–244.

Ryu, C.; Shin, D. Combined Heat and Power from Municipal Solid Waste: Current Status and Issues in South Korea. *Energies* **2013,** *6* (1), 45–57.

S, Sahni.; B, K. Sarma.; D, P. Singh; H, B. Singh.; K, P. Singh. Vermicompost Enhances Performance of Plant Growth Promoting Rhizobacteria in *Cicer arietinum* Rhizosphere against *Sclerotium rolfsii*. *Crop Prot.* **2008,** *27* (3–5), 369–376.

Sˇrut, M.; Menke, S.; Ho ̈ckner, M.; Sommer, S. Earthworms and Cadmium–Heavy Metal Resistant Gut Bacteria as Indicators for Heavy Metal Pollution in Soils? *Ecotoxicol. Environ. Saf.* **2019,** *171*, 843–853.

Sanchez-Hernandez, J. C. Bioremediation of Pesticide-Contaminated Soils by Using Earthworms. *Bioremed. Agric. Soils* **2019,** 165–192.

Scho ̈nholzer, F.; Hahn, D.; Zeyer, J. Origins and Fate of Fungi and Bacteria in the Gut of *Lumbricus terrestris* L. Studied by Image Analysis. *Microbiol. Ecol.* **1999,** *28* (3), 235–248.
Shams, S.; Sahu, J. N.; Rahman, S. S.; Ahsan, A. Sustainable Waste Management Policy in Bangladesh for Reduction of Greenhouse Gases. *Sustain. Cities Soc.* **2017,** *33*, 18–26.
Sharma, K. D.; Jain, S. Overview of Municipal Solid Waste Generation, Composition, and Management in India. *J. Environ. Eng.* **2019,** *145* (3), 04018143.
Shekdar, A. V. Sustainable Solid Waste Management: An Integrated Approach for Asian Countries. *Waste Manage.* **2009,** *29* (4), 1438–1448.
Sherman, R. *Raising Earthworms Successfully*; North Carolina Extension Service, North Carolina State University: North Carolina, 2003.
Siami, L.; Sotiyorini, T.; Janah, N. M. Municipal Solid Waste Quantification and Characterization in Banyuwangi, Indonesia. *Indones J. Urban Environ. Technol.* **2019,** *2* (2), 189–200.
Singh, R. P.; Singh, P.; Araujo, A. S. F.; Ibrahim, M. H.; Sulaiman, O. Management of Urban Solid Waste: Vermicomposting as a Sustainable Option. *Resour. Conserv. Recycl.* **2011,** *55*, 719–729.
Singh, R.; Embrandiri, A.; Ibrahim, M.; Esa, N. Management of Biomass Residues Generated from Palm Oil Mill: Vermicomposting a Sustainable Option. *Resour. Conserv. Recycl.* **2011,** *55*, 423–434.
Singh, R.; Sharma, R. R.; Kumar, S.; Gupta, R. K.; Patil, R. T. Vermicompost Substitution Influences Growth, Physiological Disorders, Fruit Yield and Quality of Strawberry (*Fragaria ananassa* Duch.). *Bioresour. Technol.* **2008,** *99* (17), 8507–8511.
Singleton, D. R.; Hendrix, P. F.; Coleman, D. C.; Whitman, W. B. Identification of Uncultured Bacteria Tightly Associated with the Intestine of the Earthworm *Lumbricus rubellus* (Lumbricidae; Oligochaeta). *Soil Biol. Biochem.* **2003,** *35* (12), 1547–1555.
Sinha, R. K.; Herat, S.; Agarwal, S.; Asadi, R.; Carretero, E. Vermiculture and Waste Management: Study of Action of Earthworms *Esinia foetida*, *Eudrilus euginae* and *Perionyx excavatus* on Biodegradation of Some Community Wastes in India and Australia. *Environmentalist* **2002,** *22* (3), 261–268.
Sivasankari, B. Study on Life Cycle of Earthworm *Eisenia foetida*. *Int. Res. J. Nat. Appl. Sci.* **2016,** *3* (5), 83–93.
Staley, B. F.; Barlaz, M. A. Composition of Municipal Solid Waste in the United States and Implications for Carbon Sequestration and Methane Yield. *J. Environ. Eng.* **2009,** *135* (10), 901–909.
Suthar, S. Production of Vermifertilizer from Guar Gum Industrial Wastes by Using Composting Earthworm *Perionyx sansibaricus* (Perrier). *Environmentalist* **2007,** *27*, 329–335.
Tchobanoglous, G.; Kreith, F. *Handbook of Solid Waste Management*, 2nd ed.; McGraw-Hill: New York, 2002.
Tripathi, G.; Bhardwaj, P. Comparative Studies on Biomass Production, Life Cycles and Composting Efficiency of *Eisenia fetida* (Savigny) and *Lampito mauritii* (Kinberg). *Bioresour. Technol.* **2004,** *92*, 275–283.
Troschinetz, A. M.; Mihelcic, J. R. Sustainable Recycling of Municipal Solid Waste in Developing Countries. *Waste Manage.* **2009,** *29*, 915–923.
Tsai, S. H.; Liu, C. P.; Yang, S. S. Microbial Conversion of Food Wastes for Biofertilizer Production with Thermophilic Lipolytic Microbes. *Renew. Energy* **2007,** *32*, 904–915.
USEPA. Facts and Figures about Materials, Waste and Recycling, 2018.
Visvanathan, C.; Trankler, J. Municipal Solid Waste Management in Asia: A Comparative Analysis. In *Workshop on Sustainable Landfill Management*; Chennai, 2003; pp 3–5.

White, P.; Franke, M.; Hindle, P. *Integrated Solid Waste Management: A Lifecycle Inventory*; Chapman & Hall: New York, 1999.

Yadav, A.; Garg, V. K. Recycling of Organic Wastes by Employing *Eisenia fetida*. *Bioresour. Technol.* **2011,** *102*, 2874–2880.

Yan, M.; Wibowo, H.; Liu, Q.; Cai, Y.; Rahim, D. A.; Yanjun, H. Municipal Solid Waste Management and Treatment in China. In *Sustainable Waste Management Challenges in Developing Countries*; IGI Global, 2020; pp 86–114.

Yang, F.; Li, G.; Zang, B.; Zhang, Z. The Maturity and CH4, N2O, NH3 Emissions from Vermicomposting with Agricultural Waste. *Compost. Sci.* **2017,** *25*, 262–271.

Yiing, Chiee. M.; Latifah A. M. Solid Waste Management Transformation and Future Challenges of Source Separation and Recycling Practice in Malaysia. *Resour. Conserv. Recycl.* **2017,** *116*, 1–14.

Yildiz, S.; Yaman, C.; Demir, G.; Ozcan, H. K.; Coban, A.; Okten, H. E.; Goren, S. Characterization of Municipal Solid Waste in Istanbul, Turkey. *Environ. Prog. Sustain. Energy*, **2013,** *32* (3), 734–739.

CHAPTER 13

Composting for Organic Wastes Management and Profitable Use

MUHAMMAD ASHIR HAMEED[1a], ABDUL QADIR[1a], ZIA UR RAHMAN FAROOQI[1], SADIA YOUNAS[2], FAZILA YOUNAS[1*], and MUHAMMAD MAHROZ HUSSAIN[1]

[1]*Institute of Soil and Environmental Sciences, University of Agriculture Faisalabad-38040, Pakistan*

[2]*Department of Chemistry, The Government Sadiq College Women University, Bahawalpur-63100, Pakistan*

[a]*These authors work as co-first authors in this chapter.*

[*]*Corresponding author. E-mail: fazila.younas@gmail.com*

ABSTRACT

Solid waste management has always been the serious problem from the last few decades. Unlike developed countries, solid waste management is a serious issue in developing countries including Pakistan. It is because of lack of waste collection strategies and processes as well as its environmental impacts. Environmental issues triggered by solid wastes are water, air, and soil contamination that poses human health impacts. From the collected wastes, one of the portions or type in solid waste is organic, which can be biodegraded and converted into compost. Compost is then used for improving soil fertility and crop production. For composting, there are numerous technologies for getting highly decomposed and nutrient-rich product (compost). This chapter highlights the mismanagement of solid wastes to the environment in developing countries focusing on environmental contamination issues. The techniques and types of composting have also been described.

Waste Problems and Management in Developing Countries. Umair Riaz, Shazia Iqbal, & Moazzam Jamil (Eds.)

13.1 INTRODUCTION

Globally uncontrolled population and improper urbanization results in the large quantities of waste generation which is an increasing problem for the society (Onwosi et al., 2017). Improper handling and dumping of this waste create pressure on the society by creating environmental pollution thus destroying natural habitat (Lin et al., 2018), although waste that is being produced is of both organic and inorganic in nature and the former can be of certain use than the later one. Nowadays, there is a growing interest in improving the management of an organic proportion of solid wastes (Wei et al., 2017). More than 50% of the waste that is produced by both the developing and developed countries is included in the category of biodegradable or recyclable waste if handled properly as it contains most of the food waste and other domestic waste (Tai et al., 2011; Thi et al., 2015; Troschinetz and Mihelcic, 2009). This much amount of organic waste caused multiple socioeconomic issues even in a worst scenario destroying the habitat and ecosystems (Awasthi et al., 2014; Sukholthaman and Sharp, 2016). To defeat these worries, efficient administration strategies for the waste must be adopted (Moh, 2017). Food energy nexus is mostly dependent upon the treatment and handling of the organic waste so as to protect the environment and to fight the global food security issue (Lin et al., 2018).

In some urban and rural regions of developing countries, solid waste disposal is a common and stagnant problem. Most of the waste is dumped directly into the main holes, due to poor and constant waste collection system, suitable landfills, open channels, and gutters. This direct disposal often results in the blockage of gutters and causing a severe sanitation issue. Most of waste, which is generated from the housing societies, residential areas, and even from offices, comprises plastics stuff, paper products/cellulose products, and a large proportion of food materials. The products of plastic and food materials sometimes are considered the main issues of chocking sanitation systems. However, such toxic substances are environmentally harmful due to the large persistent time, and its decomposition even enhances the biological oxygen demand that not only destroys the aquatic ecosystems but also plays a devastating role in the reduction of beneficial microbial count (Abdel-Shafy and Mansour, 2018).

To achieve the goal for the sustainable environment, the principle of 3Rs (recycle, reduce, and reuse) has always been the first line of defense against solid waste management. But due to increasing available technologies and food security issues methods are required that not only minimize soild waste management issues but also enhance food production. There are numerous approaches toward managing different waste types, that is, (1) recycling that can be done for inorganic waste, (2) composting for organic waste, (3) incineration, which can be used for organic and inorganic wastes, and (4) landfilling (Domingo and Nadal, 2009). Among the above-mentioned techniques, recycling is profitable but can be used in case of inorganic materials; incineration can yield energy but damages the environment by air pollution while landfilling damages soil and nearby water bodies by their leachates (Meena et al., 2019). To overcome all the possible issues an integrated approach is required. Composting is an approach whose byproducts are less hazardous and even can be used as sustainable synthetic fertilizer replacement through the use of organic wastes as well as for management solutions for organic waste (Awasthi et al., 2020). By definition composting is the process of converting organic decomposition of wastes into nutrient-rich stable humus-like product under aerobic environments (Ahmad et al., 2007). The composting process mainly involves the active microbial population that promotes the decomposition process and their population dynamics vary widely with time, space and usually involve a rise in thermophilic temperature caused by the heat generated by them (Farrell and Jones, 2009). Organic waste (food waste) usually contains much of the substances that are rich in nutrients (micro and macro nutrients) and this process gives a nutrient-rich and stable product that can be used as an organic fertilizer that can enhance soil health and productivity (Qian et al., 2014; Wang et al., 2015). UN-Sustainable Development Goals no 2 which clearly states zero hunger and it involved a major section that cover the food security issues and to provide a clean environment for life on land, that is, SDG No 15 for both these goals composting can play an important role. Potential applications of organic waste after the composting process involve its use as fertilizers and conditioners for soil, energy rehabilitation (heat, liquid-fuel, electricity), and chemical manufacturing (unstable organic acids, nitrate products, and alcohol). To increase the productivity in past few decades farmer has usually used livestock manure as a fertilizer to boast soil heath. Local biosolids and industrial organic waste have also been used to a lesser extent for this goal (Gupta et al., 2014). Energy derived from animal manure, crop waste products, and certain other environmentally friendly waste has been used for farming in various regions of the globe. The use of various organic wastes in the farming sector depends on many factors, involving waste attributes such as nutritional elements and heavy metal concentrations, energy value, waste availability, advantages for farming and regulatory issues. The significance of such factors may vary depending on the kind of organic waste, but for most organic waste use considerations are similar (Westerman and Bicudo, 2005).

Many studies have been published and state the importance of composting in sustaining the life and environment by reducing pollution and enhancing the productivity of soil by improving soil health (Diacono and Montemurro, 2011; Dotaniya et al., 2016; Farrell and Jones, 2009; Semple et al., 2001). This chapter briefly explains the types of waste and sources. It unveiled the strategies to overcome the solid waste problem through innovative and mechanistic approaches. This will provide a better understanding of the waste generation handling including the biodegradable waste that is produced from industry, domestic, and commercial levels.

13.2 POSSIBLE WAYS OF SOLID WASTE GENERATION AND ITS MANAGEMENT STRATEGIES

Solid wastes are collected from agriculture, food, and industrial areas. These wastes include animal manure, crop residues, food waste, municipal biosolids, and sludge from industries (Westerman and Bicudo, 2005). Solid waste management is considered a very important environmental problem due to wide higher generation rates and their potential for environmental contamination (Abdel-Shafy and Mansour, 2018). Solid waste is typically generated from different sources that are exposed to different human activities. Some studies report that the main sources of solid waste in emerging territories are households (55–80%), tracked by markets or commercial regions (10–30%) (Figure 13.1). The latter includes flexible quantities that are produced from industries, roads, organizations, and numerous other sectors (Nabegu, 2010).

Municipal solid wastes sources are very heterogeneous. They have different chemical and physical characteristics depending upon source of their origin. Household wastes consist of garden waste, food waste, plastic, wood, metal, paper, rubber, leather, batteries, fabrics, paint cans, demolition, and building materials and many other materials that can be included from the household on daily basis (Abdel-Shafy and Mansour, 2018). The heterogeneity of this solid waste generated is a major obstacle for its reuse. Therefore, this waste must be fractionated and classified before a significant treatment procedure. Waste disposal provides information on the quality of separated portions for their every possible use. However, the success of a solid waste classification method depends primarily on civic consciousness and active involvement of these waste producers in various communities, that is, how they manage to pursue the basic rules of waste classification (Valkenburt et al., 2008).

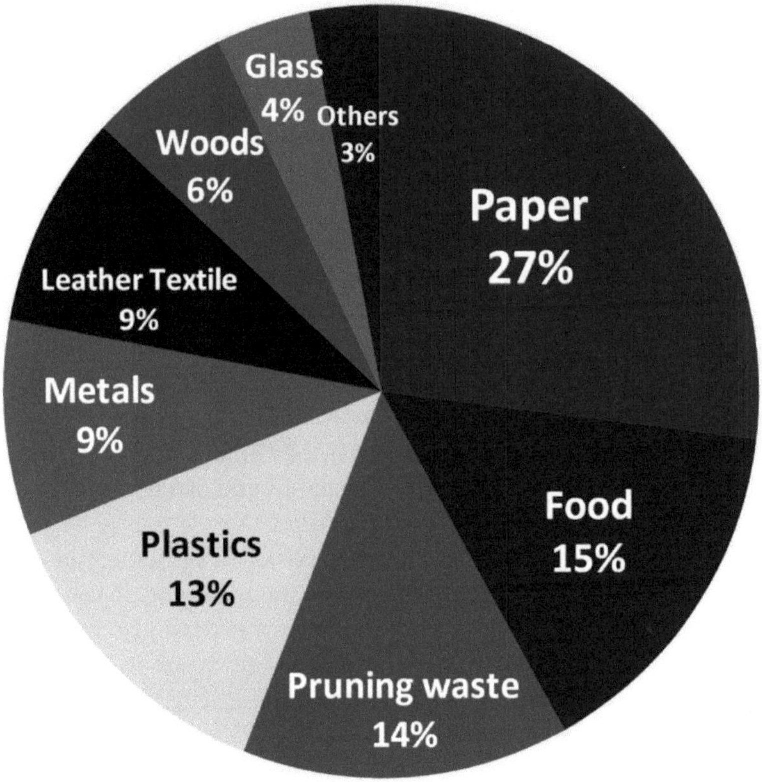

FIGURE 13.1 Types of biodegradable wastes in USA.

The Environmental Protection Agency describes food waste as "non-eaten food waste" from residential and business enterprises (comprising restaurants, food shops, agricultural product stalls, canteens, and kitchens) and industrial sources (e.g., dining rooms for employees) (Abdel-Shafy and Mansour, 2018; Zhang et al., 2007). The rise in wastes production in cities also has a significant impact on sanitation difficulties and essential services like the health, water supply, garbage disposal, and transportation infrastructure (Ferronato and Torretta, 2019; Majale-Liyala, 2011). Food commodity waste like the kitchen waste, garbage, left over of vegies, and solid food waste arises from processing of food, preparation, supply chain, and its use.

13.2.1 DOMESTIC WASTE

Some of the most important environmental issues constitute the collection, managing, and discarding of municipal waste in an urban environment (Abdel-Shafy and Mansour, 2018). The most significant environmental challenges are the compilation, administration, and the removal of municipal waste in city areas. The lack of dumping of solid waste leads to serious environmental issues. This contains soil, air, water, and aesthetic contamination (Louis, 2004). Because of the rise in greenhouse gas, environmental issues are associated with human health disorders (Weigand et al., 2003). The waste stream from industrial sources differs from dangerous substances in domestic waste (Abdel-Shafy, 2015; Abdel-Shafy et al., 2016; Abdel-Shafy and Mansour, 2017). They are not restricted by dangerous waste regulations like the European Hazardous Waste Directive 91/689/ EEC and the United States Resource Conservation and Recycling Act of 1976 (RCRA) (Abdel-Shafy and Mansour, 2018). It is generally believed that the amount of organic waste is small, so that the risk of disposal is negligible. However, the separate dumping of industrial, municipal solid sludge as well as other waste has increased the significance of the poisonous and dangerous elements that are present in the waste (Slack et al., 2005).

People are very concerned about the existence of numerous chemicals in domestic products (Blundell, 2003). The implications of the disposal of organic waste and the effects on the environment must also be considered. Consequently, the disposal of this sort of organic waste at landfills should comply with the applicable regulations to reduce environmental risks (Slack et al., 2005). Worldwide, about 71% of waste other than organic waste is settled in the landfill (Zacarias-Farah and Geyer-Allély, 2003). This type of waste mainly contains dangerous materials, including a few batteries, paints, Hg-containing waste, medicines, car repair goods, and numerous other products (Slack et al., 2004). Instead, more than 53% of the landfills contain cardboard, garden waste, paper, and food that can be broken down by anaerobic bacteria (Segbeaya et al., 2019).

13.2.2 INDUSTRIAL AND COMMERCIAL WASTE

Industrial and commercial waste mainly includes plastics, papers, steels, and other related products while plastic waste is composed off natural hydrocarbons and their calorific value is between 30 and 40 MJ/ kg. So, they can be burnt to produce electricity or used as fuel to substitute for fossil fuels in manufacturing processes like blast furnaces and brick kilns (Everaert and Baeyens, 2002). But these technologies are not used everywhere and result in air pollution unless effective waste and air pollution management is applied to reduce greenhouse gas emissions (Abdel-Shafy and Mansour, 2018; Eriksson and Finnveden, 2009). Otherwise, these plastic materials and other nonrecyclable items will continue to be managed and treated/dumped in landfills. Therefore, the burning of plastic and other nonrecyclable wastes pose environmental risks (Abdel-Shafy and Mansour, 2018). Solid wastes that are produced from the industrial and commercial sites are always of major concern as these are very critical to handle due to its large persistence time in the environment and its toxicity. Its entry into the food chain and the byproducts ability to reduce the growth and productivity is still needed to be studied critically.

13.2.3 AGRICULTURAL WASTE

Agriculture being the major contributor to the economic sector produces much more waste than expected. Agriculture sector produces large amounts of residues every year. To create a balanced nutrient waste disposal system for agricultural plants, a variety of technologies that are suitable for the processing of agricultural biomass must be considered about its long-term economic feasibility. Agricultural waste contains biodegradable hemicellulose and cellulose materials that are decomposed to provide plants with good nutrition. Establishing a farming waste disposal system requires introduction of new approaches in the agricultural and waste treatment methodologies (Barham et al., 2015; Launio et al., 2014). From the view of garbage disposal literature, the analysis of agricultural plant remains often focuses on a particular treatment approach with various waste feedback options (Baruya, 2015), or on the evaluation of the impact on the environment of a particular treatment plant (Broitman et al., 2018).

Many technologies are used in agriculture waste management, that is, baking is an anaerobic thermochemical process in which charcoal is produced (Bar-Ziv, 2012); pyrolysis is the procedure of anaerobiotic thermochemical incorporation of wood-based ingredients in the temperature spectrum of 300°C–700°C. The input is typically based on leaves and wood litter and the product is biochar and heat (Ahmad et al., 2014), mixing animal feed is the process of adding nutrient residues to cattle and sheep feed. In addition to silage, the diet of cattle and sheep can also contain 10–50% agricultural residues (Yosef et al., 2015); refused derived fuel (RDF) is a physical procedure of shredding and pressing residue into agglomerates or bits with a soaking cost for energy production within the industrial practices. The expertise uses wood scraps or leaves as an input (Srivastava et al., 2014) while composting is an aerobic procedure that uses microbes to improve the soil. The plant waste used in this technique is organic wastes (Raviv et al., 2005; Lokeshwari and Swamy, 2010).

13.3 MANAGEMENT OF SOLID WASTE

There is evidence that the high recovery rate is related to tipping at the disposal site (Wilson et al., 2010). The high waste removal price has a beneficial effect on the substantial waste generated during recycling. This will contribute to the advantages of recycling or a value chain of solid garbage. Social influence, altruism, and governing factors are essential reasons for some areas to develop solid reprocessing behaviors (Minghua et al., 2009). Some scientists also reported that people who regularly throw normal rubbish in the trash can have a greater chance to recover some products at the home. In most instances, the quantity of portions that people distinct, sort, and gather at home boosts as the distance from the trash increases. To expand the reprocessing rate, local governments want to promote the market for recycled materials and increase the number of skilled workers at recycling companies (Minghua et al., 2009).

In fact, waste management deals with the use of multidisciplinary methods such as engineering, humanities, sociology, and biology (Francois et al., 2006). A country's level of development broadly reflects the rights to waste disposal and the choice of another disposal (Cheng et al., 2007; Mahmood et al., 2017). Due to its possible heterogeneous structure, some problems can arise during the treatment of solid waste. The physical properties of solid waste are important for the selection and design of suitable disposal methods (Fudala-Ksiazek et al., 2016). Consequently, one should know the physical properties of household waste, including its composition, calorific value, and moisture content, so that you can choose a suitable management method. The moisture content varies between 5% and 40% with an average of 20% of solid waste. The scope of waste management is very broad and varies depending on the socioeconomic structures and the regional features of solid garbage (Hui et al., 2006). United Nations Environment Program (UNEP) estimated that solid waste disposal accounts for between 3% and 5% of greenhouse gas emissions. This is primarily due to the escape of methane (CH_4), carbon dioxide (CO_2), and nitrous oxide (N_2O) emissions from open landfills. Other

CO_2 emissions come from the current processes like the transportation and garbage disposal (Abdel-Shafy and Mansour, 2018; Chandak, 2010).

However, proper waste management can reduce greenhouse gas emissions in a variety of ways, involving energy generation, compost production to be used as a substitute for synthetic fertilizers, and storage of carbon in farming and landfill sites. Different internationally renowned institutions suggested a focused waste management Principle 3R (reduce, reuse, and recycle). These 3R are garbage avoidance, circular financial system inception, cleaner constructions, and enhancement of the value of the waste by conversion into a reliable source of energy and raw material. Insufficient waste treatment can lead to changes in the ecosystem, involving air, water, and soil contamination and therefore pose a real risk to human health (Rushton, 2003). Sufficient distinction between organic and inorganic waste is a necessary condition for efficient energy production. Indeed, organic waste products can affect thermal technology's effectiveness in the conditions of energy generation and greenhouse gas emissions (Abdel-Shafy and Mansour, 2018; Chandak, 2010).

13.4 WASTE MANAGEMENT STRATERGIES

Following are the major adoptive and sustainable approaches for the generated waste handling. These approaches have been used for centuries and proven useful in managing the municipal, domestic, agricultural, and industrial wastes.

13.4.1 POTENTIAL LANDFILLS AND THEIR USAGE

Accessible, unregulated, and poorly regulated dumping is widely practiced in many metropolitan cities, resulting in significant environmental devastation. In towns and cities about 90% of solid waste is disposed of directly on land through an unsatisfactory way. In several coastal cities, such dumping practice has contributed to accumulation of heavy metals leaching rapidly into coastal waters. The land availability for disposal of waste in bigger towns or cities is very small (He et al., 2019). In most urban areas, solid waste is discarded in low-lying areas far from the main city without accepting the sanitary landfill rules. Waste compaction, leveling, and landfill were hardly found at most disposal sites and these low-lying disposal sites do not have a leachate collection system or inspection and collection equipment for landfill gas (Miao et al., 2019). Since there is no isolation of solid waste at the origin, all the waste including hospital hazardous waste usually paves the way for the dumping sites. Industrial waste is also very frequently stored at dumping sites intended for household waste (Gautam et al., 2019). An appropriate and recommended approach for the ultimate management of solid waste is sanitary landfill. It is an essential aspect of the management of solid waste, since all other approaches provoke such residues that must be rid of by landfilling. However, it seems that over the next few years, landfilling in Pakistan will continue to be the most accepted method, during which some changes still must be taken to assure sanitary landfilling. Also, more work is needed to be done on the materials and lining thickness of the inner landfill sites.

13.4.2 ORGANIC WASTE RECYCLING AND REUSAGE

The natural process will begin to decay if organic waste is left untreated, resulting in smells, accommodation, and food for a wide range of insects and rodents, in turn carriers of disease, resulting in major health problems. Biological activity provides the basis of recycling in complex natural processes for integration, decomposition, and stabilization of organic waste. These cycles are sometimes very crucial to complete the biogeochemical cycling of the certain element. This recycling also helps to control and estimate about the solid waste production in the rural and urban areas.

13.4.3 AEROBIC COMPOSTING

Composting is the product of the organic waste under hot and humid conditions by the bacteria, whereas the product is of agricultural value. It is used as fertilizer and is pathogens-free and non-odorant (Yu et al., 2020). Due to the composting phase, waste volume can be reduced to 50–85%. The techniques of composting may be manual or mechanical and therefore be referred to as a manual or mechanical process. Manual composting and mechanical composting plants are built in major cities in smaller urban centers (Syafrudin et al., 2020).

In the early initiatives of the Punjab government in 2020 and the water and sanitation agency Pakistan, composting was promoted, regarding solid waste management primarily based on the encouragement of urban solid waste composting. In most cities in the developing world, the provision of public services by local officials alone cannot be maintained (Korai et al., 2020).

The range of operations ranging from itinerant manual laborers to major recycling factories are independent operators involved in waste management, procurement, segregation, repair, resale, and recycling. Kabaris are large-scale waste dealers working from stores and warehouses. Safai Kamai Bank and Waste Busters gather garbage from households and charge around $2 a month, which includes the distribution of about 30 rubbish bags. The refuse is taken to the transfer station where it is separated and loaded into trucks for recycling. Pakistan Environmental Welfare and Recycling Program (PEWARP) has built a small processing plant that manufactures three organic waste products purchased from itinerant buyers on the big vegetable market in Karachi (Korai et al., 2020). The vegetable waste is processed, and the liquid extract extracted, resulting in the selling of liquid concentrate as a pesticide, the dilution of liquid sold as fertilizer and solid residue. Shehri, a Karachi-based NGO, also known as "Citizens for a Better Environment," focuses primarily on protecting and preserving the natural and built environment. It has developed guidelines for upgraded bin designs and encourages understanding of the management of solid waste. By offering opportunities, there is a strong need to progressively involve the private sector in SWM in other small towns and cities (Azam et al., 2020).

13.4.4 VERMICOMPOSTING

Through the collective activity of earthworms and aerobic microbes, vermicomposting includes stabilization of organic waste. Biodegradable organic matter microbes originally occur through additional cellular enzyme activity (primary decomposition). Earthworms rely on partially decomposed substances, consuming organic substances five times a day. The food in the worm intestines is further decayed and hence the particle size is decreased. The worm cast is a fine, nontoxic, and granular material. It can serve as a bio-fertilizer in agriculture. The use of vermicomposting was in Hyderabad, Bangalore, Mumbai, and Faridabad. Experiments have also been performed to build vermicomposting kits for households. The area needed however is larger than dry composting (Sami-ur-Rehman et al., 2020; Srivastava et al., 2020; Yadav and Garg, 2019). This chapter revels that composting is the best solution in all circumstances (rural, urban or urban, etc.) where space is available because it avoids the load on MSW to collect and transport municipalities and therefore decreases the burden on landfills. It also offers a valuable agricultural byproduct.

13.5 INTRODUCTION TO COMPOSTING PROCESSES

Efficient planning and management approaches regarding the amount and type of waste are very important for sustainable environmental management of solid waste. Therefore, the most significant process is the quantification and description of all viable waste disposal techniques (Senzige et al., 2014). This information can of course also be helpful when planning the infrastructure, policy making, and making scale decisions in connection with the integrated waste management plan (Al-Jarallah and Aleisa, 2014). The most common and cheapest way of disposing of solid waste in ancient times was landfilling (Mahmood et al., 2017). Solid waste was disposed off in large open spaces with a low population density in these earliest periods (Cheng et al., 2007; Francois et al., 2006). From the most recent methods, composting is being done as waste management as well as substitute of chemical fertilizers to amend soils with nutrients. The use of microbes, earth worms (vermicomposting), and soldier fly larvae (*Hermetia illucens*) for organic wastes treatment has huge potential and sustainable treatment option. The added value of organic waste may be achieved through the compost and oxygen-free fermentation. The benefit of compost production is the technological effortlessness of the process (Abdel-Shafy and Mansour, 2018).

Composting is the biochemical and aerobic decay process of organic waste under controlled conditions (Kulcu and Yaldiz, 2004). It consists of three successive stages: (1) the activation stage, (2) thermophilic stage, that is, abrupt temperature increasing stage, and (3) the intermediate temperature stage, in which the organic material is cooled to ambient temperature (Abu-Khayer et al., 2013). The conversion of organic substances into stable and nutrient-rich product is caused by the enzyme activity of specific microbial communities (Agnolucci et al., 2013; Federici et al., 2011). The metabolic activity of microorganisms generates heat, which means that the organic substance is subject to physical and chemical changes and becomes CO_2 and humus-like final products (Cooperband, 2002).

The whole composting process is affected by various factors, that is, the nutrient balance, moisture, particle size, porosity and pH, as well as oxygen level, temperature, and compression (Zhang et al., 2013). It may be important to check parameters like the temperature, C/N ratio, pH, nutrient content, bulk density, moisture, porosity, particle size, and oxygen source to understand the required optimal process conditions. During the composting process, microorganisms need C, P, N, and K for their food requirements (Iqbal et al., 2015). Furthermore, Bernal et al. (2009) reported that decomposition of organic waste can vary throughout the composting process, depending upon chemical arrangement, natural load on the compost matrix, and the microbial efficiency in waste. Similarly, environmental conditions have a direct impact on bacterial activity as well as on the rate of decay of organic matter through the composting procedure. Composting is seen as an oxygen-consuming procedure in which organic substances are hydrolyzed to a stable product called compost (humus) (Białobrzewski et al., 2015). For this, anaerobic organism *Clostridium* spp. is involved in the development process (Franke-Whittle et al., 2014).

13.6 COMPOSTING AND ENVIRONMENTAL FACTOR AFFECTING IT

Composting is currently used as the main method of stabilizing agricultural organic waste by breaking down biodegradable components in the microbial community in a controlled environment. Microbes and fungi play a significant role in the breakdown and mineralized the agricultural waste (Zhu et al., 2020). There are numerous factors that affect composting, like the water content, C: N ratio, temperature, and airflow. Working with a nonoptimal C/N ratio increases the processing time. When the humidity is high, the compost cannot move through the air so the water will fill the necessary air space for anaerobic fermentation. On the other hand, microorganisms cannot develop and alter the normal biopharmaceuticals during composting if the water content is too poor (Sánchez et al., 2017). If the air supply is improper, the process slows down as microorganisms that are unsuitable for composting with poor growth even the composting is fermented anaerobically. Certain lignocellulosic materials such as wood scraps or wood chips can also slow down the composting process because they are resistant to microbial enzymatic attack (Zayed and Abdel-Motaal, 2005) (Figure 13.2).

Regarding the connection between the optimal humidity levels and the compost moisture limit, in the advanced concept of moisture control, it is concluded that a 60% to 80% limit is best for composting. The main contribution of the composting process to pollution includes the exhaust of gases or volatile substances into the air and washing out of pollutants into the groundwater. Soil pollution like heavy metals is more related to the use of compost in the field and has nothing to do with the composting process (Zhang et al., 2011).

Several composting technologies exist like windrows, aerated static dumps, bunkers, and in-vessel systems (Füleky and Benedek, 2010; Wang and Li, 2009). The solids are stacked in windrow composting in long parallel rows (conventional system) (conventional system). For better supply and porosity, the windrows are rotated, mixing moisture, or removing, and

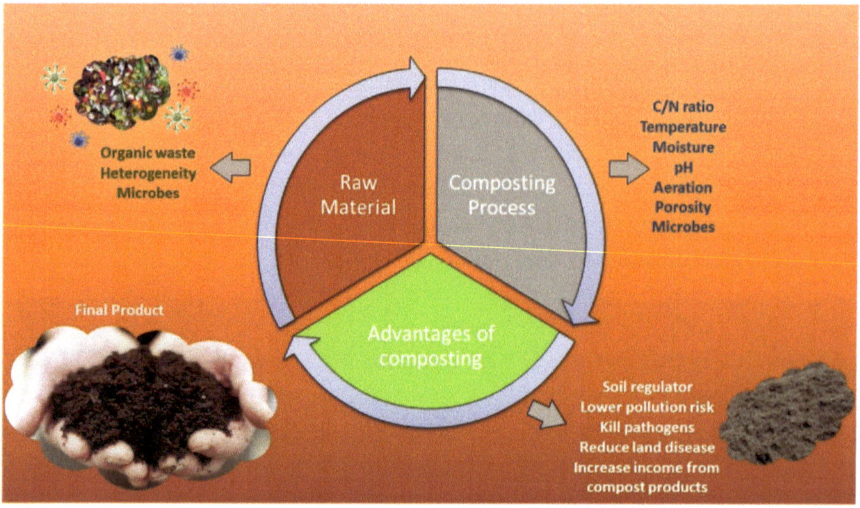

FIGURE 13.2 Environmental factors affecting the composting process.

redistributing cooler and hotter pile pieces. The windrows may be trapezoidal or triangular according to the characteristics of the machinery used to blend the stacks. The scale of the windrows depends on the number of residues to be processed and the environment appropriate for work (Wang and Li, 2009). The windrows can be 3 m in width, 2 m in height, and 25 m in length (Vargas et al., 2005). Under these conditions, about 1–1.2 m^2 of compost is required for one ton of initially formulated material. Aerated static piles use an air distribution system that consists of the perforated pipe at the bottom of the pile; this system decreases the area needed to be composted because the waste cannot be transmitted to another ventilation area. Moreover, the temperature does not drop sharply during this replacement (Wang and Li, 2009). For the processing of large quantities of residues, bunker compost is sufficient. This system includes durable materials and a raised floor with a PVC pipe network for the air distribution in a rectangular space enclosed by walls. Odors may be managed with finished composition or combined with enclosed buildings with external biological filters on the material in the bunker (Sánchez et al., 2017). The in-vessel method is ultimately based on the use of mainly cylindrical solid-state bioreactors. Temperature, humidity, and oxygen supply can be regulated theoretically (Mohee and Mudhoo, 2005); however, it is very difficult for large bioreactors to control these operating parameters due to the fluid dynamic constraints that arise with increasing size (De Guardia et al., 2012); therefore, the system is best used for laboratory and pilot scale research. In principle, the use of bunkers, systems in vessels, and aerated static piles may reduce costs compared to conventional systems, as the area used is reduced and time taken to work can be reduced by better ventilation. However, there has not been a sufficiently detailed report on the economic feasibility assessment of this system, so further research is needed, especially using modeling and simulation tools and plant data. Semi-industrial trials or commercial installations are usually rare or not available.

The composting process is divided into four stages. The first stage or mesophilic stage starts with the initial organic decomposition of mesophilic microorganisms, such as warm microorganisms that grow between 15°C and 35°C. These bacteria belong to the family *Pseudomonas, Rhodobacteraceae, Comamonadaceae, Enterobacteriaceae, Streptomyces*, *Bulbecaceae*, etc. These microorganisms use soluble and easily assimilated compounds that are present in the raw materials used to make compost, such as sugars, amino acids, and lipids (Bernal et al., 2009). The metabolic activity of microorganisms produces an exothermic reaction, which raises the compost temperature up to 65°C–85°C. The competition of mesophilic micro-organisms is reduced and replaced with thermal micro-organisms under these conditions (Insam and De Bertoldi, 2007). During the period known as the thermophilic process, the propagation of actinobacteria (mostly in the *Thermoactinomycetaceae*, *Thermomonosporaceae*, and *Pseudonocardiacaceae* families) and other Thermophile occurs. Enzymes that degrade complex molecules such as cellulose, lignin, hemicellulose, and proteins are present in these microorganisms (Franke-Whittle et al., 2014; Insam and De Bertoldi, 2007; Probst et al., 2015). Similarly, the removal of germinating infectious agents and seeds occurs as temperatures rise (Neklyudov et al., 2008; Vélez-Sánchez et al., 2008). Then, energy was used up and the temperature of the compost reached 15°C–35°C, causing mesophilic bacteria to colonize the compost for the second time. The remaining sugar, cellulose, and hemicellulose are broken down by mesophilic microorganisms in this process called cooling phase. The catalysts of the humic substances are produced in the final step or maturation (Sánchez et al., 2017; Zeng et al., 2010).

13.7 COMMERCIAL PRODUCTS DERIVED FROM BIODEGRADABLE WASTES

Exhaustive agricultural output typically leads to a loss of soil fertility, water pollution, soil erosion, compaction, and a reduced content of organic matter.

A low level of organic matter is a common characteristic of Mediterranean soils. Since it directly influences the physical, chemical, and biological characteristics of the soil, it correlates strongly with potential productivity and fertility (Alburquerque et al., 2009). Hence, the usage of compost on farmland can increase the organic content of the soil, and compost can perform an important role in contemporary sustainable agricultural practices (Chowdhury et al., 2013).

13.7.1 USES OF ORGANIC WASTES-DERIVED COMPOSTS

The potential use of organic wastes as animal feed was assessed in a rather extreme study (Garcıa et al., 2005). The author concluded that after the heat treatment (65°C, 20 min) all waste was effectively disinfected and suitable for animal consumption. In addition, the lead and cadmium contents in the domestic waste tested in their research are higher than those permitted in European Union animal feed. Regardless of whether it is theoretically possible to use household waste lawfully and securely in the manufacture of animal feed, this is not a generally acceptable procedure (Farrell and Jones, 2009). Compost is used as complementary fertilizer with numerous benefits according to Chowdhury et al. (2013): (1) to improve the aggregate stability and water retention capacity, (2) to enhance cation exchange capacity, (3) to enhance microorganism activity, and (4) to promote decomposition of synthetic fertilizer and pesticides residue.

13.7.2 AGRICULTURE AND HORTICULTURE

Agricultural and horticultural uses represent a tremendous potential marketplace for the composting of municipal waste. To disposal of waste, the soil can therefore represent an almost endless collection of environmental friendly residues (Farrell and Jones, 2009). The nutritional value of household waste compost and its potential to improve soil quality make it an ideal choice for farming. That provides the right safety measures that are taken to reduce damage to the environment and gain community recognition. In soil addition of compost unreasonably enhances the heavy metal content (Ramos and López-Acevedo, 2004). If these residues are applied to the land, it is possible to significantly increase the SOM content, which is decreasing sharply in various parts of the earth, especially in drier conditions (Bellamy et al., 2005). Politically, the conservation and increase of the organic matter reserves in the soil (a key indicator of soil quality) is viewed politically as the key to maintaining ecosystem functions. Many experiments have demonstrated that the addition of municipal waste compost to the soil can at best temporarily increase the SOM content and promote the biological activity of the soil. Extra benefits of adding compost involve reducing erosion loss; bulk density is reduced and improving structural constancy (Tejada and Gonzalez, 2007).

As similar amounts of P deliver to mineral fertilizers as well as MSW compost application in some soils (Mkhabela and Warman, 2005). At standard recommended rates 10–100 t/ha, due to the insecurity about the availability of nutrients during a growing season, it is also challenging to predict exactly the harvesting needs. From the perspective of horticulture and gardening, there have been recently increasing concerns regarding peat mining and the demolition of peatlands by the horticultural industry, which is why substitutes to peat are instantly required (Farrell and Jones, 2009). Organic waste compost contains organic substances and nutrients that becomes more beneficial after adding with peat (Silva et al., 2007). The main barrier to its application is the huge quantity of high-quality green waste compost that is saturating the market. The ability to use household waste compost in containers of horticulture remains extremely limited.

13.8 SUSTAINABLE AND ENVIRONMENTAL REMEDIATION PRODUCTS

13.8.1 MINERAL SITES WASTE

This waste spikes are widespread globally and their geographical weakening and recovery can eventually take thousands of years (Soudejani et al., 2019). The main barriers to repairing these areas are the absence of a finely structured matrix that can retain water and provide nutrients. Waste composting may be an inexpensive solution to this issue, if it could be purchased on site and used safely without running the risk of wind and water erosion. Although green waste has been successfully applied to remediate these sites, there are no reports of the usage and achievement of composting municipal waste in large-scale remediation programs. This is primarily due to the current legal restrictions that prevent the spread of composting of municipal waste. Because inorganic fertilizers are easily lost through leaching, composting offers a better solution because it provides immediate nutrients (which are kept in solution and in exchange places, etc.) and at the same time creates a long-term supply with the decomposition of the compost. A potential problem is that too much paper-based material should not be used in the initial organic waste raw materials, as this can temporarily include nutrients (e.g., N and P), which restricts the plant growth over the short term.

13.8.2 BIOCHAR

Biochar is commonly used as a soil conditioner and is converted from residue through an anoxic, high-temperature thermal decay procedure (called pyrolysis) (Li and Chen, 2018). Also, different types of raw materials can be used in the production of biochar, involving cultivated waste (Cea et al., 2012), wood (Creamer et al., 2014), animal dung (Zhang et al., 2018), and biosolids (Hossain et al., 2010). Therefore, the manufacturing of biochar can be a promising alternative method of soil improvement for sustainable farming, even though its economic feasibility is still the most feasible (Clare et al., 2014; Li et al., 2019).

The two key pyrolysis procedures that generate biochar, such as slow and fast pyrolysis, vary in terms of heat transmission rate, temperature limit, and time of residence. The most conventional method is slow pyrolysis to produce biochar (Qambrani et al., 2017). In the slow pyrolysis process, the temperature limit is usually 300°C–700°C, and several hours to several days is the residence time (Amen et al., 2020a; Duku et al., 2011). Though slow pyrolysis generally leads to a better yield of biochar and improved homogeneity, it is time consuming and so not suitable for large scale. The rapid pyrolysis has newly invited wide attention, since the ensuing liquid can be used in high yield (up to 75%) for the production of biofuels or other fuels (Qambrani et al., 2017).

Biochar is too valuable to be used directly in the ground without being used at least one time for more useful objectives, regardless of whether it is used as a storage capacity for unstable nutrients, as an absorbent. Batteries are used as filters in sewage treatment plants, silage agents, or feed supplements (Amen et al., 2020a; Yu et al., 2018). While you use the same raw material, distinct temperatures have a significant impact on obtained biochar. Its physical and chemical properties are comparable to those of pecan shell biochar, which is manufactured at 700 °C (Novak et al., 2009). Many laboratory and field researches have demonstrated that biochar increases water retention capacity (Amoakwah et al., 2017), increases uptake of nutrient (Malakootian et al., 2018), facilitating soil pH (Teutscherova et al., 2017), improvement of the CEC (de Andrade et al., 2015), and enrichment of the bacterial diversity (Edenborn et al., 2018). The scientists also consider that addition of biochar to agricultural soils will efficiently decrease greenhouse gas emissions (e.g., CO_2 and N_2O) (Cayuela et al., 2014; Li et al., 2019).

Before composting, these applications can be used in sludge pits or wastewater treatment plants for farmers. It must only be processed in the soil toward the end of this "cascade" to produce *terra preta* (Schmidt and Wilson, 2014). The poured use of biochar in animal farming: silage agent, feed additive/supplement, water treatment in fish farming, litter additive, slurry treatment, dung composting (Gerlach and Schmidt, 2012); use as a soil conditioner: carbon source, compost, substitute for peat in potting soil, plant protection, fertilizer for trace elements; use in the building sector: insulation, air cleansing, moisture management, shield against electromagnetic pollution (Schmidt, 2013). Use as a soil stabilizer for remediation, soil substrates (highly adsorbing, use in cleaning wastewater with heavy metals) (Hussain et al., 2019), production of biogas, biogas slurry treatment, micro-filters, macro-filters in developing countries (Schmidt, 2012); exhaust filters for other purposes for divers: (emission control, indoor air filter) industrial materials, carbon fiber, plastics) electronics, semiconductor, battery, cosmetics, (soap, skin cream, therapeutic bath additives) color and coloring (food coloring, industrial coatings), energy and biogas production, granulate, alternatives to brown coal (Inthapanya et al., 2012); medicines: detoxification, carrier of active substances; textiles, health filled mattress, fill pillow, shield against electromagnetic radiation (Amen et al., 2020b; Lin and Chang, 2008; Schmidt and Wilson, 2014).

13.8.3 GREEN FUELS

With increasing the population, energy consumption also increasing that becomes a great interest in developing countries as complementary energy resources to complement the massive global demand for oil and gas. Biogas is a by-product of the anaerobic decay of organic substances and it has been viewed as an alternate energy resource. Biogas can be used for cooking, heating, and lighting in small household units, while larger facilities can be used for heating system or generating electricity. Another prospective source of energy from renewable sources is alcohol-based, a fluid biofuel that could be produced by fermenting organic waste like the cassava, sugarcane, molasses, and corn. In addition to its present utilization as a fuel, ethanol has been used extensively in the chemicals industry, pharmaceutical, and cosmetic industries. The three main raw materials for biomass that are appropriate for ethanol manufacture are: (1) raw materials comprising sugar (like the sweet sorghum, sugarcane, molasses, etc.); (2) starchy substances (like the potatoes, cassava, corn, etc.); (3) cellulosic materials (like the timber and agricultural waste products, etc.). The biological reaction throughout anaerobic absorption in a biogas fermenter reduces the organic content by 30–60% and creates a stable slurry that can be used as a compost or soil improver. The production of biogas in the rural areas may have many benefits, for example, reducing demand for electricity, coal, oil, and firewood, as well as related to administration/distribution networks. The organic substance necessary for biogas manufacturing is plentiful and easy to obtain. Reduce the need for firewood, leave the forest idle, and continue reforestation (Polprasert, 2007).

13.9 SUMMARY

Solid waste is the most significant environmental challenge. Inadequate facilities of waste handling can lead to changes in the ecosystem, including pollution of air, water, and soil, and therefore pose a real threat to human health. Various researches indicate that residents are born near low birth weight municipal waste facilities, have congenital abnormalities, and have few cancers. The increasing production of solid waste is burdening the elevated costs of the municipal financial plan. The rise in population, the acceleration of urbanization, the booming economy, and the improvement of living standards have significantly accelerated the speed, quantity, and quality of solid municipal waste generation. Over time, the biodegradation of solid municipal waste has become a significant factor in determining the quantity of decomposable materials (especially the organic content). Municipal waste from emerging countries is very diverse in nature. Most countries around the world face major challenges in household food waste disposal. Composting is usually a time-consuming procedure, but advances in composting technology have shortened the length of the composting development.

ACKNOWLEDGMENTS

The authors want to thank the Institute of Soil and Environmental Sciences, University of Agriculture Faisalabad, Pakistan. Higher Education Commission, Pakistan for awarding fellowship under International Research Support Initiative Program (IRSIP) (Miss Fazila Younas No: *1-8/HEC/HRD/2020/10831*;

Mr Muhammad Mahroz Hussain No: *1-8/HEC/HRD/2020/10831*; Mr Zia Ur Rehman Farooqi No: *1-8/HEC/HRD/2020/10610*).

CONFLICT OF INTEREST

The authors declared no conflict of interest for this chapter's publishing.

KEYWORDS

- **environmental sustainability**
- **soil fertility**
- **solid waste management**
- **composting**
- **UN-sustainable development goals**

REFERENCES

Abdel-Shafy, H. I. Chemical Treatment for Removal of Heavy Metals from Industrial Wastewater. *Egypt. J. Chem.* **2015,** *58,* 1–12.

Abdel-Shafy, H. I.; El-Khateeb, M. A.; Mansour, M. S. Treatment of Leather Industrial Wastewater via Combined Advanced Oxidation and Membrane Filtration. *Water Sci. Technol.* **2016,** *74,* 586–594.

Abdel-Shafy, H. I.; Mansour, M. S. Treatment of Pharmaceutical Industrial Wastewater via Anaerobic/Aerobic System For Unrestricted Reuse, 2017.

Abdel-Shafy, H. I.; Mansour, M. S. Solid Waste Issue: Sources, Composition, Disposal, Recycling, and Valorization. *Egypt J. Petrol.* **2018,** *27,* 1275–1290.

Abu-Khayer, M.; Chowdhury, M.; Akratos, C.; Vayenas, D.; Pavlou, S. Olive Mill Waste Composting: A Review. *Int. Biodeter. Biodegr.* **2013,** *85,* 108–119.

Agnolucci, M.; Cristani, C.; Battini, F.; Palla, M.; Cardelli, R.; Saviozzi, A. et al. Microbially-Enhanced Composting of Olive Mill Solid Waste (Wet Husk): Bacterial and Fungal Community Dynamics at Industrial Pilot and Farm Level. *Bioresour. Technol.* **2013,** *134,* 10–16.

Ahmad, M.; Rajapaksha, A. U.; Lim, J. E.; Zhang, M.; Bolan, N.; Mohan, D. et al. Biochar as a Sorbent for Contaminant Management in Soil and Water: A Review. *Chemosphere* **2014,** *99,* 19–33.

Ahmad, R.; Jilani, G.; Arshad, M.; Zahir, Z. A.; Khalid, A. Bio-conversion of Organic Wastes for Their Recycling in Agriculture: An Overview of Perspectives and Prospects. *Ann. Microbiol.* **2007,** *57,* 471–479.

Al-Jarallah, R.; Aleisa, E. A Baseline Study Characterizing the Municipal Solid Waste in the State of Kuwait. *Waste Manage.* **2014,** *34*, 952–960.

Alburquerque, J. A.; Gonzálvez, J.; Tortosa, G.; Baddi, G. A.; Cegarra, J. Evaluation of "alperujo" Composting Based on Organic Matter Degradation, Humification and Compost Quality. *Biodegradation* **2009,** *20*, 257–270.

Amen, R.; Bashir, H.; Bibi, I.; Hussain, M. M.; Shaheen, S. M.; Shahid, M. et al. *Arsenic Removal from Water Using Biochar-Based Sorbents: Production, Characterization, and Sequestration Mechanisms. Soil and Groundwater Remediation Technologies*; CRC Press, 2020a; pp 63–80.

Amen, R.; Bashir, H.; Bibi, I.; Shaheen, S. M.; Khan Niazi, N.; Shahid, M. et al. A Critical Review on Arsenic Removal from Water Using Biochar-Based Sorbents: The Significance of Modification and Redox Reactions. *Chem. Eng. J.* **2020b,** 125195.

Amoakwah, E.; Frimpong, K. A.; Okae-Anti, D.; Arthur, E. Soil Water Retention, Air Flow and Pore Structure Characteristics after Corn Cob Biochar Application to a Tropical Sandy Loam. *Geoderma* **2017,** *307*, 189–197.

Awasthi, M. K.; Pandey, A. K.; Khan, J.; Bundela, P. S.; Wong, J. W.; Selvam, A. Evaluation of Thermophilic Fungal Consortium for Organic Municipal Solid Waste Composting. *Bioresour. Technol.* **2014,** *168*, 214–221.

Awasthi, S. K.; Sarsaiya, S.; Awasthi, M. K.; Liu, T.; Zhao, J.; Kumar, S. et al. Changes in Global Trends in Food Waste Composting: Research Challenges and Opportunities. *Bioresour. Technol.* **2020,** *299*, 122555.

Azam, M.; Jahromy, S. S.; Raza, W.; Raza, N.; Lee, S. S.; Kim K-H. et al. Status, Characterization, and Potential Utilization of Municipal Solid Waste as Renewable Energy Source: Lahore Case Study in Pakistan. *Environ. Intern.* **2020,** *134*, 105291.

Bar-Ziv, E. Production of Bio-coal (Torrefaction) from Agricultural Waste. A Report Submitted to the Ministry of Environmental Protection, 2012.

Barham, B. L.; Chavas, J. P.; Fitz, D.; Ríos-Salas, V.; Schechter, L. Risk, Learning, and Technology Adoption. *Agric. Econ.* **2015,** *46*, 11–24.

Baruya, P. *World Forest and Agricultural Crop Residue Resources for Cofiring*; IEA Clean Coal Centre: London, 2015.

Bellamy, P. H.; Loveland, P. J.; Bradley, R. I.; Lark, R. M.; Kirk, G. J. Carbon Losses from All Soils across England and Wales 1978–2003. *Nature* **2005,** *437*, 245–248.

Bernal, M. P.; Alburquerque, J.; Moral, R. Composting of Animal Manures and Chemical Criteria for Compost Maturity Assessment. A Review. *Bioresour. Technol.* **2009,** *100*, 5444–5453.

Białobrzewski, I.; Mikš-Krajnik, M.; Dach, J.; Markowski, M.; Czekała, W.; Głuchowska, K. Model of the Sewage Sludge-Straw Composting Process Integrating Different Heat Generation Capacities of Mesophilic and Thermophilic Microorganisms. *Waste Manage.* **2015,** *43*, 72–83.

Blundell, T. *Chemicals in Products Safeguarding the Environment and Human. Health Twenty-Fourth Report*; Royal Commission On Environmental Pollution: London, 2003.

Broitman, D.; Raviv, O.; Ayalon, O.; Kan, I. Designing an Agricultural Vegetative Waste-Management System under Uncertain Prices of Treatment-Technology Output Products. *Waste Manage.* **2018,** *75*, 37–43.

Cayuela, M.; Van Zwieten, L.; Singh, B.; Jeffery, S.; Roig, A.; Sánchez-Monedero, M. Biochar's Role in Mitigating Soil Nitrous Oxide Emissions: A Review and Meta-Analysis. *Agric. Ecosys. Environ* **2014,** *191*, 5–16.

Cea, M.; Gonzalez, M.; Sangaletti, N.; Gonzalez, A.; Toro, C.; Navia, R. Biochar Derived from Agricultural and Forestry Wastes: Characterization and Potential Application for Enzyme Immobilization. *New Biotechnol. 2012*, S60.

Chandak, S. Trends in Solid Waste Management: Issues, challenges and Opportunities. International Consultative Meeting on Expanding Waste Management Services in Developing Countries, 2010; pp 18–19.

Cheng, H.; Zhang, Y.; Meng, A.; Li, Q. Municipal Solid Waste Fueled Power Generation in China: A Case Study of Waste-to-Energy in Changchun City. *Environ. Sci. Technol.* **2007,** *41*, 7509–7515.

Chowdhury AKMMB, Akratos, C. S.; Vayenas, D. V.; Pavlou, S. Olive Mill Waste Composting: A Review. *Int. Biodeteri. Biodegrad.* **2013,** *85*, 108–119.

Clare, A.; Barnes, A.; McDonagh, J.; Shackley, S. From Rhetoric to Reality: Farmer Perspectives on the Economic Potential of Biochar in China. *Int. J. Agric. Sustain.* **2014,** *12*, 440–458.

Cooperband, L. The Art and Science of Composting. Center for Integrated Agricultural Systems, 2002.

Creamer, A. E.; Gao, B.; Zhang, M. Carbon Dioxide Capture Using Biochar Produced from Sugarcane Bagasse and Hickory Wood. *Chem. Eng. J.* **2014,** *249*, 174–179.

de Andrade, C. A.; Silveira Bibar, M. P.; Coscione, A. R.; Moreno Pires, A. M.; Soares, A. G. Mineralization and Effects of Poultry Litter Biochar on Soil Cation Exchange Capacity. *Pesquisa Agropecuária Brasileira* **2015,** *50*, 407–416.

De Guardia, A.; Petiot, C.; Benoist, J.; Druilhe, C. Characterization and Modelling of the Heat Transfers in a Pilot-Scale Reactor during Composting under Forced Aeration. *Waste Manage.* **2012,** *32*, 1091–1105.

Diacono, M.; Montemurro, F. *Long-term Effects of Organic Amendments on Soil Fertility. Sustainable Agriculture*, Vol. 2; Springer, 2011; pp 761–786.

Domingo, J. L.; Nadal, M. Domestic Waste Composting Facilities: A Review of Human Health Risks. *Environ. Int.* **2009,** *35*, 382–389.

Dotaniya, M.; Datta, S.; Biswas, D.; Dotaniya, C.; Meena, B.; Rajendiran, S. et al. Use of Sugarcane Industrial By-products for Improving Sugarcane Productivity and Soil Health. *Int. J. Recycl. Organ. Waste Agric.* **2016,** *5*, 185–194.

Duku, M. H.; Gu, S.; Hagan, E. B. Biochar Production Potential in Ghana—A Review. *Renew. Sustain. Energy Rev.* **2011,** *15*, 3539–3551.

Edenborn, S. L.; Johnson, L. M.; Edenborn, H. M.; Albarran-Jack, M. R.; Demetrion, L. D. Amendment of a Hardwood Biochar with Compost Tea: Effects on Plant Growth, Insect Damage and the Functional Diversity of Soil Microbial Communities. *Biol. Agric. Horti.* **2018,** *34*, 88–106.

Eriksson, O.; Finnveden, G. Plastic Waste as a Fuel-CO 2-Neutral or Not? *Ener. Environ. Sci.* **2009,** *2*, 907–914.

Everaert, K.; Baeyens, J. The Formation and Emission of Dioxins in Large Scale Thermal Processes. *Chemosphere* **2002,** *46*, 439–448.

Farrell, M.; Jones, D. Critical Evaluation of Municipal Solid Waste Composting and Potential Compost Markets. *Bioresour. Technol.* **2009,** *100*, 4301–4310.

Federici, E.; Pepi, M.; Esposito, A.; Scargetta, S.; Fidati, L.; Gasperini, S. et al. Two-Phase Olive Mill Waste Composting: Community Dynamics and Functional Role of the Resident Microbiota. *Bioresour. Technol.* **2011,** *102*, 10965–10972.

Ferronato, N.; Torretta, V. Waste Mismanagement in Developing Countries: A Review of Global Issues. *Int. J. Environ. Res. Public Health* **2019,** *16*, 1060.

Francois, V.; Feuillade, G.; Skhiri, N.; Lagier, T.; Matejka, G. Indicating the Parameters of the State of Degradation of Municipal Solid Waste. *J. Hazard. Mater.* **2006,** *137*, 1008–1015.

Franke-Whittle, I. H.; Confalonieri, A.; Insam, H.; Schlegelmilch, M.; Körner, I. Changes in the Microbial Communities during Co-composting of Digestates. *Waste Manage.* **2014,** *34*, 632–641.

Fudala-Ksiazek, S.; Pierpaoli, M.; Kulbat, E.; Luczkiewicz, A. A Modern Solid Waste Management Strategy–the Generation of New By-products. *Waste Manage.* **2016,** *49*, 516–529.

Füleky, G.; Benedek, S. *Composting to Recycle Biowaste. Sociology, Organic Farming, Climate Change and Soil Science*; Springer, 2010; pp 319–346.

Garcıa, A.; Esteban, M.; Marquez, M.; Ramos, P. Biodegradable Municipal Solid Waste: Characterization and Potential Use as Animal Feedstuffs. *Waste Manage.* **2005,** *25*, 780–787.

Gautam, P.; Kumar, S.; Lokhandwala, S. Advanced Oxidation Processes for Treatment of Leachate from Hazardous Waste Landfill: A Critical Review. *J Clean Prod* **2019,** *237*, 117639.

Gerlach, A.; Schmidt H-P. The Use of Biochar in Cattle Farming. *Ithaka J.* **2012,** *2012*, 281–285.

Gupta, V. K.; Potumarthi, R.; O'Donovan, A.; Kubicek, C. P.; Sharma, G. D.; Tuohy, M. G. Bioenergy Research: An Overview on Technological Developments and Bioresources. In *Bioenergy Research: Advances and Applications*; Elsevier, 2014; pp 23–47.

He, P.; Chen, L.; Shao, L.; Zhang, H.; Lü, F. Municipal Solid Waste (MSW) Landfill: A Source of Microplastics?-Evidence of Microplastics in Landfill Leachate. *Water Res.* **2019,** *159*, 38–45.

Hossain, M. K.; Strezov, V.; Chan, K. Y.; Nelson, P. F. Agronomic Properties of Wastewater Sludge Biochar and Bioavailability of Metals in Production of Cherry Tomato (*Lycopersicon esculentum*). *Chemosphere* **2010,** *78*, 1167–1171.

Hui, Y.; Li'ao, W.; Fenwei, S.; Gang, H. Urban Solid Waste Management in Chongqing: Challenges and Opportunities. *Waste Manage.* **2006,** *26*, 1052–1062.

Hussain, M. M.; Bibi, I.; Shahid, M.; Shaheen, S. M.; Shakoor, M. B.; Bashir, S. et al. Chapter Two—Biogeochemical Cycling, Speciation and Transformation Pathways of Arsenic in Aquatic Environments with the Emphasis on Algae. In *Comprehensive Analytical Chemistry*; Duarte, A. C., Reis, V., Eds., Vol. 85; Elsevier, 2019; pp 15–51.

Insam, H.; De Bertoldi, M. Microbiology of the Composting Process. *Compost Sci. Technol.* **2007,** *8*, 25–48.

Inthapanya, S.; Preston, T.; Leng, R. Biochar Increases Biogas Production in a Batch Digester Charged with Cattle Manure. *Livest. Res. Rural Dev.* **2012,** *24*, 212.

Iqbal, M.; Nadeem, A.; Sherazi, F.; Khan, R. Optimization of Process Parameters for Kitchen Waste Composting by Response Surface Methodology. *Int. J. Environ. Sci. Technol.* **2015,** *12*, 1759–1768.

Korai, M. S.; Ali, M.; Lei, C.; Mahar, R. B.; Yue, D. Comparison of MSW Management Practices in Pakistan and China. *J. Mater. Cycl. Waste Manage.* **2020,** 1–11.

Kulcu, R.; Yaldiz, O. Determination of Aeration Rate and Kinetics of Composting Some Agricultural Wastes. *Bioresour. Technol.* **2004,** *93*, 49–57.

Launio, C. C.; Asis, Jr. C. A.; Manalili, R. G.; Javier, E. F.; Belizario, A. F. What Factors Influence Choice of Waste Management Practice? Evidence from Rice Straw Management in the Philippines. *Waste Manage. Res.* **2014,** *32*, 140–148.

Li, S.; Chen, G. Thermogravimetric, Thermochemical, and Infrared Spectral Characterization of Feedstocks and Biochar Derived at Different Pyrolysis Temperatures. *Waste Manage.* **2018,** *78*, 198–207.

Li, S.; Harris, S.; Anandhi, A.; Chen, G. Predicting Biochar Properties and Functions Based on Feedstock and Pyrolysis Temperature: A Review and Data Syntheses. *J. Clean. Prod.* **2019,** *215*, 890–902.

Lin, C.; Chang, C. Production of Thermal Insulation Composites Containing Bamboo Charcoal. *Tex. Res. J.* **2008,** *78*, 555–560.

Lin, L.; Xu, F.; Ge, X.; Li, Y. Improving the Sustainability of Organic Waste Management Practices in the Food-Energy-Water Nexus: A Comparative Review of Anaerobic Digestion and Composting. *Renew. Sustain. Ener. Rev.* **2018,** *89*, 151–167.

Lokeshwari, M.; Swamy, C. N. Waste to Wealth-Agriculture Solid Waste Management Study. *Pol. Res.* **2010,** *29*, 129–133.

Louis, G. E. A Historical Context of Municipal Solid Waste Management in the United States. *Waste Manage. Res.* **2004,** *22*, 306–322.

Mahmood, K.; Batool, A.; Faizi, F.; Chaudhry, M. N.; Ul-Haq, Z.; Rana, A. D. et al. Bio-Thermal Effects of Open Dumps on Surroundings Detected by Remote Sensing—Influence of Geographical Conditions. *Ecol. Indica* **2017,** *82*, 131–142.

Majale-Liyala, C. Modernising Solid Waste Management at Municipal Level: Institutional Arrangements in Urban Centres of East Africa. *Environ. Pol. Ser.* **2011,** 3.

Malakootian, M.; Mahdizadeh, H.; Nasiri, A.; Mirzaienia, F.; Hajhoseini, M.; Amirmahani, N. Investigation of the Efficiency of Microbial Desalination Cell in Removal of Arsenic from Aqueous Solutions. *Desalination* **2018,** *438*, 19–23.

Meena, M. D.; Yadav, R. K.; Narjary, B.; Yadav, G.; Jat, H.; Sheoran, P. et al. Municipal Solid Waste (MSW): Strategies to Improve Salt Affected Soil Sustainability: A Review. *Waste Manage.* **2019,** *84*, 38–53.

Miao, L.; Yang, G.; Tao, T.; Peng, Y. Recent Advances in Nitrogen Removal from Landfill Leachate Using Biological Treatments–A Review. *J. Environ. Manage.* **2019,** *235*, 178–185.

Minghua, Z.; Xiumin, F.; Rovetta, A.; Qichang, H.; Vicentini, F.; Bingkai, L. et al. Municipal Solid Waste Management in Pudong New Area, China. *Waste Manage.* **2009,** *29*, 1227–1233.

Mkhabela, M.; Warman, P. The Influence of Municipal Solid Waste Compost on Yield, Soil Phosphorus Availability and Uptake by Two Vegetable Crops Grown in a Pugwash Sandy Loam Soil in Nova Scotia. *Agric. Ecosys. Environ.* **2005,** *106*, 57–67.

Moh, Y. Solid Waste Management Transformation and Future Challenges of Source Separation and Recycling Practice in Malaysia. *Resour. Conserv. Recyc.* **2017,** *116*, 1–14.

Mohee, R.; Mudhoo, A. Analysis of the Physical Properties of an In-vessel Composting Matrix. *Power Technol.* **2005,** *155*, 92–99.

Nabegu, A. B. An Analysis of Municipal Solid Waste in Kano Metropolis, Nigeria. *J. Human Ecol.* **2010,** *31*, 111–119.

Neklyudov, A.; Fedotov, G.; Ivankin, A. Intensification of Composting Processes by Aerobic Microorganisms: A Review. *App. Biochem. Microbiol.* **2008,** *44*, 6–18.

Novak, J. M.; Lima, I.; Xing, B.; Gaskin, J. W.; Steiner, C.; Das, K. et al. Characterization of Designer Biochar Produced at Different Temperatures and Their Effects on a Loamy Sand. *Ann. Environ. Sci.* **2009**.

Onwosi, C. O.; Igbokwe, V. C.; Odimba, J. N.; Eke, I. E.; Nwankwoala, M. O.; Iroh, I. N. et al. Composting Technology in Waste Stabilization: On the Methods, Challenges and Future Prospects. *J. Environ. Manage.* **2017,** *190*, 140–157.

Polprasert, C. *Organic Waste Recycling: Technology and Management*; IWA Publishing, 2007.

Probst, M.; Walde, J.; Pümpel, T.; Wagner, A. O.; Insam, H. A Closed Loop for Municipal Organic Solid Waste by Lactic Acid Fermentation. *Bioresour. Technol.* **2015,** *175*, 142–151.

Qambrani, N. A.; Rahman, M. M.; Won, S.; Shim, S.; Ra, C. Biochar Properties and Eco-Friendly Applications for Climate Change Mitigation, Waste Management, and Wastewater Treatment: A Review. *Renew. Sustain. Ener. Rev.* **2017,** *79*, 255–273.

Qian, X.; Shen, G.; Wang, Z.; Guo, C.; Liu, Y.; Lei, Z. et al. Co-composting of Livestock Manure with Rice Straw: Characterization and Establishment of Maturity Evaluation System. *Waste Manage.* **2014,** *34*, 530–535.

Ramos, M.; López-Acevedo, M. Zinc Levels in Vineyard Soils from the Alt Penedès-Anoia Region (NE Spain) after Compost Application. *Adv. Environ. Res.* **2004,** *8*, 687–696.

Raviv, M.; Oka, Y.; Katan, J.; Hadar, Y.; Yogev, A.; Medina, S. et al. High-Nitrogen Compost as a Medium for Organic Container-Grown Crops. *Bioresour. Technol.* **2005,** *96*, 419–427.

Rushton, L. Health Hazards and Waste Management. *Br. Med. Bull.* **2003,** *68*, 183–197.

Sami-ur-Rehman, Z. A.; Bellitürk, K.; Ahmad, A.; Nadeem, M.; Waqas, M. Vermicomposting in Pakistan: Current Scenario and Future Prospectives, 2020.

Sánchez, Ó. J.; Ospina, D. A.; Montoya, S. Compost Supplementation with Nutrients and Microorganisms in Composting Process. *Waste Manage.* **2017,** *69*, 136–153.

Schmidt, H-P. Treating Liquid Manure with Biochar. *Ithaka J.* **2012,** *1*, 273–276.

Schmidt, H-P.; Wilson, K. 55 Uses of Biochar. *Biochar J.* **2014,** 286–288.

Schmidt, H. P. Biochar as Building Material for an Optimal Indoor Climate. *Ithaka J.* **2013,** *2013*, x–y.

Segbeaya, K. N.; Koledzi, E. K.; Baba, G.; Feuilade-Cathalifaud, G. The Impact of Household and Similar Solid Wastes on Kara River Quality Due to Their Potential to Release Nitrogen. *J. Environ. Chem. Ecotoxicol.* **2019,** *11*, 29–42.

Semple, K. T.; Reid, B. J.; Fermor, T. Impact of Composting Strategies on the Treatment of Soils Contaminated with Organic Pollutants. *Environ. Pol.* **2001,** *112*, 269–283.

Senzige, J. P.; Nkansah-Gyeke, Y.; Makinde, D. O.; Njau, K. N. The Potential for Solid Waste Recycling in Urban Area of Tanzania: The Case of Dar Es Salaam. *I. J. Environ. Prot. Pol.* **2014,** *2*, 147–152.

Silva MTB, Menduina, A. M.; Seijo, Y. C.; Viqueira FD-F. Assessment of Municipal Solid Waste Compost Quality Using Standardized Methods before Preparation of Plant Growth Media. *Waste Manage. Res.* **2007,** *25*, 99–108.

Slack, R.; Gronow, J.; Voulvoulis, N. Hazardous Components of Household Waste. *Crit. Rev. Environ. Sci. Technol.* **2004,** *34*, 419–445.

Slack, R.; Gronow, J.; Voulvoulis, N. Household Hazardous Waste in Municipal Landfills: Contaminants in Leachate. *Sci. Total Environ.* **2005,** *337*, 119–137.

Soudejani, H. T.; Kazemian, H.; Inglezakis, V. J.; Zorpas, A. A. Application of Zeolites in Organic Waste Composting: A Review. *Biocata Agric. Biotechnol.* **2019,** *22*, 101396.

Srivastava, N.; Narnaware, S.; Makwana, J.; Singh, S.; Vahora, S. Investigating the Energy Use of Vegetable Market Waste by Briquetting. *Renew. Ener.* **2014,** *68*, 270–275.

Srivastava, V.; Vaish, B.; Singh, R. P.; Singh, P. An Insight to Municipal Solid Waste Management of Varanasi City, India, and Appraisal of Vermicomposting as Its Efficient Management Approach. *Environ. Monit. Asses.* **2020,** *192*, 1–23.

Sukholthaman, P.; Sharp, A. A System Dynamics Model to Evaluate Effects of Source Separation of Municipal Solid Waste Management: A Case of Bangkok, Thailand. *Waste Manage.* **2016,** *52*, 50–61.

Syafrudin, S.; Samadikun, B.; Wardhana, I.; Rizaldianto, A. Biogas Generation During Anaerobic Composting of Organic Waste. *E&ES* **2020,** *448*, 012131.

Tai, J.; Zhang, W.; Che, Y.; Feng, D. Municipal Solid Waste Source-Separated Collection in China: A Comparative Analysis. *Waste Manage.* **2011,** *31*, 1673–1682.

Tejada, M.; Gonzalez, J. Influence of Organic Amendments on Soil Structure and Soil Loss under Simulated Rain. *Soil Til. Res.* **2007,** *93*, 197–205.

Teutscherova, N.; Vazquez, E.; Masaguer, A.; Navas, M.; Scow, K. M.; Schmidt, R. et al. Comparison of Lime-and Biochar-Mediated pH Changes in Nitrification and Ammonia Oxidizers in Degraded Acid Soil. *Biol. Fert. Soil* **2017,** *53*, 811–821.

Thi, N. B. D.; Kumar, G.; Lin, C-Y. An Overview of Food Waste Management in Developing Countries: Current Status and Future Perspective. *J. Environ. Manage.* **2015,** *157*, 220–229.

Troschinetz, A. M.; Mihelcic, J. R. Sustainable Recycling of Municipal Solid Waste in Developing Countries. *Waste Manage.* **2009,** *29*, 915–923.

Valkenburt, C.; Walton, C. W.; Thompson, B. L.; Gerber, M. A.; Jones, S. B.; Stevens, D. J. Municipal Solid Waste (MSW) to Liquid Fuels Synthesis, Volume 1: Availability of Feedstock and Technology. Pacific Northwest National Lab.(PNNL), Richland, WA (United States), 2008.

Vargas, D. C.; Monedero, M. S.; Urpilainen, S.; Kamilaki, A.; Stentiford, E. Assessing the Stability and Maturity of Compost at Large-Scale Plants. *Ingeniería* **2005,** *9*, 25–30.

Vélez-Sánchez, C.; Pinedo-Álvarez, C.; Viramontes-Oliva, O.; Ortega, C.; Melgoza-Castillo, A. Bio-tecnologías ambientales para el tratamiento de residuos ganaderos (Environmental Biotechnologies for Treatment of Livestock Waste, in Spanish). *Creativ. Desarr. Tecnol.* **2008,** *2*, 131–144.

Wang, K.; He, C.; You, S.; Liu, W.; Wang, W.; Zhang, R. et al. Transformation of Organic Matters in Animal Wastes during Composting. *J. Hazard. Mater.* **2015,** *300*, 745–753.

Wang, L.; Li, Y. *Sequencing Batch Reactors. Biological Treatment Processes, Handbook of Environmental Engineering Series*, Editors. Wang, L. K., Pereira, N. C., Hung, Y. T. Humana Press: New York, 2009.

Wei, Y.; Li, J.; Shi, D.; Liu, G.; Zhao, Y.; Shimaoka, T. Environmental Challenges Impeding the Composting of Biodegradable Municipal Solid Waste: A Critical Review. *Resour. Conserv. Recyc.* **2017,** *122*, 51–65.

Weigand, H.; Fripan, J.; Przybilla, I.; Marb, C. Composition and Contaminant Loads of Household Waste in Bavaria, Germany: Investigating Effects of Settlement Structure and Waste Management Practice. In *Proceedings of the 9th International Waste Management and Landfill Symposium*, Cagliari, CD-ROM, 2003.

Westerman, P.; Bicudo, J. Management Considerations for Organic Waste Use in Agriculture. *Bioresour. Technol.* **2005,** *96*, 215–221.

Wilson, D. C.; Rodic-Wiersma, L.; Scheinberg, A.; Alabaster, G. Comparative Analysis of Solid Waste Management in Cities around the World, 2010.

Yadav, A; Garg, V. Biotransformation of Bakery Industry Sludge Into Valuable Product Using Vermicomposting. *Bioresour. Technol.***2019,** *274*, 512–517.

Yosef, E.; Weinberg, Z.; Elyahu, D.; Nakbahat, M.; He, Y.; Solomon, R. et al. The Potential of Agricultural Residuals in Animal Feed. The Israeli Agricultural Research Organization-Volcani Center, 2015.

Yu, H.; Zhao, Q.; Meng, F.; Ruan, L.; Sun, T.; Liu, X. et al. Deciphering the Role of Calcium Peroxide on the Fate of Antibiotic Resistance Genes and Mobile Genetic Elements during Bioelectrochemically-assisted Anaerobic Composting of Excess Dewatered Sludge. *Chem. Eng. J.* **2020,** 125355.

Yu, Z.; Chen, L.; Pan, S.; Li, Y.; Kuzyakov, Y.; Xu, J. et al. Feedstock Determines Biochar-Induced Soil Priming Effects by Stimulating the Activity of Specific Microorganisms. *Eur. J. Soil Sci.* **2018,** *69*, 521–534.

Zacarias-Farah, A.; Geyer-Allély, E. Household Consumption Patterns in OECD Countries: Trends and Figures. *J. Clean. Prod.* **2003,** *8*, 819–827.

Zayed, G.; Abdel-Motaal, H. Bio-Active Composts from Rice Straw Enriched with Rock Phosphate and Their Effect on the Phosphorous Nutrition and Microbial Community in Rhizosphere of Cowpea. *Bioresour. Technol.* **2005,** *96*, 929–935.

Zeng, G.; Yu, M.; Chen, Y.; Huang, D.; Zhang, J.; Huang, H. et al. Effects of Inoculation with Phanerochaete Chrysosporium at Various Time Points on Enzyme Activities during Agricultural Waste Composting. *Bioresour. Technol.* **2010,** *101*, 222–227.

Zhang, H.; Schuchardt, F.; Li, G.; Yang, J.; Yang, Q. Emission of Volatile Sulfur Compounds during Composting of Municipal Solid Waste (MSW). *Waste Manage.* **2013,** *33*, 957–963.

Zhang, J.; Zeng, G.; Chen, Y.; Yu, M.; Yu, Z.; Li, H. et al. Effects of Physico-chemical Parameters on the Bacterial and Fungal Communities during Agricultural Waste Composting. *Bioresour. Technol.* **2011,** *102*, 2950–2956.

Zhang, R.; El-Mashad, H. M.; Hartman, K.; Wang, F.; Liu, G.; Choate, C. et al. Characterization of Food Waste as Feedstock for Anaerobic Digestion. *Bioresour. Technol.***2007,** *98*, 929–935.

Zhang, T.; Xu, H.; Li, H.; He, X.; Shi, Y.; Kruse, A. Microwave Digestion-assisted HFO/Biochar Adsorption to Recover Phosphorus from Swine Manure. *Sci. Total Environ.* **2018,** *621*, 1512–1526.

Zhu, L.; Wei, Z.; Yang, T.; Zhao, X.; Dang, Q.; Chen, X. et al. Core Microorganisms Promote the Transformation of DOM Fractions with Different Molecular Weights to Improve the Stability during Composting. *Bioresour. Technol.* **2020,** *299*, 122575.

CHAPTER 14

Role of Modern Technologies and Legislation Measures to Manage Biosolid Waste

TABINDA ATHAR[1*], ANAMIKA PANDEY[2], MOHD. KAMRAN KHAN[2], ZULFIQAR AHMAD SAQIB[1], MUHAMMAD SALMAN SARWAR[3], AQSA[1], UMER FAROOQ[1], MEHMET HAMURCU[2], and SAIT GEIZGIN[2]

[1]*Institute of Soil and Environmental Sciences, University of Agriculture, Faisalabad, Faisalabad 38000, Pakistan*

[2]*Department of Soil Science and Plant Nutrition, Selcuk University, Konya 42130, Turkey*

[3]*School of Chemical and Materials Engineering, National University of Science and Technology, Islamabad 44000, Pakistan*

**Corresponding author. E-mail: athartabinda@gmail.com*

ABSTRACT

With the latest advancement in science and technology, huge amount of biosolid waste is produced throughout the globe and posing serious threats to human beings, environment, and agricultural lands. Biosolids are rich sources of energy and nutrients, and their proper management can help to reduce burden on landfills, along with various other advantages. In this chapter, characteristics, handling, legislation, different disposal, management technologies, latest trends, and prospects of biosolid waste management are discussed. This chapter will also review various reports published by reputable international organizations about frameworks of biosolid waste management around the globe, to outline the sustainable methods and options

Waste Problems and Management in Developing Countries. Umair Riaz, Shazia Iqbal, & Moazzam Jamil (Eds.)

along with strategies to produce value-added products from the biosolid waste. These management practices can significantly reduce the magnitude of health and environmental concerns.

14.1 INTRODUCTION

Rapid development and industrialization has greatly contaminated water sources and environment. Production of biosolids from the municipal wastewater is of a major concern, as its direct disposal is creating environmental complications and health concerns (Selvarajan et al., 2020). Biosolids are complex mixtures of biological, organic, and inorganic pollutants generated from commercial establishments, households, health care systems, and industrial facilitations. Moreover, more complex compounds can be created during treatment of wastewater and disposal practices.

Sewage sludge contains different pollutants such as trace elements, metals, polychlorinated biphenyls, surfactants, pharmaceuticals, dioxins, pesticides, viruses, and bacteria. The treatment of sewage sludge is intended to reduce organic contents of pollutants and volume of waste. Furthermore, the presence of pathogens is minimized, whereas useful bacteria are retained. Resultant biosolids are being used for noncultivated and agricultural lands as they have good potential to improve the biological, physical, and chemical properties of the soil. They also boost crop growth and development by supplying trace elements and nutrients (Meyer et al., 2019).

Based on the presence of pathogenic contents, biosolids are categorized into two main classes, Class A and Class B. Biosolids in Class A are specifically well treated and have no or very low pathogenic populations, so they can be used in the application site without any restriction about the pathogens. These biosolids can be bagged and sold to general public for specific purpose. Class B biosolids are also well treated, but they have low levels of pathogenic populations (Khiari et al., 2020). So, their application to different sites is restricted due to detectable level of pathogenic populations. These biosolids cannot be sold to public; however, pathogen populations in these biosolids can be reduced by the environmental factors such as sunlight, desiccation, and heat. Sustainable and appropriate technologies and legislation measures can significantly help to create excellent opportunities to maximize food production and environmental protection by the management of biosolids (Torri et al., 2017).

Effective and safe management of sewage sludge and biosolids is also helpful to reduce the adversative effects of various pollutants on human healthiness and metabolic activities (Akash et al., 2018). Preventing the production of biosolid waste is the most important strategy to reduce the hazards associated with biosolids. It is the basic goal of waste management authorities, and citizens of various countries are giving due importance to this concept. Various options and wise practices ranging from manufacturing, usage, and recycling can significantly reduce the generation levels of biosolid wastes (Oruko et al., 2020). There are many cases when biosolid waste cannot be directly eliminated; however, there could be implementation of some strategies to minimize and reduce the production of biosolid waste (Brown et al., 1998).

At present management of biosolids and sewage sludge are gaining significant importance in the developed countries due to significant modification of quality of sludge, evolutionary approaches for the legislation measurements, technological options for the material, and energy recovery. Sustainable and cost-effective treatment of biological sewage sludge is significantly helpful to produce stabilized biosolid's residues. The quality of biosolids is being assured in these countries to offer friendly and sustainable use of biosolids in the agricultural soils (Katsoyiannis et al., 2019).

14.2 IMPORTANCE OF BIOSOLIDS FOR LAND APPLICATION AND ENVIRONMENTAL PROTECTION

Municipal sludge is now treated as a valuable resource and is being referred to as biosolids. It is carefully processed by using diverse efficient unit operations, leading to a sought-after, aesthetically fair, and safe biosolid product. Engineering and management policies and practices are greatly concerned with the safe utilization and disposal of biosolids, thus providing great opportunities for volume reduction, reduced vector attraction, stabilization of waste, and pathogenic inactivation (Schulte et al., 2014).

Although uncertainty about the quality of biosolids and consequences after application are complicating the decision-making capabilities, the perception about the use of biosolids is changing, and people are accepting its beneficial aspects for land application. Biosolids are a source of good quality and volatile fatty acids for the carbon-limited systems. Moreover, it also acts as a soil conditioner and provides a slow and required release of micronutrients and macronutrients. However, many farmers are still reluctant to use biosolids in their fields due to risks related to the use of pathogen-contaminated biosolids (Al-Khatib et al., 2017).

Some negative factors are attempting to limit the application of biosolid to land. These efforts are the labeling biosolids as synthetic products, whereas manures are labeled as organic by some agricultural communities in the USA. These practices combined with the poor education and negative reporting have led to idiosyncratic and negative behavior like banned application of biosolids to agricultural lands. There are many scientifically justified evidences advising the biosolids application to agricultural lands (Harrison et al., 1999). A year of negotiations was taken in the Colorado to lift ban for applying sewage sludge and biosolids to the agricultural lands. Keeping in view the negativity associated with the application of biosolids a National Biosolids Partnership (NBP) was built in USA in 1999. This partnership had the responsibilities of combining municipalities association, agricultural communities, sludge processors, and developers of value-added products for marketing.

14.3 ROLE OF MODERN TECHNOLOGIES FOR SUSTAINABLE UTILIZATION OF BIOSOLIDS

There is an overwhelming array about the availability of processing units for engineering approaches and designs for the management of biosolids. Right processes and technologies for the management of biosolid are selected based on cost analysis, cost-effectiveness, and sustainability to achieve desired goals for the production of quality biosolids (Zhang, 2020). Mainly there are five parameters that influence the selection of process train and right technologies. These parameters include total solids or total dissolved solids, concentration of organic and hazardous compounds, pathogenic contents, suitability of available land for disposal, and concentration of metals. Other than the availability of land, all parameters can be changed or modified by using specialized technological options (Miino et al., 2019).

Removal of organic materials and metals is not possible from the economic point of view, and the only way to reduce the presence of these hazardous materials is to minimize their production and entry through the prevention approaches (Paz-Ferreiro et al., 2018). Even if the entry of these toxic compounds is reduced, the significance of wastewater treatment is still there. The huge cost of overdesigning associated with the aeration compartments for the suppressed nitrification rates can be prevented by the sustainable implementation of source control and rigorous management programs. Due to these efforts the contents of hazardous substances are being eliminated to some extent, but metal toxicity of cadmium, zinc, and mercury is still exhibited at the elevated levels (Singh et al., 2017).

Deep shaft high pressure stabilization or pyrolysis is being used at full scale, but its cost-effectiveness in the context of other technologies has not been properly accessed through formally valued engineering approaches. Utilization of various technologies in multiprocessing unit operation is also possible. For example, biological stabilization can be achieved in only single-step process by using principle of alkaline composting that can also cause drying and disinfection to accomplish high quality and safe preparation of value-added biosolid product (McNamara et al., 2020). Drying centrifuge can also be used for drying and dewatering along with disinfection. This is an attractive opportunity for smaller and resource-limited installations where utilization of manpower is cost prohibitive for unit operations consisting of complex series (Koukoulakis et al., 2020). Technologies such as enzymatic alkaline hydrolysis, high-energy-based electric arc disinfection, dewatering, electro-osmotic, and acoustic conditioning are in experimental phases, but scientists are expecting good and high-quality results for the management of biosolids.

Biosolids are the nutrient-loaded organic materials arising due to treatment of wastewater residuals and their processing. The problem related to maintainable management of biosolids is the significant problem in the whole developing world. Throughout the previous era there has been a key alteration in the method sewage sludge is disposed (Ødegaard et al., 2002). The agricultural practice has become the primary discarding way of biosolids. Thermal processing of sewage sludge is one of the highly straightforward methods for discarding risk and giving some energy (Capodaglio et al., 2016). The latest developments for the sewage sludge management such as gasification, pyrolysis, wet-oxidation, and co-combustion of sludge, along with other substances for additional use of energy, have developed important technical intention (Fytili and Anastasia, 2008). Pretreatment has been used for the management of biological sludge and also used for mixed sludge. Applied power causes the breakdown of bacterial cell wall which causes release of organic material coming out from the cell (Wang et al., 1999).

There is significant utilization of modern concepts in the biological nutrient removal plants for the management of sewage sludge. Separate thickening of biological waste-activated sludge (WAS) and primary sludge solid should be regularly practiced. There is the use of primary clarifiers to reduce nitrification basin's volume and the use of primary solids in the activated sludge is leading to the reduced fractions of smaller nitrifier population and active biomass (Garrido-Baserba et al., 2019).

Primary solids can also be utilized for producing short chain and the volatile fatty acids. Mechanical or aerobic thickening of biological waste-activated sludge with minimum residence timing for limiting the release of phosphorus, dissolved air flotation, and nitrogen are being used in Canada. Processing and treatment of returning the sludge liquors for nitrogen, phosphorus elimination by using ammonia stripping, and chemical precipitation for separation of nitrified products is recommended for ammonia oxidation. The mass of biologically waste-activated sludge is reliant on the various factors such as kind of wastewater to be treated, inherent biomass, residence time of solids, temperature, and type of technological process to be used (Sandino et al., 2018).

14.3.1 VOLUME REDUCTION

Particles of sludge and biosolids contain water contents in different forms. Four forms have been defined at present, and these forms can also be determined by using analytical practices. Nonbonded or free water can be easily removed by the simple thickening techniques that can cause a good increase in the total solid contents ranging from 1 to 8%. In this way, 75% of water contents can be removed which can cause an approximately 5% volume reduction (Joo et al., 2015).

Capillary or bonded water can be used by using the mechanical dewatering technique. This dewatering leads to 20% of total solid sludge cake, where 95% water contents have efficiently been removed from sludge. Colloidal or adsorbed layer of water can also be easily removed by using the mechanical dewatering technique. This removal of water is favored by the addition of conditioning polymers. Typically, 22–35% of total solids can be removed by the dewatering machines. Moreover, advanced dewatering machines are more efficient and can take 40–45% of total solids from sludge cake (McNamara et al., 2020).

Intracellular water contents can also be removed by using advanced and drastic actions such as sonification and thermal drying. Most of the intracellular water contents can be removed by the thermal drying. Biosolids treated in this way will have only 2–8% of water contents. The benefits of thickening can never be over emphasized regarding the performance of dry stream processing (Liu et al., 2020). Typically, the process of separate gravity thickening of the primary solids can produce 5–6% concentration of total solids. But mechanical thickening or DAF can achieve 4–5% concentration of total solids. There are numerous alternative mechanical thickeners, but the use of cost-effective technologies is a sustainable approach (Zhang et al., 2020).

14.3.2 THERMAL TECHNOLOGIES

Through the previous years, studies disclosed that the presence of raw sludge in the situation of terrestrial use can be an important cause of unwanted ingredient in the water and soil. The major pollutants of sewage sludge include organic pollutants, some heavy metals, and pathogens (Dichtl et al., 2007). Mostly used procedures applied to sewage sludge comprise thermal, chemical, biological, and drying processes.

Thermal processes:

Thermal processes comprise the following applications.

1. Incineration
2. Pyrolysis
3. Gasification and mineralization of organic matter.

The direct practice of sewage sludge in crop growing is provocative because it creates pollution with organic matter and heavy metal. The basic purpose of thermal processing is the consumption of energy and decrease in the environmental problems at the same time (Fytili and Anastasia, 2008). This process produces such products that can be used in agriculture, such as biochar and sewage ashes.

14.3.2.1 INCINERATION

It is one of the trendy thermal treatments used to manage the sludge. These waste ashes are used in agricultural land which provide the chance for the improvement of essential nutrients. Although the number of heavy metals in ash is more than in soil (Zhang et al., 2002), the sludge ash contains higher amount of phosphorus, varying from 4% to 9%. Sewage sludge ash could exchange the phosphate rocked-based item, but its application is not conceivable as of its heavy metals. For this purpose, various kinds of technologies for the recovery of phosphorus from sludge ash have been developed such as RuePa-Process, SEPHOS-Process, BioCon-Process, and ASH DEC Umwelt AG (Dichtl et al., 2007).

14.3.2.2 PYROLYSIS

Pyrolysis is a specific method through which the biosolids are decayed in the absenteeism of oxygen resulting in the formation of black carbon-rich substance known as char and in gas. Latest studies demonstrate that pyrolysis is also beneficial to eliminate organic micro pollutants of sludge. Pyrolysis of contaminated residues at 800°C eradicates more than 99% of the dioxins and polychlorinated biphenyl (PCBs), and the procedure carried out at 450°C eliminates 75% PCBs from sludge (McNamara et al., 2016).

14.3.2.3 GASIFICATION

Gasification is a special process in which biosolids or sludge is converted into explosive gas known as synthesis gas or syngas. In this thermochemical process input waste is converted into ash and char (carbon-rich compound) known as biochar. Biochar can be burned for high temperature and power, for adoption application, and is also applied to soil for the improvement of soil. Biochar plays an important role in carbon sequestration (Mayer et al., 2016). Moreover, it increases the plant growth and yield, holding nutrients improving the biological and physical properties. In this way it possibly changes the water-holding capacity, aeration, nutrient-holding capacity, and microbial activities of the soil in the rooting zone (Colón et al.,2017).

14.3.3 CHEMICAL PROCESSES

Sewage sludge causes the increased acidity of soil because of proton discharge from organic content. This increase in acidity can cause the solubility of metals and as a result increased the leaching potential, specifically in soil with low buffer capacity. More attention is paid toward stabilization of sludge; the objective is to decrease the heavy metal mobility by using different preservatives (Samaras et al., 2008). Use of lime for the maintenance of biosolid or sewage sludge relies on various kinds of restrictions, such as the related cost, lime availability, and the time required for stabilization.

Lime is considered as mostly used material for stabilization of sludge, as it plays an important role by increasing agricultural advantages, reducing the environmental problems, and reducing the heavy metals and microbes such as pathogens of sludge (Wong and Selvam, 2006). Moreover, the alternative substance such as fly ash could be used for stabilization of biosolids.

Studies showed that the biosolid or sewage sludge alteration with fly ash and lime first offend the phytotoxic effects on different plant varieties (Samaras et al., 2008).

14.3.4 BIOLOGICAL PROCESSES

Biological processes comprising of different methods.

14.3.4.1 ANAEROBIC DIGESTION

Anaerobic digestion is a modern technological advancement that is cost effective and environmentally friendly. Anaerobic treatment is the standard technology used to deal with primary sludge and secondary sewage sludge (Lee et al., 2020). Anaerobic digestion is a biotic procedure in which sludge matter is specifically decomposed in an oxygen-free environment by using microbes that generate biogas consisting of carbon dioxide and methane. It also decreases the quantity of solid for disposal. The biochemical component of gas is 30–35% CO_2 and 60–65% methane and also has little amount of N_2, H_2, and H_2S. Generally, mesophilic digestion is more extensively used than thermophilic (Schwartzbrod et al., 2001). Through the anaerobic digestion, the organic content changes into biogas. The sewage sludge is dissolved by gas movement, forcing, or tub blender and then it heats up to get an optimum digestion rate.

The method begins in a closed digester. The solid material depends upon the input contents. The digested waste is removed, water is extracted and then stabilized in the presence of oxygen for about two weeks. This process fulfils the stabilization of content; the final item is nontoxic and used for the amendment of soil. Studies showed that biosolid that is anaerobically digested is less toxic than the raw sludge (Ramírez et al., 2008).

14.3.4.2 BIOLOGICALLY CELL BREAKDOWN

Activated sludge is transferred to a vessel in which some type of chemicals diverted. These chemicals have strong power of oxidation, which weakens the cell wall of bacteria and eventually causes their breakdown. The prepared bacteria is sent back to the activated sludge component and then disintegrated into water and CO_2. This biological cell destruction procedure deliberates for only organic sewage sludge.

14.3.4.3 COMPOSTING

Composting is a specific process in which the organic matter goes through biological decomposition, under the situation that creates a thermophilic condition to produce the final stable product. This resulting product should be free from crop seeds and pathogens that can provide benefits to land. Composting lessens the transporting cost and reduces its volume. Furthermore, in agriculture soil it has certain effects as compost has a more binding ability for both the organic and inorganic substances, improves the water-holding capacity, and helps in the formation of soil aggregates that provide air to roots and enhance infiltration; it also decreases erosion and runoff (Sánchez et al., 2015).

Mostly composting consists of these basic steps:

i. Mixing of sewage sludge with bulking agent.
ii. Increasing the degradation by the mechanical method or by the addition of air.
iii. Stabilization and cooling of compost.
iv. Sieving for the removal of non-decomposable substances such as glass, plastic, and metal
v. Final disposition and preserving.

Studies show that before applying to agricultural soil it decreases the phytotoxicity of sludge (Ramírez et al., 2008).

14.3.4.4 VERMICOMPOSTING

Vermicomposting is a simple technology and the mostly used process for sewage sludge stabilization. It is biochemical and mechanical modification of sludge done by the microbial action. Earthworms have power of enhancing the degradation rate, stabilizing the organic waste, composting sludge, and extracting the heavy metals (Sinha et al., 2010). In the vermicomposting process microorganisms and earthworms work together. Earthworms add to pile which degrade the organic matter to generate a fine product. The earthworms live in the bed and increase its population by using food and space. Solid should be adjusted according to condition suitable for microbial growth. The resulting product contains plant growth regulators and it is also used as bio-fertilizers. Various studies show that the earthworms have capability to bio-accumulate heavy metals. Earthworms may not be capable of eradicating the harmful toxic substances, but it can modify the sludge by changing its chemical makeup to reduce its harmful impact on soil (Bajsa et al., 2004). The most suitable worms for vermicomposting are Red Tiger Worm (*Eisenia andrei*), Red Worm (*Lumbricus rubellus*), Tiger Worm (*E. foetida*), and the Indian Blue Worm (*Perionyx excavatus*), under all the environmental conditions (Sinha et al., 2010).

14.3.4.5 BIOLEACHING

Bio-acidification or bioleaching is the mostly used method for metal leaching. Chemiolithotrophic bacteria are specifically used for the oxidation of iron and sulfur. *Thiobacillus ferrooxidans* and *Acidithiobacillus ferrooxidans* are mostly used microbes for metal leaching. By using bioleaching methods, researchers were capable of removing 38.96% of Pb, 83.96% of Cu, and 24.73% of Fe from biosolids. Furthermore, this process of bioleaching can decrease the nutrient matter of biosolids (Pathak et al., 2009).

14.3.4.6 ENZYMES CONDITIONING

Enzymes play a major role in the breakdown of oils and fats, and also reduce chemical oxygen demand (COD) in biological digester and aeration tanks. Enzymatic mixture, some nutrients, and some bacteria are inserted to the system that generates enzyme which decomposes organic substance and converts them into water and carbon dioxide. The process is cost effective and has very little harm to the system as the enzymatic conditioning will occur for some days (Dentel et al., 2006).

14.3.5 DRYING PROCEDURE

Drying is a comparatively simple process in which water is evaporated from sludge by applying thermal energy. The drying process can also be attained by using belt compress, open drying bed, centrifuge, and other procedures. The process is cost effective, most easily transferable, environment friendly, and advantageous for agricultural use. The most usual method of the drying process is thermal treatment. The use of this technology can be difficult as it needs high amount of energy. The recently introduced method is bio-drying to decrease the content of sludge by producing heat biologically (Dufour, 2006).

Drying decreases the volume of sludge and makes it easy for conveyance, storage, and packing. Drying of sludge as well deactivates pathogens and reactive chemicals and then sanitizes the product (Fernández et al., 2007). When a thermal-dried product is applied to soil it increases the concentration of nitrogen as well as increases the leaching of nitrate; it also increases the carbon mineralization. Some studies showed that the thermally treated sludge has a toxic effect on microbes and plants. Energy required for drying can be given through radiation, electricity, conduction, convention, and other methods to apply temperature and aeration to extract water vapors from sludge. The major thermal drying treatments are rotatory drying (Chen et al., 2015; Treviño and Rodríguez, 2015).

14.3.5.1 SOLAR DRYING

The process of solar drying is using solar energy for sludge drying. Solar drying is a useful way of applying the sun radiations with the purpose of eradicating the action of microorganisms and humidity content of sludge. There are various types of solar drying, but the greenhouse drier has more benefits than other methods use for processing of sewage sludge (Singh et al., 2018). Solar and bio-drying are two different processes, another method for water extraction by using heat produced by aerobic microorganisms (Bennamoun, 2012).

14.3.5.2 GREEN HOUSE SCHEME

Floor is made up of concrete which is covered by the transparent material such as plastic foil and Plexiglas which permit the high conductance of incoming solar radiation. This impervious floor contains some small channels that collect the excessive water, further collected to well from where water can return back to fluid process units. A mechanical mixer is used to transfer the liquid from the bottom to the surface for air drying. The structure consists of exhaust fans for circulation of air, increasing the drying power of inside atmosphere. These horizontal fans increase the rate of humidity removal (Krawczyk, 2016).

Solar drying process has some effects on sludge properties and makeup by enhancing humid-like products, reducing amount of lipids, and using this method for long-time amendment in soil. The two major issues related to solar drying are pathogen and odor. This method has many advantages as it reduces the energy need and also decreases surface requirements (Lemée et al., 2017).

TABLE 14.1 Technologies for Efficient Management of Sewage Sludge and Biosolids.

Technologies	Descriptions	References
Thermal processes: 1. incineration	The thermal process to convert sewage sludge into sludge ash which contains essential nutrients.	Zhang et al. (2002); Dichtl et al. (2007)
2. pyrolysis	The anaerobic method in which biosolids are decomposed to char and gas.	McNamara et al. (2016)
3. gasification	The process used to convert sludge into explosive gas and syngas.	Mayer et al. (2016); Colón et al. (2017)
Chemical processes	Lime is used for the maintenance of biosolids and also has a major role in environmental issues.	Wong and Selvam (2006); Samaras et al. (2008)
Biological processes: 1. Anaerobic digestion	The process used to decompose matter and convert it into carbon dioxide and methane gas.	Gantzer et al. (2001); Ramírez; Lee et al. (2020)
2. Biologically cell breakdown	Some kind of chemicals use for the destruction of bacterial cell wall.	Lee et al. (2020)
3. Composting	It is the biological decomposition of matter to produce stable product.	Ramírez et al. (2007; Sánchez et al. (2015)
4. Vermicomposting	Modification of sludge by microbes.	Bajsa et al. (2004); Sinha et al. (2010)
5. Bioleaching	The process uses for metal leaching.	Pathak et al. (2009)
6. Enzymes conditioning	Enzymes are used for the decomposition of matter into water and carbon dioxide.	Dursun et al. (2006)
Drying processes: 1. Solar drying	Use of solar energy for drying sewage sludge	Bennamoun (2012); Singh et al. (2018)
2. Green house scheme	Green house method used to increase the rate of drying.	Krawczyk (2016); Lemée et al. (2017)

14.4 REGULATION FOR PRODUCTION AND UTILIZATION OF BIOSOLIDS

Regulation to define the quality of biosolids is being developed in the European Union and Canada. Almost all of their regulatory authorities and management teams are focused on capitalization to pioneer the development of regulation in the USA (Kim et al., 2020). United States Environmental Protection Agency (USEPA) has developed standards about the pathogenic presence in the biosolids under the Resources Conservation Recovery Act (RCRA), in the specific subtitle D, 40 CFR, Part 503 for pathogenic requirements (USEPA-1994). The specifications of pathogens are developed on the basis of performance-related standards for two specific classes of pathogenic reduction. Class A biosolids are well suited for land application, whereas class B biosolids have some restrictions or limitations. All the biosolids for land application must meet these standards (Mohajerani and Bojana, 2020). The processing criteria for disinfecting the biosolids are specific for each type of disinfecting, and the stabilization process under the rule of 503 regulations.

Regulators in Canada such as Canadian Council of Ministers of Environment are stepping forward more cautiously and are trying their utmost to develop more stringent standards for the management and application of biosolid waste. Almost 70% of the sludge was incinerated in the Toronto, but now authorities are planning to utilize it 100% for the application to land in the form of stable dried pelletized materials and wet solids. There are larger ongoing efforts to educate and convince public about potential benefits of land application. They are focusing on guiding community that there should be no tolerance about the negativities associated with the land use of biosolids (Badgett and Anelia, 2020).

14.5 LEGISLATION MEASURES FOR THE MANAGEMENT OF BIOSOLIDS

The safest possible management of biosolids has always been the concern of the people dealing with these beneficial byproducts of sewage sludge treatment, because these can equally harm our environment if some rational steps are not taken regarding their management during all the stages involved from their production to their end use (Katsoyiannis et al., 2019). Some legislation strategies have been devised by the concerned authorities from all over the world, to ensure the implementation of the best possible methods and procedures to effectively manage biosolids. But, with the passage of time, the quality of biosolids remained altering because of the continuous addition of new contaminants at intervals, in the sewage being discharged from houses, hotels, and industries, resulting in the necessary amendments in the legislation strategies and measures each time, to avoid the potential harm from any contaminant being discharged along with the sewage.

Also, the technological advancements, diversified usages of biosolids, and their increasing demand for various purposes lead to such necessary amendments in these legislation measures and strategies. Sewage being discharged contains many potentially harmful contaminants, whose treatment, if managed inappropriately, can pose harm to the lives or wealth of the people using the biosolids produced through this treatment. So, not even a single of the measures for the management of biosolids can be compromised (Christodoulou, A.; Stamatelatou, 2016).

Biosolids have a vast application in the field of agriculture. One of the largest uses of biosolids is their application as fertilizers for various crops or as amendments to reclaim soils. They are being used prevalently in agricultural field to fertilize crops and reclaim soils. Their application has helped a lot in improving soil characteristics, crop production, and economic returns. Also, the agricultural application of biosolids has helped in the conservation of space required for landfilling and has also helped to improve air quality by reducing biosolids incineration load (Yurgel et al., 2020).

Governments of different states have allowed their vast application in agricultural field; as a result of which the agricultural application has become the top-of-the-line solution to manage sewage sludge and biosolids in many of these states. But their application in this field demands efficient risk assessment procedures and tools, because the quality of biosolids produced as a result of sewage sludge treatment may sometimes be below the standard required to declare them safe for agricultural application. Because of diverse sources of sewage and those of wastewater, biosolids may contain a large number of heavy metals, pharmaceuticals, disease-causing germs, and other life-threatening and environment-degrading substances, which, if not addressed timely, can lead to fatal consequences.

The agricultural application of such biosolids may pose potential risks to human and animal health, environment, and food safety. So, the legislation measures devised and amended by different governments as per the need felt by the concerned authorities at different time intervals have performed a substantial role in controlling the quality of biosolids being applied in agricultural field (Abdon, 2019). So, the application in these soils of the biosolids produced due to inappropriate handling of sewage sludge may cause bioaccumulation and biomagnification of heavy metals in various food chains originating from the vegetation growing in such soils. Other areas, where biosolids are being used, include incineration, landfilling, and thermal procedures yielding the products that are used as the components of various construction materials. So different states of the world have designed necessary legislations in this regard to ensure, as much as possible, the sustainability of the environment.

Following are the legislations and underlying regulations of varying natures, which are being implemented by different countries of the world to manage biosolids.

14.5.1 EMINENT LEGISLATION MEASURES SET BY USA TO MANAGE BIOSOLIDS

In USA, the population density is low and biosolids produced are applied in agriculture on a much larger scale, as compared with all other countries of the world. Since the 1970s, a number of legislation measures have been devised to encourage the valuable agricultural use of biosolids. Other means of biosolids management include incineration, landscaping, and forestry. Following are the regulations and acts implemented in USA to manage biosolids.

14.5.2 THE MARINE PROTECTION, RESEARCH AND SANCTUARIES ACT (MRSA)

Marine Protection, Research and Sanctuaries Act, sometimes also called ocean dumping act (ODA), was passed in 1972. This act sets regulation on the dumping of all kinds of material into oceans, and specifically forbids the dumping of those materials into oceans that are disastrous for human health, marine life, environment, ecology, and economic capabilities. Although this act is not directly concerned to the management of biosolids, but under its provisions biosolid managers have shown inclination for other biosolids management practices like land application, landfilling, and incineration, altogether raising the need for the formulation of new legislations to regulate these practices.

14.5.3 THE FEDERAL WATER POLLUTION CONTROL ACT AMENDMENT OF 1972

Standards set by this act are concerned with the regulation of waste contents of water. This law was called Clean Water Act after amendments made in it in 1972. The act was structured to control the addition of pollutants into water, and also sets important regulations to preserve the quality of surface water. Under this act EPA has launched many projects to control pollution by regulating the addition of discharge into waters of the state. Clean water act prohibited the addition of the discharge into navigable water, originating from a point source unless a permit was granted under its terms for this purpose. Under the provisions set by clean water act, EPA assigned the task to National Research Council of assessing the methods and procedures adopted to set the standards concerning the presence of cadmium, PCBs, and pathogens in biosolids used to apply to lands for agricultural purposes (Gadhamshetty et al., 2020).

14.5.4 WATER QUALITY ACT OF 1987

This set of regulations was originally enacted in 1948 but was amended in 1972. The act is composed of two discrete parts, one of which permits financial assistance for the construction of the plants to treat sewage. The other one is focused on the regulation of discharges of industrial and municipal origin into water. Till 1987 it was focused on regulating the point sources of pollution, but in 1987 it was amended, and some inclusions were made to regulate the nonpoint sources of pollution (Hackett, 1987).

14.5.5 STANDARDS FOR EFFICIENT USE AND DISPOSAL OF SEWAGE SLUDGE (40CFR PART 503 RULE)

The legislation devised by USA has provided many risk assessments tools and has introduced a set of standards known as "standards for the efficient utilization or disposal of sewage sludge" (40CFR Part 503 rule). This rule is applicable to all the handling works producing sewage sludge, and not for the people carrying out the agricultural application of biosolids. This rule provides us with certain tools for the identification of the chemical pollutants existing in sewage sludge and for the regulation of the presence of pathogens in biosolids. Several management practices and sewage sludge treatment technologies for the production of biosolids have also been standardized through the implementation of this rule. According to this rule, biosolids are not considered waste clearly anymore, rather they are considered a potential source of nutrients. Hence in USA the practice of biosolids management through agricultural application has been reinforced during the last two decades (USEPA, 1995).

14.5.6 LEGISLATION MEASURES TO MANAGE BIOSOLIDS IN JAPAN

In Japan, thermal methods of sewage sludge management like incineration, gasification melting, and carbonization are being implemented on much larger scales since long. But many projects have been initiated to aggrandize the concept of biosolids reuse, and to reinforce the credibility of their agricultural application, altogether ensuring health and environmental conservation. Policies and legislations have been in accordance with Japan's national strategy to develop a community that is focused on the reuse and the recycling of wastes including biosolids. For instance, many components recovered from sewage sludge such as Magnesium ammonium phosphate are being used on the commercial scale for the industrial production of organic fertilizers. Following are the eminent legislations devised by Japan to manage biosolids (Christodoulou and Stamatelatou, 2016).

14.5.6.1 WASTE MANAGEMENT AND PUBLIC CLEANING LAW

The purpose of the enactment of this law was to conserve the living environment and make joint efforts to uplift health standards by regulating industrial and municipal discharge and appropriating various tasks associated with biosolids management. Through this law, citizens were directed to reduce the waste production and discharge and were persuaded to use the recycled goods. Businesses were directed to manage waste originating from their business activities by themselves as far as possible. It also set some additional regulations regarding the application of biosolids on agricultural lands and in green areas.

14.5.6.2 SOIL CONTAMINATION COUNTER MEASUREMENT LAW

Soil often gets contaminated as a result of the hazardous substances applied either directly to the soil or from the landfill or disposal sites as a result of leaching or runoff. So, this law was enacted to curb the contamination of soil by numerous kinds of hazardous substances, through the formulation of important regulations, and to ensure their imposition where and when needed. The quality of biosolids applied to land is regulated through the standards based on the provisions of this law (Luo et al., 2014).

14.5.6.3 ENVIRONMENTAL POLLUTION PREVENTION ACT

Thermal methods of sewage sludge management such as gasification, incineration, drying, melting, and carbonization have flourished to a great extent during the last three decades. The residues of these thermal procedures are used for different purposes such as the production of construction materials, compost, and soil amendment products. So, this act has been imposed for controlling the quality of biosolids and their residues obtained after thermal procedures used for aforementioned purposes.

14.5.6.4 FERTILIZER MANAGEMENT ACT

In Japan, the agricultural application is the second largest practice to manage biosolids, and the Japanese government has also devised many legislations measures to avoid the hazardous consequences of agricultural application of inappropriately managed biosolids. Fertilizer management act that was amended in the year 2000 provides the criteria for heavy metals to be used as fertilizers in sewage sludge, and regulates the necessary tasks to be performed, like reporting and recording the data about various aspects of production and processing of biosolids (Makino et al., 2019).

14.5.7 LEGISLATION MEASURES TO MANAGE BIOSOLIDS IN EUROPEAN UNION

Several member states of the European Union favor the use of biosolids to agricultural lands, and it is the commonly practiced method of biosolids management, but some developed member states discourage this practice because of the public concerns and other social issues. The European Union's legislation concerning the agricultural application of biosolids is more rigorous than EPA's, although the sustainable application of phosphorus is being fostered and favored by the policies forged by the European union, but phosphorus extracted from sewage sludge is not yet recommended to be applied on land because of its origination from waste. Sweden has amended their legislation to recover phosphorus from wastewater, and many other European countries are working on the formulation of a legislation to permit the recovery of phosphorus from sewage and its safe agricultural application (Katsoyiannis et al., 2019).

Despite all these legislation measures and permissions from the governments of these and other developed states, many people still have doubts concerning the agricultural application of biosolids because of their waste origin. Although other methods of biosolids management like incineration, land filling, and ocean disposal are also in practice in some member states, these methods are less prominent. There is a need to create awareness among the people to persuade them to reuse these beneficial and cheap fertilizer replacements and soil amendments. All the member states of European Union have together formulated some legislations that are discussed here as follows (Christodoulou and Stamatelatou, 2016).

14.5.7.1 URBAN WASTEWATER TREATMENT DIRECTIVE

This directive is focused on the protection of the people and environment from the hazardous impacts of wastewater discharges from residential and industrial areas. According to Joint Research Center and European Environment Agency, a great reduction in the emissions of nitrogen and phosphorus to rivers and seas has been observed after the imposition of this directive. Even though the practicality of the directive looks costly, the excellent effectiveness compensates the expenses. The imposition of this directive has also supported the other EU legislations, including Water Framework Directive and Marine Strategy Framework Directive (Marek et al., 2017). But a number of gaps have been identified which need to be addressed to maintain the effectiveness of this directive. For example, it does not deal appropriately with upcoming issues like pharmaceuticals and microplastics getting access to the water reserves. These contaminants can be a source of pathogens, polychlorinated biphenyls, and polycyclic aromatic hydrocarbons.

14.5.7.2 SEWAGE SLUDGE DIRECTIVE

The fundamental legislation regarding biosolids management in European Union has been sewage sludge directive (directive/278/EEC) since 1986. It emphasizes on the reinforcement of agricultural application of biosolids, all together fulfilling the conditions forged to avoid its disastrous impacts on humans, animals, land, and other living and nonliving entities of interest. The Regulations designed under this directive set limitations for the components of biosolids that are potentially toxic, such as heavy metals. This directive has at several times been subjected to amendments by the European commission as per the recommendations of research and monitoring agencies and a deliberation process among the states of the union (Hudcová et al., 2019).

14.5.8 LEGISLATION MEASURES DEVISED BY AUSTRALIA TO MANAGE BIOSOLIDS

Agricultural application of biosolids is highest in Australia as compared with that in USA and European countries. In Australia, the low nutrient value of soils along with the low population density make the agricultural application of biosolids one of the inevitable practices of fertilizing crops and reclaiming soils. Following are the legislations that Australia has devised to manage biosolids. Legislation regarding biosolids management in Australia focuses on agricultural application of biosolids. And the regulations thus set have been supported and favored by the environmental acts of the past and the existing environmental conservation laws also authorize these regulations (Mohajerani et al., 2019).

These regulations have been designed in accordance with EPA's 503 rule, following the same principles of categorization and operation. But some of these regulations are such that they thwart the presence of some contaminants of organic nature along with some heavy metals and pathogen content. Removal of these contaminants from sewage sludge being treated is costly. The cost of this removal is to be borne by the sewage treatment plant customers and the end consumers. So Australian Water Association has recommended some amendments in these regulations to permit the presence of these contaminants up to safe limits. In Australia, biosolids are still placed in the category of wastes, but after the amendments made by South Wales in some regulations, in addition to the legislation revised by the Queensland, some restrictions have been abrogated to permit the application of biosolids on land, and biosolids are now considered a resource.

14.5.8.1 ENVIRONMENTAL GUIDELINES: USE AND DISPOSAL OF BIOSOLIDS PRODUCTS

The imposition of these guidelines is restricted to the application of biosolids and/or their disposal to land, but these guidelines are not applied on the other methods of biosolids management like ocean disposal and thermal processing. Classification of biosolids on the basis of their qualities is done under the provisions of these guidelines. A classification framework has been designed for the recognition of the quality of biosolids to categorize it in the respective class determining its fate, regarding its use in soil reclamation or its landfilling. After the recognition of biosolid quality, it is classified as either for unrestricted use, restricted use, or unsuitable for use. It sets regulations for the application procedure of biosolids on the basis of their qualities. A discrete section of these guidelines has been designed to regulate the application rates of the biosolids, and to recognize the site suitability, management practices, and statutory requirements for their application (Omidghane et al., 2020).

A separate section covers the regulatory measures for the disposal of biosolids, which regulate the disposal procedures by prioritizing environment safety, public health, and promotion of a sustainable process. Summarization of these guidelines reveals that these guidelines are imposed on sewage treatment plant operators, biosolids processers carrying out biosolids treatment to further obtain products like composters and pelletizers, biosolids appliers to land or to those carrying out their landfilling and disposal, and the end users like famers and foresters etc. (Delibacak et al., 2020).

14.5.8.2 PROTECTION OF THE ENVIRONMENT OPERATIONS ACT 1997

The purpose of the imposition of this act is to protect, restore, and promote the environmental quality of New South Wales. An allocation of diverse duties between the EPA, local councils, and other public authorities has been made under the provisions of this act. The act consists of a set of policies to regulate the injection of wastes into the environment, their proper management through appropriate disposal, landfilling, incineration, reuse, and recycling. The act has also been designed to encounter the offences and mismanagements that may lead to the hazardous consequences, if not tackled (Greenstone, 2002).

14.5.9 LEGISLATION MEASURES DEVISED BY NEW ZEALAND TO MANAGE BIOSOLIDS

New Zealand also manages biosolids through their agricultural application because of the sufficient availability of agricultural land. Even though a number of strategies have been made and multiple projects have been launched during the last two decades, to improve agricultural lands, forestry, and to reclaim soils including the application of biosolids for these purposes, the efficient reuse of these beneficial products of sewage sludge treatment has not been seen as yet (Mohajerani et al., 2019). A recently conducted survey revealed that the lack of awareness among general public regarding the beneficial use of biosolids and other social hindrances thwart the agricultural application of biosolids. People still practice the age-old methods of biosolids management, like their ocean disposal. So, there is a dire need to create awareness among the people and to improve the markets in order to ensure the sustainable management of biosolids (Fisher et al., 2019).

14.5.9.1 GUIDELINES AND LEGISLATION MEASURES FOR SAFE APPLICATION OF BIOSOLIDS TO AGRICULTURAL LAND IN NEW ZEALAND

Although biosolids contain a number of essential nutrients in sufficient amounts and their application can be very beneficial for the health of soils and crops, this fact can also not be denied that biosolids may contain a number of harmful pathogens and hazardous compounds, which can pose serious hazards to humans, animals, and environment. These guidelines have been designed to manage the agricultural application of biosolids. These guidelines safeguard the protection of health and environment by regulating agricultural application of biosolids. Also, the guidelines are applicable on biosolids manufacturers and distributors, biosolids consumers, and concerned authorities (Schofield, 2010).

The purpose of the application of these guidelines on the aforementioned groups of people is to minimize the associated risks with the existence of such pathogens and hazardous compounds as a result of inappropriate treatment, management, and land application of biosolids. The guidelines provide a grading scheme to categorize biosolids on the basis of their quality (Lumsden and Charles, 2000). According to this scheme, biosolids are given a grade "A" or "B" on the basis of presence or absence of pathogens and/or vector attracting volatile solids. Grade "A" biosolids are those in which pathogens or vector-attracting compounds have been removed or reduced to standardized limits. Also, a biosolid is given a grade "a" if all the contaminants like metals and organic chlorides present in it, are at, or below the standardized limits, and a biosolid is given a grade "b" if any of these contaminants exceeds the limits set for the presence of that contaminant in a biosolid to be called an "a" grade biosolid (Yuan, 2009).

14.6 FUTURE PROSPECTS

With due consideration to policy making, and regulations safe disposal, use, and management of biosolids is a scientific concern and has a good research focus from reputable international organizations and research centers. Even though there are good regulations and technologies available to deal with the management of biosolids, still more research is needed to know about the nature of biological, physical, and chemical pollutants in the sewage sludge for odor research, sudden increasing level of concerns, effects of processing, reduced availability of substrate, inhibition of growth indicators, and regrowth research.

Findings of scientific studies have also reported that there is a good relationship between climate change and the strategies for the management of biosolid waste. Reduced emissions of greenhouse gases and less acceleration of global warming can reduce the carbon footprints resulting from the operations of biosolids management and handling and emphasized corresponding on the availability of renewable energy sources.

There should be a detailed and comprehensive survey to check the associated risks with the use of biosolids. Biosolids are classified into two classes, such as class A and class B, according to the presence of pathogens, so their role toward toxicities should be properly defined before selling to the public. Any mishandling can greatly affect the soil quality, soil fertility, and soil productivity by affecting the biological, physical, and chemical properties of soils. Moreover, potential entry of pollutants to water sources, and environment from class B should be reduced. There should be proper testing to classify biosolids in classes A and B. Sometimes, testing standards are not managed efficiently, so these may exert negative influences on human and aquatic life by contaminating food sources and environment.

EPA should pay more attention to evaluate and develop efficient means for monitoring of contaminants in the biosolids. There is an exigent need to modify the existing regulations for removal of harmful chemicals and pollutants from biosolids. Although use of biosolids for agricultural lands is a good approach, all biosolids should be properly tested before marketing and only exceptional quality biosolids without any contaminants should be sold to general public. Standards should be developed and monitored for checking the biosolids application to agricultural lands so that the problem of groundwater contamination can be avoided on a sustainable basis. Moreover, potential long-term impacts of all biosolids should be accessed on all application sites and groundwater quality.

Despite the fact that biosolids contain very less to no pathogens, and potentially harmful contaminants, their use can cause significant health concerns for human beings, livestock, and aquatic life. Exposure to biosolids may cause death by causing different kinds of sickness such as gastrointestinal illness, respiratory problems, headaches, and cardiovascular problems. These health issues are quite variable in different localities, so quality and application rates of biosolids should be carefully regulated. Utilization of biosolids also produces specific odor in the environment and attracts various vectors such as birds and insects. As a result, there is a significant damage to the roads and property and the value of property is also reduced. Other than this, vector attraction may cause health concerns, so this problem should be managed on sustainable basis, and strict measures should be undertaken to reduce the odors.

Despite the availability of good resources and technological options for biosolid waste management, there is an exigent need to efficiently manage the sewage sludge in the most efficient and sustainable way. The present measures are not enough to handle the problem and citizens of almost all countries across globe are suffering from serious health concerns and environmental contamination problems. Ineffective management of sewage sludge and biosolids is greatly contaminating our food, water resources, air, and surroundings. Resultantly, people are eating, inhaling, and consuming toxic compounds responsible for various severe and acute diseases. Effective management will not only be helpful to improve soil health and environmental quality, but also help to reduce the burden of diseases and pharmaceuticals on poor people. This in turn is essentially an excellent approach to improve cultural and socioeconomical conditions of communities and deprived societies.

Cost-effective and friendly management of biosolids is a burning issue around the globe, and there are ongoing efforts of practitioners, researchers, and manufacturers for improving the performances of various processes and equipment, but these technologies cannot be used by the poor countries and societies. So, their efforts should be more focused toward creating economical feasible options for all nations. Furthermore, extraction and cleaning efficiencies of these technologies are also questionable, so good quality must be ensured to overcome the management-related challenges. Performances of processes and equipment are directly related to public perception, regulatory factors, and technological constraints. Thus, performance improvements should be of prime importance for all the biosolids management and handling processes.

Moreover, there should be community awareness campaigns to improve the perception of farmers for applications of biosolids to their crop lands. Biosolids can greatly improve soil fertility status and soil health along with significant improvement of rhizospheric activities. But there is a constraint that both classes of biosolids may not be good for all food crops, so different criteria should be made for the application of biosolids to food crops, cash crops, and commercial crops. Application of dewatered and liquid biosolids is a proven cost-effective and reliable management practice for different purposes, but creation of value-added products is an ultimate requirement to get maximum benefits.

Development of the value-added products can open up new outlets and marketing opportunities to generate more revenue from product scale. It is also helpful to potentially reduce the weight of total biosolids to be transported that in turn can reduce the cost associated with fuel consumption and truck loads. Increased public acceptance and perception about the use of biosolids as biofuel, amendment, and fertilizer can reduce the environmental concern along with maximum restoration of soil fertility and health. Furthermore, there should be opening up of different private and public ventures for facilitation of cooperative operations and financing.

Even though management of biosolids is being evolved as a major response to the diversifying majors and policy shifts, regulatory measures are driving shifts related to both management and technical trends at rapid pace. Urbanization and industrialization has greatly changed the world as a whole, but it has created significantly negative consequences on human beings, aquatic life, soil quality, and environment. Based on this fact, regulatory consultancy and operations are required to respond to negative effects of all kinds of pollution. These management tools, practices, and strategies can provide excellent opportunities and abilities for prediction and preparation about the omnipresent opportunities in the near future. So, there should be a strong collaboration between researchers, international organizations, governments, communities, students, farmers, scientists, and extension workers to promote the concept of biosolids management for restoring soil health and protecting environment, and all living creatures.

14.7 CONCLUSION

Regulatory practices, policy drivers, and technological prospects can offer potential opportunities to dispose, use, and manage biosolids in the most sustainable way. Biosolids are an exceptional source of energy, water, organic matter, and nutrients. Cultural shifts and perceptions of farmers and local communities about the safe and efficient use of biosolids can provide valuable resources to the communities for enhanced food security, environmental protection, and urban sustainability. Improved focus for the development of sustainable approaches and technologies should be viewed within the specialized context of complex regulatory landscaping and limited budget resources. The conflicting pressure about legislation measures, sustainable technologies, and efficient management of biosolids can be resolved with the help of innovative technologies and creative approaches. Furthermore, biosolids management programs can be sustained and promoted by productive, efficient, and flexible strategies related to regionalized operations, community awareness, need-based management drivers, operations, recognition of business mission, expanding vision beyond boundaries, considering leveraging private, and public partnership, and focused knowledge management according to training needs and communication tools.

KEYWORDS

- **biosolids**
- **management technologies**
- **disposal**
- **legislation measures**

REFERENCES

Abdon, J. D. A Systematic Review of Management Practices for Biosolids Land Application, 2019.

Badgett, A.; Anelia, M. A Summary of Standards and Practices for Wet Waste Streams Used in Waste-to-energy Technologies in the United States. *Renew. Sust. Energ. Rev.* **2020,** *117*, 109425.

Bajsa, O., Nair, J.; Mathew, K.; Ho, G. E. Vermiculture as aTool for Domestic Wastewater Management. *Water Sci. Technol.* **2004,** *48*, 125–132.

Bennamoun, L. Solar Drying of Wastewater Sludge: A Review. *Renew. Sust. Energ. Rev.* **2012,** *16*, 1061–1073.

Brown, S. L.; Rufus, L. C.; Scott, A. J.; James, A. R. The Phyto Availability of Cadmium to Lettuce in Long-term Biosolids-Amended Soils. *J. Environ. Qual.* **1998,** *27*, 1071–1078.

Capodaglio, A. G.; Callegari, A.; Dondi, D. Microwave-induced Pyrolysis for Production of Sustainable Biodiesel from Waste Sludges. *Waste Biomass Valorization* **2016,** *7*, 703–709.

Chen, G.; Yue, P. L.; Mujumdar, A. S. Dewatering and Drying of Wastewater Treatment Sludge. In *Handbook of Industrial Drying*; Mujumdar, A. S., Ed., 4th ed.; CRC Press: Boca Raton, 2015; pp 867–882.

Christodoulou, A.; StamatelatouK. Overview of Legislation on Sewage Sludge Management in Developed Countries Worldwide. *Water Sci. Technol.* **2016,** *73*, 453–462.

Collard, M.; Benoit, T.; Laurent, L. Comparison of Three Different Wastewater Sludge and Their Respective Drying Processes: Solar, Thermal and Reed Beds–Impact on Organic Matter Characteristics. *J. Environ. Manag.* **2017,** *203*, 760–767.

Collivignarelli, M. C.; Alessandro, A.; Andrea, F.; Marco, C. M.; Sergio, P.; Ioannis, K.; Vincenzo, T. Legislation for the Reuse of Biosolids on Agricultural Land in Europe: Overview. *Sustain.* **2019a,** *21*, 6015.

Collivignarelli, M. C.; Alessandro, A.; Massimiliano, B.; Barbara, M. C. Marco, C. M. Electrolytic Recovery of Nickel and Copper From Acid Pickling Solutions Used to Treat Metal Surfaces. *Water Air Soil Pollut.* **2019b,** *5*, 101.

Colón, J.; Manuel, A.; Mark, H.; Ayten, N.; Sergio, P. F.; Dilek, S.; and Carlota, T. Producing Sludge for Agricultural Applications. In *Innovative Wastewater Treatment & Resource Recovery Technologies: Impacts on Energy, Economy and Environment*; IWA Publishing London, 2017.

Delibacak, S.; Lyudmila, V.; Ekaterina, M.; Ali, R. O. Use of Sewage Sludge in Agricultural Soils: Useful or Harmful. *Eurasian J. Soil Sci.* **2020,** *9*, 126–139.

Diaz-Elsayed, N.; Nader, R.; Tianjiao, G.; Shima, M.; Qiong, Z. Wastewater-based Resource Recovery Technologies Across Scale: A Review. *Resour. Conserv. Recycl.* **2019,** *145*, 94–112.

Dichtl, N.; Svenja, R.; Katrin, B. Novel Strategies in Sewage Sludge Treatment. *CLEAN–Soil Air Water* **2007,** *35*, 473–479.

Dufour, P. Control Engineering in Drying Technology: Review and Trends. *Dry. Technol.* **2006,** *24*, 889–904.

Dursun, D.; Azize, A.; Steven, K. D. Pretreatment of Biosolids by Multi-enzyme Mixtures Leads to Dramatic Improvements in Dewaterability. *Proc. Water Environ. Feder. 2006,* **2006,** *2*, 392–402.

Eriksen, S. N. U.; Per, H. N.; Ib, P.; Mikkel, H. A.; Sandino, J. Side stream Management as Part of an Energy Optimization Roadmap for an Advanced BNP Facility. *Proc. Water Environ. Feder. 2018*, **2018,** *4*, 1087–1094.

Fernández, J. M.; Plaza, C.; Hernández, D.; Polo, A. Carbon Mineralization in An Arid Soil Amended with Thermally Dried and Composted Sewage Sludges. *Geoderma.* **2007,** 137, 497–503.

Fisher, R. M.; Juan, P. A.; Richard, M. S. Review of the Effects of Wastewater Biosolids Stabilization Processes on Odor Emissions. *Crit. Rev. Environ. Sci. Technol.* **2019,** *49*, 1515–1586.

Fytili, D.; Anastasia, Z. Utilization of Sewage Sludge in EU Application of Old and New Methods—AReview. *Renew. Sustain. Energy Rev.* **2008,** 116–140.

Gadhamshetty, V.; Mark, J. k.; Catherine, A. Peters. AEESP Spotlight: Mid 2020. *Environ. Eng. Sci.* **2020.**

Gantzer, C.; Gaspard, P.; Galvez, L.; Huyard, A.; Dumouthier, N.; Schwartzbrod, J. Monitoring of Bacterial and Parasitological Contamination during Various Treatment of Sludge. *Water Res.* **2001,** *35*, 3763–3770.

Greenstone, M. The Impacts of Environmental Regulations on Industrial Activity: Evidence from the 1970 and 1977 Clean Air Act Amendments and the Census of Manufactures. *J. Polit. Econ.* **2002,** *110*, 1175–1219.

Hackett, M. J. Remining and the Water Quality Act of 1987: Operators Beware. *Colum. J. Environ. L.* **1987,** *13*, 99.

Harrison, E. Z., Murray, B. McBride, D.; Bouldin, R. Land Application of Sewage Sludges: An Appraisal of the US Regulations. *Int. J. Environ. Pollut.* **1999,** *1*, 1–36.

Hudcová, H.; Jan, V.; Miloš, R. Present Restrictions of Sewage Sludge Application in Agriculture within the European Union. *Soil Water Res.* **2019,** *14*, 104–120.

Iloms, E.; Olusolao, O.; Ololade, H.; Ogola, J. O.; Ramganesh, S. Investigating Industrial Effluent Impact on Municipal Wastewater Treatment Plant in Vaal, South Africa. *Int. J. Environ. Res. Public Health.* **2020,** *3*, 1096.

Joo, S. H.; Francesca, D. M.; Eric, A.; Philip, C. Sustainable Approaches for Minimizing Biosolids Production and Maximizing Reuse Options in Sludge Management: A Review. *J. Environ. Manag.* **2015,** *158*, 133–145.

Khiari, L.; Claude-Alla, J.; Jacques, G.; Ned, B.; Sidki, B. Classification and Assessment Models of First Year Biosolids Phosphorus Bioavailability. *Waste Biomass Valorization* **2020,** *11*, 1443–1452.

Kim, L.; Rooney, Y. J. C.; Michael, L. M.; Linda, S. L. "Characterizing and Comparing Per-and Polyfluoroalkyl Substances in Commercially Available Biosolid and Organic Non-Biosolid-Based Products. *Environ. Sci. Technol.* **2020,** *54*, 8640–8648.

Koukoulakis, P. H.; Kyritsis, S. S.; Kalavrouziotis, I. K. The Contribution of Decision Support System (DSS) to the Approach of the Safe Wastewater and Biosolid Reuse. *Wastewater Biosolids Manag.* **2020,** *159*.

Krawczyk, P. Control Strategy for Ventilation System of Sewage Sludge Solar Dryer. *J. Power Technol.* **2016,** *96*, 145–148.

Lee, E.; Deborah, S. B. L. O.; Luiza, S. B. L. O.; Eduardo, J.; Youngwoon, K. Meng, W.; Sarina, J. E. Qiong, Z. Comparative Environmental and Economic Life Cycle Assessment of High Solids Anaerobic Co-digestion for Biosolids and Organic Waste Management. *Water Res.* **2020,** *171*, 115443.

Liu, Z.; Brooke, K.; Mayer, K.; Venkiteshwaran, S. S.; Arun, S. K.; Raju, D. Z.; Patrick, J. M. The State of Technologies and Research for Energy Recovery from Municipal Wastewater Sludge and Biosolids. *Curr. Opin. Environ. Sci. Health.* **2020,** *14*, 31–36.

Lumsden, J.; Charles, H. Landfill Guidelines: Towards Sustainable Waste Management in New Zealand. **2000.**

Makino, T.; Masaharu, M.; Satoru, I.; Tadashi, A. Regulations for Cadmium in Rice and Soil in Japan and Countermeasures to Reduce the Concentrations. In *Cadmium Toxicity*; Springer: Singapore, 2019; pp 103–114.

Marek, D.; Michael, B.; Marcin, D. The challenge of Implementing European Union Environmental Law in the New Member States: The Urban Wastewater Treatment Directive in the Czech Republic and Poland. *Environ. Plan. C: Polit. Sp.* **2017,** *35*, 1117–1135.

Mayer, P.; Hilber, I.; Hale, S. How to Determine the Environmental Exposure of PAHs Originating from Biochar. *Environ. Sci. Technol.* **2016,** *50*, 1941–1948.
Meyer, A. M.; Christina, K.; Elisabeth, F.; Ralf, K.; Horst, P. Beck. Real-time Monitoring of Water Quality to Identify Pollution Pathways in Small and Middle Scale Rivers. *Sci. Total Environ.* **2019,** *651*, 2323–2333.
Mohajerani, A.; Aruna, U.; Tristan, J.; Michael, S. Mohtashim, A.; Glen, R.; Simon, B.; Nicky, E.; Sujeeva, S. A Proposal for Recycling the World's Unused Stockpiles of Treated Wastewater Sludge (Biosolids) in Fired-Clay Bricks. *Building.* **2019,** *9*, 14.
Mohajerani, A.; Bojana, K. Microplastics and Pollutants in Biosolids Have Contaminated agricultural soils: An Analytical Study and AProposal to Cease the Use of Biosolids in Farmlands and Utilize Them in Sustainable Bricks. *Waste Manag.* **2020,** *107*, 252–265.
Ødegaard, H.; Paulsrud, B.; Karlsson, I. Wastewater Sludge as AResource: Sludge Disposal Strategies and Corresponding Treatment Technologies Aimed at Sustainable Handling of Wastewater Sludge. *Water Sci. Technol.* **2002,** 46, 295–303.
Omidghane, M.; Mattia, B.; Justice, A. Lin, X.; Michael, C.; David, C. Bressler. Pyrolysis of Fatty Acids Derived from Hydrolysis of Brown Grease with Biosolids. *Environ. Sci. Pollut. Res.* **2020,** 1–11.
Oruko, R. O.; Selvarajan, R.; Ogola, H. J. O.; Edokpayi, J. N.; Odiyo, J. O. Contemporary and Future Direction of Chromium Tanning and Management in Sub-Saharan Africa Tanneries. *Process Saf. Environ. Prot.* **2020,** *133*, 369–386.
Pathak, A.; Dastidar, M. G.; Sreekrishnan, T. R. Bioleaching of Heavy Metals from Sewage Sludge: AReview. *J. Environ. Manag.* **2009,** *90*, 2343–2353.
Paz-Ferreiro, J.; Aurora, N.; Ana, M.; Matthew, P. J. A.; Gabriel, G. Biochar from Biosolids Pyrolysis: A Review. *Int. J. Environ. Res. Public Health.* **2018,** *15*, 956.
Price, G. W.; Morgan, G. I. L.; Svetlana, N. Y. Microbial Co-occurrence Network Analysis of Soils Receiving Short-and Long-term Applications of Alkaline Treated Biosolids. *Sci. Total Environ.* **2020,** *751*, 141687.
Ramírez, W. A.; Domene, X.; Ortiz, O.; Alcañiz, J. M. Toxic Effects of Digested, Composted and Thermally Dried Sewage Sludge on Three Plants. *Bioresour. Technol.* **2008,** *99*, 7168–7175.
Rashid, M.; Mary, G.; Kattou'a, I.; Al-Khatib, A.; Chikashi, S. Farmers' Attitude Toward Treated Sludge Use in the Villages of West Bank, Palestine. *Environ. Monit. Assess.* **2017,** *7*, 353.
Rehman, K.; Fiza, F.; Iqra, W.; Muhammad, S. H. A. Prevalence of Exposure of Heavy Metals and Their Impact on Health Consequences. *J. Cell. Biochem.* **2018,** 119, 157–184.
Ross, J. J.; Zitomer, D. H.; Miller, T. R.; Weirich, C. A.; Patrick, J. M. Emerging Investigators Series: Pyrolysis Removes Common Microconstituents Triclocarban, Triclosan, and Nonylphenol from Biosolids. *Environ. Sci.: Water Res. Technol.* **2016,** *2*, 282–289.
Samaras, P.; Papadimitriou, C. A.; Haritou, I.; Zouboulis, A. I. Investigation of Sewage Sludge Stabilization Potential by the Addition of Fly Ash and Lime. *J. Hazard. Mater.* **2008,** *154*, 1052–1059.
Sánchez, A.; Xavier, G.; Adriana, A.; Raquel, B.; Joan, C.; Xavier, F. Dimitrios, K. Composting of Wastes. Resource Recovery to Approach Zero Municipal Waste. *Green Chemistry and Chemical Engineering Series*; CRC Press, Taylor & Francis Group, 2015; pp 77–106.
Schofield, S. A. Waste Management Law in New Zealand. *NZ J. Environ. L.* **2010,** *14*, 223.
Schulte, R. P. O.; Rachel, E. C.; Trevor, D.; Niall, F.; Reamonn, F.; Cathal, O'D.; Daire O'H. Functional Land Management: A Framework for Managing Soil-Based Ecosystem Services for the Sustainable Intensification of Agriculture. *Environ. Sci. Policy* **2014,** *38*, 45–58.

Sharma, B.; Abhijit, S.; Pooja, S.; Rajeev, P. S. Agricultural Utilization of Biosolids: A Review on Potential Effects on Soil and Plant Grown. *Waste Manag.* **2017,** *64*, 117–132.

Singh, P.; Shrivastava, V.; Kumar, A. Recent Developments in Greenhouse Solar Drying: A Review. *Renew. Sust. Energ. Rev.* **2018,** 82, 3250–3262.

Sinha, R. K.; Herat, S.; Bharambe, G.; Brahambhatt, A. Verm Stabilization of Sewage Sludge (biosolids) by Earthworms: Converting APotential Biohazard Destined for Landfill Disposal into A Pathogen-free, Nutritive and Safe Biofertilizer for Farms. *Waste Manag. Res.* **2010,** 28, 872–881.

Torri, S. I.; Rodrigo, S. C.; Giancarlo, R. Biosolid Application to Agricultural Land—A Contribution to Global Phosphorus Recycle: A Review. *Pedosphere* **2017,** *1*, 1–16.

Treviño, A. B.; Rodríguez, C. J. Analysis of Drying Technologies for Wastewater Treatment Plant Sludge As An Alternative Source of Energy. Third LACCEI International Latin American and Caribbean Conference for Engineering and Technology (LACCET'2005) *Advances in Engineering and Technology: A Global Perspective*, Cartagena de Indias, Colombia, 2015.

United States. Environmental Protection Agency. Office of Wastewater Management. A Guide to the Biosolids Risk Assessments for the EPA Part 503 rule. US Environmental Protection Agency, Office of Wastewater Management, 1995.

Wang, Q.; Masaaki, K.; Kohji, K.; Hiroaki, I.; Yasuhiko, K. Upgrading of Anaerobic Digestion of Waste Activated Sludge by Ultrasonic Pretreatment. *Bioresour. Technol.* **1999,** *68*, 309–313.

Wen, X.; Qingming, L.; Hualong, H.; Na, W.; Ying, C.; Jing, J.; Yongli, H.; Guanying, Xu.; Fengming, Li.; Wenjie, F. Comparison Research on Waste Classification Between China and the EU, Japan, and the USA. *J. Mater. Cycl. Waste Manag.* **2014,** *16*, 321–334.

Wong, J. W. C.; Selvam, A. Speciation of Heavy Metals during Co-composting of Sewage Sludge with Lime. *Chemosphere*. **2006,** *63*, 980–986.

Yin, Y.; Ya-Juan, L.; Shu-Juan, M.; Esra, U. K.; Yu, L. Enzymatic Pretreatment of Activated Sludge, Food Waste and Their Mixture for Enhanced Bioenergy Recovery and Waste Volume Reduction Via Anaerobic Digestion. *Appl. Energy*. **2016,** *179*, 1131–1137.

Yuan, G. Copper, Zinc, and Nickel in Soil Solution Affected by Biosolids Amendment and Soil Management. *Soil Res.* **2009,** *47*, 305–310.

Zhang, D. Effect of Process Intensification Techniques on Biosolids Management. PhD diss., Virginia Tech, **2020**.

Zhang, F. S.; Yamasaki, S.; Kimura, K. Waste Ashes for Use in Agricultural Production: II. Contents of Minor and Trace Metals. *Sci. Total Environ.* **2002,** *286*, 111–118.

Zhao, G., Manel, G. B.; Samuel, R.; Jin-Chemg, Xu.; annfd, D. R. Comparative Energy and Carbon Footprint Analysis of Biosolids Management Strategies in Water Resource Recovery Facilities. *Sci. Total Environ.* **2019,** *665*, 762–773.

Index

N

O

P

Q

R